Specimen Handling, Preparation, and Treatments in Surface Characterization

METHODS OF SURFACE CHARACTERIZATION

Series Editors:

Cedric J. Powell, *National Institute of Standards and Technology, Gaithersburg, Maryland*
Alvin W. Czanderna, *National Renewable Energy Laboratory, Golden, Colorado*
David M. Hercules, *Vanderbilt University, Nashville, Tennessee*
Theodore E. Madey, *Rutgers, The State University of New Jersey,*
Piscataway, New Jersey
John T. Yates, Jr., *University of Pittsburgh, Pittsburgh, Pennsylvania*

A Continuation Order Plan is available for this series. A continuation order will bring delivery of each new volume immediately upon publication. Volumes are billed only upon actual shipment. For further information please contact the publisher.

Specimen Handling, Preparation, and Treatments in Surface Characterization

Edited by

Alvin W. Czanderna

National Renewable Energy Laboratory
Golden, Colorado

Cedric J. Powell

National Institute of Standards and Technology
Gaithersburg, Maryland

and

Theodore E. Madey

Rutgers, The State University of New Jersey
Piscataway, New Jersey

KLUWER ACADEMIC / PLENUM PUBLISHERS
NEW YORK, BOSTON, DORDRECHT, LONDON, MOSCOW

Library of Congress Cataloging-in-Publication Data

Specimen handling, preparation, and treatments in surface
  characterization / edited by Alvin W. Czanderna, Cedric J. Powell
  and Theodore E. Madey.
      p.   cm. -- (Methods of surface characterization ; v. 4)
  Includes bibliographical references and index.
  ISBN 0-306-45887-X
  1. Surfaces (Physics)--Methodology. 2. Surface chemistry-
-Methodology. 3. Surfaces (Technology)   I. Czanderna, Alvin
Warren, 1930-   . II. Powell, C. J. (Cedric John) III. Madey,
Theodore E.   IV. Series.
QC173.4.S94S65   1998
530.4'17--dc21
                                                           98-45737
                                                                CIP

ISBN 0-306-45887-X

© 1998 Kluwer Academic / Plenum Publishers, New York
233 Spring Street, New York, N.Y. 10013

10 9 8 7 6 5 4 3 2 1

A C.I.P. record for this book is available from the Library of Congress.

Printed in the United States of America

Contributors

John R. Arthur, Oregon State University, Corvallis, OR 97331
David G. Castner, University of Washington, Seattle, WA 98195
C. A. Colemenares, Lawrence Livermore National Laboratory, Livermore, CA
    94551
Ulrike Diebold, Tulane University, New Orleans, LA 70118
Leonard B. Hazell, CSMA (South) Ltd., Middlesex TW16 5DB, UK
Paul A. Lindfors, Mankato State University, Mankato, MN 56002
W. McLean, Lawrence Livermore National Laboratory, Livermore, CA 94551
R. G. Musket, Lawrence Livermore National Laboratory, Livermore, CA 94551
Piero A. Pianetta, Stanford Synchrotron Radiation Laboratory, Stanford
    University, Stanford, CA 94309
W. J. Siekhaus, Lawrence Livermore National Laboratory, Livermore, CA 94551
Harland G. Tompkins, Motorola Incorporated, Tempe, AZ 85284

About the Series

Many techniques are now being used to determine different surface properties, particularly surface composition, and improvements in these techniques are reported each year. While the techniques are generally relatively simple in concept, their successful application involves the use of complex instrumentation, adequate understanding of the physical principles, avoidance of many problems, identification of possible artifacts, and careful analysis of the data.

We have designed this series to assist newcomers to the field of surface characterization, although we hope that the series will also be of value to more experienced workers. The approach is pedagogical or tutorial. Our main objective is to describe the principles, techniques, and methods that are considered important for surface characterization, with emphasis on how important surface characterization measurements are made and how to ensure that the measurements and interpretations are satisfactory to the greatest extent possible. At this time, we have planned six volumes, but others may follow.

The first volume provides a description of methods for vibrational spectroscopy of molecules on surfaces. Most of these techniques are still under active development; commercial instrumentation is not yet available for some techniques, but this situation could change in the next few years. The current state-of-the-art of each technique is described as are the relative capabilities. An important component of the first volume is a summary of the relevant theory.

The second volume contains descriptions of ion spectroscopies for surface analysis and is the first of two volumes on the techniques and methods of electron and ion spectroscopies that are in widespread use for surface analysis. These two volumes address techniques for which commercial instrumentation is available. The books are intended to fill the gap between a manufacturer's handbook and review articles that highlight the latest scientific developments.

The third volume is concerned with infrared (IR) and Raman spectroscopies that provide detailed molecular-level information about the species present on surfaces in ultrahigh vacuum as well as in various absorbing media. This book is written for those who are not spectroscopists but who are beginning to use IR or Raman spectroscopy to investigate surfaces of practical materials.

This fourth volume is about techniques for specimen handling, preparation, and treatments in surface analysis and other surface studies. It provides a compilation of methods that have proven useful for those engaged in surface science, applied surface science, thin-film deposition, surface analysis, and related areas.

The fifth volume will address common artifacts and problems associated with the bombardment of solid surfaces by photons, electrons, and ions. Descriptions will also be given of methods for characterizing the surface topography and of sputter depth profiling for measuring the near-surface composition of a solid.

Surface characterization measurements are being used increasingly in diverse areas of science and technology. We are confident that this series will be useful in ensuring that these measurements can be made as efficiently and reliably as possible. Comments on the series are welcomed, as are suggestions for volumes on additional topics.

C.J . Powell
Gaithersburg, Maryland

A. W. Czanderna
Golden, Colorado

D. M. Hercules
Nashville, Tennessee

T. E. Madey
Piscataway, New Jersey

J. T. Yates, Jr.
Pittsburgh, Pennsylvania

Foreword

With the development in the 1960s of ultrahigh vacuum equipment and techniques and electron, X-ray, and ion beam techniques to determine the structure and composition of interfaces, activities in the field of surface science grew nearly exponentially. Today surface science impacts all major fields of study from physical to biological sciences, from physics to chemistry, and all engineering disciplines. The materials and phenomena characterized by surface science range from semiconductors, where the impact of surface science has been critical to progress, to metals and ceramics, where selected contributions have been important, to biomaterials, where contributions are just beginning to impact the field, to textiles, where the impact has been marginal. With such a range of fields and applications, questions about sample selection, preparation, treatment, and handling are difficult to cover completely in one review article or one chapter. Therefore, the editors of this book have assembled a range of experts with experience in the major fields impacted by surface characterization. It is *the only* book which treats the subject of sample handling, preparation, and treatment for surface characterization. It is full of tricks, cautions, and handy tips to make the laboratory scientist's life easier.

With respect to organization of the book, the topics range from discussion of vacuum to discussion of biological, organic, elemental or compound samples, to samples prepared *ex situ* or *in situ* to the vacuum, to deposition of thin films. Generic considerations of sample preparation are also given.

In Chapter 1, Harland Tompkins explains that a vacuum is necessary for the surface characterization techniques using charged particles (electrons and ions) because of the probability of collision with a gas molecule at a pressure $>10^{-5}$ torr. However, to keep the surface from being contaminated by residual gases, a vacuum of better than 10^{-9} torr is required. He also describes how the system conductance, materials selection and treatment, vacuum pumps, and vacuum pressure gauges affect the surface and its preparation and maintenance.

To illustrate that surface characterization has advanced to the point where it can be used to analyze interfaces not created in a vacuum, Leonard Hazell describes in Chapter 2 the technique of SPLINT (sample preservation using liquid nitrogen transfer). True to the intent of the book, he describes in great detail the use of SPLINT for surface preservation. The proper application of the technique to studies of corrosion is illustrated for mild steel and aluminum. Persons using surface characterization for biological materials should find this chapter very interesting reading.

In a more generic discussion of factors critical to preservation of the original composition and structure of surfaces, Paul Lindfors discusses the critical issues involved with analysis of solid surfaces in Chapter 3. Practical details are provided on sample handling, cleaning, shipping, charging, and storage. The comments apply to electron and optical microscopy, rather than just to the normal compositional analysis techniques. Specific hints are given for depth profiling using chemical removal of overlayers in *in situ* as well as *ex situ*, cross section, angle lapping, and fracturing. In addition, techniques for analysis of wires, fibers, powders, and other unusual geometries are discussed.

Continuing with detailed recipes for sample preparation, Musket, McLean, Colemenares, and Siekhaus describe in Chapter 4 the literature on how to produce atomically clean surfaces of elemental samples. In all cases, the results of the cleaning procedure(s) have been verified using element-specific surface-sensitive analytical techniques. Of the 75 solid elements, only 10 have no report of cleaning procedures. Several of the 65 have reports both for polycrystalline and single crystals of various orientations, all of which are discussed.

Ulrike Diebold, in Chapter 5, gives a general discussion of how cleaning techniques have been used to prepare surfaces of metal compounds, with particular emphasis on surface preparation of oxides. In the case of compounds versus elements, the surface morphology and stoichiometry must be preserved during preparation. In addition to discussing *ex situ* and *in situ* chemical reaction techniques, she discusses cleaving, polishing, annealing, and sputtering. Problems with sample mounting and temperature measurement are also discussed, subjects rarely found in books on surface characterization.

Piero Pianetta reviews in Chapter 6 the details behind surface dosing with gases or alkali metals dispensers, and creating surfaces by cleaving. Excellent discussion is given on the fundamentals of gas adsorption and ideal gas behavior with respect to a surface. Seldom-reported aspects of the operation of alkali dispenser sources are given, especially the uniformity of dose across a surface and getter pumps to avoid reactions with gases. Some critical aspects and tricks in cleaving semiconductors are given.

In Chapter 7, David Castner describes the modification of surfaces with self-assembled monolayers (SAMs), glow discharge polymerization, and chemical reactions on catalytic materials. The application of surface characterization to SAMs and polymerized surface films is known to relatively few surface scientists, and this description will allow the neophyte to quickly gauge the field. Application of surface analysis to catalytic materials is better known, but this chapter provides a detailed look at how surface characterization has been used in conjunction with isolation and high-pressure reaction cells to correlate surface composition with catalytic activity.

The most extensive use of surface characterization has been in the area of thin films, the task left for John Arthur in Chapter 8. To meet this challenge, a brief description is given of the physical vapor and chemical vapor classes of thin-film deposition techniques. However, detailed descriptions of the impact of surface characterization techniques are given for molecular beam epitaxy. The use of compositional and structural analytical techniques is described. It is a fitting final chapter for a book which provides the reader with detailed hints and direct recipes for the handling, preparation, and treatment of surfaces for characterization of their composition, structure, and properties. I recommend the entire book to the novice and to the experienced surface scientist.

Paul H. Holloway
University of Florida
Gainesville, FL

Preface

The determination of surface composition is essential in many areas of science and technology, and many approaches may be used for this purpose. Each method has strengths and limitations that often are directly connected to the physical processes involved. Typically, atoms on the surface and in the near-surface region may be excited by photons, electrons, ions, or neutral species, and the detected particles are usually ions, electrons, or photons that have been emitted, ejected, or scattered. A major part of any surface analysis or other surface characterization involves using various useful methods for specimen handling, preparation, and treatments that are rarely discussed in detail in the literature; the purpose of this book is to present a discussion of these methods. Characterization of the resulting surface is usually necessary for assessing the effectiveness of the specimen handling or process.

The first chapter deals with the use of ultrahigh vacuum and the vacuum compatibility of materials. The concepts of monolayer formation time, pressure regimes, pumps, and pressure measurement are considered. The second chapter is concerned with a successful cryogenic sample transfer technique designed to preserve surface chemistry from a liquid environment, and gives examples about mild steel, aluminum, and other materials corroded in solution prior to transfer for surface analysis by X-ray photoelectron spectroscopy. The third chapter is an overview of cleaning and processing different types of specimens and of miscellaneous procedures in specimen handling. The fourth chapter is a detailed review of surface cleaning procedures for preparing atomically clean elemental solids; procedures are described for 65 elements and for different low-Miller-index surfaces of crystalline solids. The fifth chapter provides an overview of surface preparation for compound materials (mainly oxides) and discusses sample handling, mounting, sputtering, and annealing. The sixth chapter considers *in situ* processing of solid surfaces by gas and alkali metal dosing as well as by cleaving a solid in vacuum to expose a "clean" surface. In the seventh chapter, the modification of surfaces by various chemical methods is discussed. These methods include the use of chemical reactions at organic surfaces, a radio-frequency glow discharge for etching or deposition processes, self-assembled monolayers, and modifications of catalytic surfaces. In the final chapter, techniques for depositing thin films by physical and chemical processes are described. Molecular beam epitaxial growth is a major focus of Chapter 8.

Each of the eight chapters contains extensive references to prior work and sufficient figures to illustrate the concepts being described. Tables are used to summarize information about a concept, technique, method, or process.

The editors are deeply grateful to the authors, whose work made this book possible, for taking the time from their active research programs to prepare their contributions.

Alvin W. Czanderna
Golden, Colorado
Cedric J. Powell
Gaithersburg, Maryland
Theodore E. Madey
Piscataway, New Jersey

Contents

2. Cryogenic Sample Transfer for Preservation of Surface Chemistry
Leonard B. Hazell

3. Specimen Handling: Cleaning and Processing
Paul A. Lindfors

7. Chemical Modification of Surfaces

David G. Castner

8. Physical and Chemical Methods for Thin-Film Deposition and Epitaxial Growth

John R. Arthur

Ultrahigh Vacuum and Vacuum Compatibility of Materials

Harland G. Tompkins

1. Why Vacuum?

Aristotle coined the phrase "nature abhors a vacuum." Why is it then that scientists and engineers who study surfaces and thin films insist so vehemently that the measurements be done in such an inhospitable environment? We shall see that there are two primary reasons.

We note that the probes which surface scientists use are often charged particles and/or photons. In some techniques (e.g., transmission electron microscopy and Fourier transform infrared spectroscopy), we examine the primary probes which pass through the sample. In some techniques, (e.g., x-ray diffraction, Rutherford backscattering spectrometry, ellipsometry) we observe the primary probing particle as it is reflected. In some cases, (e.g., Auger electron spectroscopy, X-ray photo-electron spectroscopy, secondary-ion mass spectroscopy, scanning electron micros-copy) we observe secondary particles which are ejected.

Since charged particles interact strongly with gas-phase molecules, it is essential that a technique which uses charged particles be done in a vacuum. The electrons or ions must be able to move through their trajectory without interference from gas-phase molecules. We shall see that this requires that the gas density be reduced to roughly 10^{-7} that of the atmosphere. From the point of view of collisions with the gas phase, a pressure of 5×10^{-5} Torr would be sufficient. If fact, some of the

Harland G. Tompkins • Motorola, Inc., Phoenix Corporate Research Laboratory, Tempe, AZ 85284.

Specimen Handling, Preparation, and Treatments in Surface Characterization, edited by Czanderna *et al*. Kluwer Academic / Plenum Publishers, New York, 1998.

older Auger electron spectroscopy systems operate at such a pressure of argon during depth profiling. This pressure falls into what is often called "high vacuum."

At this pressure, however, the arrival rate of molecules at the surface is such that if all of the molecules which strike the surface stick, a monolayer can form in less than a tenth of a second. When studying oxides, for example, it is essential that we not be inadvertently measuring oxygen from gas-phase species such as O_2, CO_2, or H_2O which have arrived at the surface and stayed there. Indeed, often our probing particle will interact with adsorbed species and the fragments will be more reactive with the surface than the original molecule. We shall find that from a cleanliness point of view, pressures of 10^{-9} Torr are routinely required for most surface analysis techniques. Pressures below 10^{-8} Torr are called "Ultrahigh Vacuum."

When the probing particle is a photon, the interaction of the probe with the gas-phase species is significantly less than for charged particles. Techniques such as total reflection X-ray fluorescence, ellipsometry, and Fourier transform infrared spectroscopy are routinely done in air. The interaction of the ambient with the surface, however, is no less important than when using charged particles. In many cases with these kinds of techniques, the sample is kept in vacuum in order to control interactions of the ambient with the surfaces of interest.

2. Definitions and Equations

In this section, we give various definitions and equations which will be useful in the following sections. In addition, brief comments are included to put the subject matter in perspective. These definitions and equations are treated in detail in monographs[1-3] and textbooks.[4-6]

2.1. Ideal Gas Law

Since it is not possible to remove all gases from an experimental chamber, "vacuum" generally means a subatmospheric pressure. The parameters for describing a low-pressure gas are pressure P, volume V, absolute temperature T, and number of moles present. The ideal gas law describes the interactions between these parameters. The ideal gas law is $PV = nRT$, where R is the proportionality constant, called the "gas constant." When pressure is expressed in Torr, volume is in liters, n is in moles, and T is in kelvins, R has a numerical value of 62.4. The ideal gas law is sometimes given as $P = NkT$ where k is Boltzmann's constant and N is the gas density in molecules per cubic centimeter.

The ideal gas law is appropriate in situations where the gas molecules do not interact at a distance and where the molecular volume is negligible compared to the total volume occupied by the gas.

2.2. Throughput and Volumetric Flow (Pumping Speed)

Gas flow is expressed in two different ways. The mass flow rate, Q, is called the *throughput*, and the volumetric flow, S, is sometimes called *pumping speed*. The choice of terms for the volumetric flow is somewhat misleading, since it sometimes confuses volumetric flow with the ability to induce volumetric flow.

The throughput is essentially the net number of molecules passing a plane per unit time. The units are pressure–volume per unit time, e.g., Torr-liters/second (Torr-L/s). Note that PV is proportional to n for a constant temperature, hence the use of the term "essentially."

Volumetric flow, S, is easy to understand for pressures where the flow is laminar, and more difficult to visualize when the flow is molecular (see below). Suppose we have a gas moving down a tube of cross-sectional area A, and that the average flow velocity is such that an imaginary plane moving with the fluid moves a distance d in 1 s. If the volume swept out in 1 s is $V = Ad$ (and the units of A and d are such that V can be expressed in liters), then the volumetric flow is V L/s, regardless of the pressure. The actual amount of substance which has moved a distance d is not specified, since that depends on the pressure P.

With a somewhat lower gas density where the gas molecules collide with the walls but not with each other, the visualization of the "flow" velocity is not as straightforward. With the random motion of the molecules, there will still be more molecules passing the plane in one direction than in the opposite direction and there will therefore be a net "flow." If one were to count the net number of molecules passing the plane per second and then consider the volume V necessary to contain this number of molecules at the pressure P, then the volumetric flow is again V L/s. The key to clarity is to mentally substitute for a volumetric flow of "15 liters per second" the words "the amount of gas contained in 15 liters per second."

2.3. Conductance

If the pressure difference between two ends of a component is $P_1 - P_2$ and the induced flow (throughput) is Q, then we might call the proportionality constant Z the "resistance." Instead, we normally use the inverse of the resistance, a term which refers to ease of flow rather than difficulty of flow. The *conductance*, C, is defined as $C = Q/(P_1 - P_2)$. For high and ultrahigh vacuums, the equation which describes the conductance of a long tube (length greater than five times the diameter) is $C = 80D^3/L$ where C is in L/s, and D and L are in inches. Clearly, large diameter and short length improve ease of flow, and small diameter and long length allow for large pressure gradients to be built up.

2.4. Pumping Speed at Locations away from the Pump

The volumetric flow at the opening of the pump is traditionally called the *pumping speed* of the pump, and we denote it S_p. At a location which is separated from the pump by a component with conductance C, the volumetric flow S (or "pumping speed") is can be calculated from the equation

$$\frac{1}{S} = \frac{1}{S_p} + \frac{1}{C} \tag{1}$$

One can readily see that the pumping speed at any given point is no larger than the smallest conductance between the point and the pump.

2.5. Velocity of a Gas Molecule

The gas molecules have a range of velocities which is described by the Maxwell–Boltzmann distribution. The average velocity is $v_{av} = (8kT/\pi m)^{1/2}$, where m is the mass of the molecule, T is the absolute temperature, and k is Boltzmann's constant. Note that molecules leaving a hot surface travel faster than molecules leaving a cold surface. Also, in general, small molecules move faster than large molecules. At room temperature, the average speed of N_2 is about 450 m/s, whereas the average speed of H_2 is about 1700 m/s.

2.6. Mean-Free Path

Molecules travel various distances between collisions. The average distance is called the *mean-free path*, L, and is given by $L = 1/(\sqrt{2}\pi N d^2)$ where N is the gas density in molecules/cm^3 and d is the molecular diameter in cm. A useful adaptation of the equation for air at about 20°C is $L = 0.005/P$, where P is the pressure in Torr and the units of L are cm. Note that the mean-free path at a pressure of 5 mTorr is about 1 cm.

2.7. Monolayer Formation Time

If every molecule which strikes a surface were to stick to it (sticking coefficient equals unity), then the time, t_m, to form a monolayer is $t_m = 4/Nvd^2$, where v is the average velocity. Again, at room temperature and with nitrogen, this equation reduces to $t_m = 1.9 \times 10^{-6}/P$ where P is in Torr. Note that for pressures in the low 10^{-6} Torr range, monolayers are formed in about a second. For pressures in the low 10^{-9} Torr range, one has about 30 min before a monolayer is formed. The value of the sticking coefficient varies from near unity for reactive gases on reactive surfaces (e.g., CO on Zr) to near zero (e.g., N_2 on SiO_2).

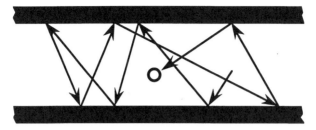

Figure 1. The motion of a gas molecule, shown schematically, where the mean free path is greater than the dimensions of the container. (After Tompkins[1])

2.8. Molecular Flow

In a typical pumpdown from atmospheric pressure, the system passes through several flow regimes. Initially, the gas leaves the system in a somewhat turbulent manner. After the pressure is reduced somewhat, the gas flows through the plumbing in a laminar manner. As the pressure continues to be reduced and the mean-free path increases, a transition occurs to a regime where the gas molecules do not interact with each other to any extent, but simply move from one wall surface to another, as suggested by Fig. 1. We call this random motion of the gas molecules *molecular flow*.

The interactions of the gas molecules with the surface can conveniently be visualized as adsorption, residence, and desorption. Specifically, the direction of motion when the molecule leaves the surface is independent of the direction of motion when it arrived at the surface. A molecule which leaves a surface is equally likely to go in the "wrong" direction (i.e., away from the pump) as in the "right" direction (toward the pump). The reason that "flow" occurs is that more molecules arrive at a given location from a higher-pressure region than from a lower-pressure region; hence flow is caused by pressure gradients.

3. Pumping Regimes

3.1. The Volume Gas

When we start a pumpdown from atmospheric pressure, the amount of gas inside the vacuum envelope can be represented as the product of pressure and volume, PV. The throughput, Q_p, going into the plumbing leading to the pump is $Q_p = SP$, where S is the pumping speed at the opening of the volume V. The source of this gas is the gas contained in the volume of interest. The differential equation which describes the change in the amount of gas in the volume of interest (the volume gas) is

$$\frac{d(PV)}{dt} = -SP \tag{2}$$

The solution is $P = P_0 \exp(-St/V)$ and is shown as the solid line in Fig. 2.

3.2. Gas from Walls

Equation (2) deals with the volume gas, but does not consider any gas sources or additional sinks. A more realistic version of Eq. (2) is $d(PV)/dt = -SP + Q_w$, where Q_w represents a gas source, specifically the walls of the chamber. The solution of this equation is represented by the broken line in Fig. 2. Generally, Q_w is a slowly varying function of time. Early in the pumpdown, the change in pressure is dominated by the removal of the volume gas, and Q_w is negligible. Eventually, however, Q_w becomes dominant, and the solution to the equation becomes simply $Q_w(t) = SP(t)$, which is referred to as the *gas load*. For a small chamber such as a turbopumped specimen introduction chamber on an Auger system, the volume–gas domination lasts for only a few seconds. For a pumpdown after opening the entire chamber, the volume gas domination may last a few minutes. We spend most of our vacuum–technology life waiting for the gas to come off (or out of) the walls!

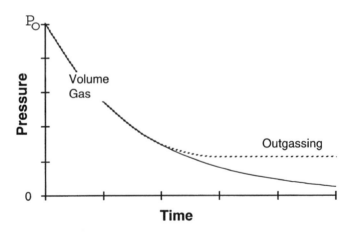

Figure 2. Pressure versus time, plotted linearly. The solid curve represents only the pressure change for the volume gas (i.e., the gas originally in the volume of the chamber being evacuated). The dotted line represents the pressure change associated with the volume gas along with gas desorbed from the chamber walls.

Suppose that we were in the gas-source-dominant mode and the pressure is in the 10^{-6} Torr range. If we were able to abruptly increase the size of our pump by 10-fold, we would only reduce the pressure by one order of magnitude to the 10^{-7} Torr range. To reduce the pressure to the ultrahigh vacuum range, it is necessary to reduce the effect of the gas source, i.e., Q_w.

3.3. Outgassing

Outgassing is a generic term involving natural evolution of species contributing to the gas load from one of the following:

- *Sublimation* of materials inside the vacuum envelope. This may be due to improper choice of materials or improper treatment of components or other materials introduced into the vacuum system.
- *Desorption* from species chemisorbed on the walls (most physically adsorbed molecules are removed during the roughing).
- *Diffusion* of gas out of the vacuum envelope (e.g., out of grain boundaries, defects, etc.)
- *Permeation* of gases through the vacuum envelope (hydrogen through the walls, helium through the windows, and air through O-rings, etc.).

A significant contribution to outgassing due to sublimation is usually the result of a mistake. Desorption, diffusion, and permeation, on the other hand, are natural phenomena which have to be dealt with routinely. The various contributions are shown graphically in Fig. 3.

The four regimes in Fig. 3 represent pressure changes associated with removal of the volume gas, desorption, diffusion, and permeation. Note that time is shown on a logarithmic scale. The volume gas is normally pumped out in a few minutes or tens of minutes, after which surface desorption dominates. Although Fig. 2 shows generally that outgassing is constant, the amount of desorption decreases as a function of time as the adsorption sites are depopulated, and the pressure varies inversely as time. This dependence represents an average of several mechanisms and desorption energies. Appendices in O'Hanlon[4] list the outgassing rates of unbaked metals after 1 h in vacuum and after 10 h. He also lists outgassing rates for ceramics, glasses, and elastomers at various times, and for vacuum-baked metals. The outgassing rates are given as a throughput per unit area (i.e., Torr-L/s cm^2).

The volume of a system determines the time required to pump out the volume gas. The surface area determines the pumpdown time in the desorption regime. A large volume does not necessarily mean a long pumpdown. The reason that it usually takes a long time to pump down a large volume is that we tend to put many components in a large volume and these have a large surface area.

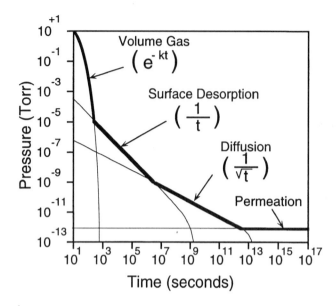

Figure 3. Characteristic behavior of pressure versus time during the pumpdown of a UHV chamber showing the limits associated with different sources. (After Tompkins,[1] data from O'Hanlon[4])

Figure 3 suggests that for pressures below 10^{-9} Torr, we move into the diffusion regime, where we must wait for gas molecules to come out of the bulk of the materials rather than simply desorb from the surfaces. Permeation is placed at the bottom of Fig. 3, and this is the case for a copper-gasket-sealed system. For an O-ring-sealed system, the permeation regime would be just below the desorption regime and the diffusion would not appear on a display such as Fig. 3. The ultimate pressure of a cryopump might be 5×10^{-10} Torr when metal gasket seals were used, but only 1×10^{-8} Torr when O-ring seals were used.

Early in the pumpdown, the volume gas dominates the pressure–time relation. During this period, the residual gas will be made up of N_2, O_2, Ar, and the other major atmospheric gases. When the pressure has been reduced to a few tenths or hundredths of a mTorr, where desorption from the walls predominates, the residual gas will consist mainly of water molecules. At pressures below the 10^{-8} Torr range, the main residual gas species are H_2, CO, and CO_2.

It is not uncommon to bake a UHV system during pumpdown. Baking increases the desorption rate (and diffusion rate). It would at first seem that this would be the wrong procedure because it increases the pressure (increases Q_w), as suggested by Fig. 4. As the adsorption sites are depopulated, however, Q_w decreases and the pressure eventually drops below the level where it would have been if the system had not been baked.

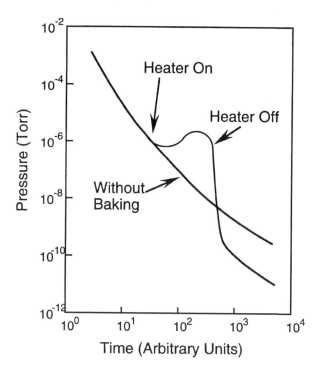

Figure 4. The effect of baking a vacuum chamber on the pressure as a function of time. (After Tompkins[2])

4. Vacuum Materials

4.1. Properties

From the point of view of vacuum technology, the properties of materials which are of interest[5–6] are

- Mechanical strength
- Magnetic properties
- Corrosion properties
- Machinability
- Vapor pressure
- Outgassing properties
- Gas diffusion
- Thermal properties
- Electrical properties

- Availability
- Cost

We shall, rather arbitrarily, divide the materials into

- Metals
- Glass and ceramics
- Polymers, epoxy, and grease

4.2. Metals

Metals are now nearly always used for the vacuum envelope and for structural support inside the vacuum envelope. Metals are also used for conducting electricity and heat and for making the vacuum seal.

From the point of view of strength, stainless steel and aluminum are the primary metals used for the vacuum envelope. In the past, glass vacuum systems were common, but now are used only for special-purpose ultrahigh vacuum systems (e.g., some field-ion microscopes).

Metal alloys which contain zinc, lead, cadmium, selenium, and sulfur have unsuitably high vapor pressure and should not be used for ultrahigh vacuum applications. Brass is commonly used in valves and other applications where no heating is expected, but for UHV applications it should be avoided. In the stainless steels, types 304 and 316 are commonly used. Type 303 stainless steel contains sulfur and selenium to make it more machinable, and should be avoided for UHV applications. Aluminum is used by some for the vacuum envelope, but its use makes joining and making vacuum seals more difficult. Possible advantages (near the ultimate pressure) are that hydrogen does not permeate through aluminum as readily as through stainless steel. This is only an issue for systems which will be expected to achieve pressures below the 10^{-10} Torr range.

Another positive attribute of 304 stainless steel is that it is nonmagnetic. This is important when using magnets outside the vacuum envelope to affect charged particles or other magnets inside the chamber.

The refractory metals such as W, Ta, Mo, and Re are used as electron sources and in other applications where high-temperature stability is required. Copper is used as an electrical conductor and as a gasket and, along with Au and Ag, for brazing alloys. The very reactive metals such as Ti and Zr are used for pumping by chemical reaction with the gas species (gettering).

Gas is dissolved in a metal during its initial fabrication, and the diffusion of this gas contributes to the gas load throughout the service lifetime. This contribution can be reduced by high-temperature treatment of the materials and components during their initial fabrication. This treatment is normally done by the manufacturer and is often not a concern of the end user.

There are two issues associated with adsorbed gas. First, the adsorbed material which is present because of fabrication must be removed. Various surface treatments, both chemical and physical, remove most of the contamination which results from fabrication methods and materials, but the final removal occurs in the completed system and is usually thermal.

The second issue is that even if all the adsorbed gas was removed from the walls, water vapor readsorbs on the empty adsorption sites when the system is opened and must be again removed during subsequent pumpdown. If the system is to be open for 15 min or less, backfilling with a dry gas will reduce the time needed to achieve UHV on the subsequent pumpdown. Various surface treatments[5] will decrease surface roughness, thereby decreasing the available surface area for adsorption.

4.3. Glasses and Ceramics

Although entire systems were made of glass and a vacuum technologist was required to be a glassblower in the past, glass is today primarily used for windows. Structural strength is normally not a problem since glass can easily withstand the required 1 atm pressure from the outside. When letting a system up to atmospheric pressure with a pressurized dry gas, one should be careful to avoid pressures much greater than 1 atm on the inside.

When reducing the pressure below the 10^{-10} Torr range, one should be aware that helium permeates glass to some extent. For ultimate pressures above this level, this phenomenon is normally not an issue.

Ceramics are used as electrical insulators and for structural strength. The primary problem with ceramics is porosity (adsorption of water and other contaminants). Most manufacturers use a glaze which reduces this problem significantly. Most ceramics are extremely hard and accordingly cannot be custom machined or fabricated by the end user. However, Macor, a registered trademark of Corning Inc., is designed specifically for this purpose. This is useful for fabricating small items, but the properties which make it machinable are expected to add to the gas load to some extent, so it should be used with discretion in a system which is to go below the 10^{-10} Torr range.

4.4. Polymers, Elastomers, Epoxy, and Grease

Because of outgassing, polymers, elastomers, epoxy, and grease are generally to be avoided. Electrical conductors should be insulated with ceramic "fish spine" insulators rather than with plastic coatings. Viton, a registered trademark of DuPont Inc., was developed as a gasket material with emphasis on low outgassing properties. This material can be baked up to 250°C when uncompressed and to 125°C when compressed. Permeation through elastomer gaskets usually minimizes the use of O-ring seals in UHV systems. Epoxy is used in cryopumps (see below). Outgassing of this material is not a problem during operation when this material is at temperatures of the order of 20 K or below Torr Seal, is a low-outgassing two-part epoxy and a registered trademark of Varian Associates, Inc. It is sometimes used to repair broken ceramic parts on electrical feedthroughs. This should be considered an interim solution, used to complete a project in progress, and not a normal operation.

When using O-rings, one should recall that the function of O-ring grease is not to seal but to lubricate the O-ring so that it will move to its optimum position. The optimal amount of grease is none if lubrication is not required.

5. Pumps

5.1. Historical Aspects

The first type of pump used regularly to obtain high vacuum, and eventually ultrahigh vacuum, was the diffusion pump. The working fluid was normally comprised of large organic molecules (or sometimes mercury). Diffusion pumps were not inherently contaminating; however, they were inherently unforgiving, and any deviation from proper operation would result in oil or mercury contamination in the chamber. Hence, when other options were made available, the UHV community chose to use them.

5.2. Ion Pumps

The ion pump (sometimes called sputter-ion or getter-ion pump) was the first UHV pump where contamination was not a problem. This type of pump uses a combination of burial and gettering to remove molecules from the gas phase. For burial, a Penning discharge formed with a modification of the basic Penning ionization gauge is used.

Figure 5 shows the configuration of the Penning gauge. The anode ring is biased a few kilovolts with respect to the cathode plates. Electrons in the discharge region are attracted to the ring plane and most pass through, to again be attracted back. A magnetic field significantly lengthens the electron path before the electron eventually reaches the anode ring and is removed from the process. Gas molecules are ionized along the electron trajectory to create additional electron–ion pairs. The ions are attracted to the cathode plates and strike the surface with enough energy to bury most of them (which are, of course, neutralized). Secondary electrons are generated on ion impact with the surface, and the discharge continues. Burial of gas molecules is one of the two pumping mechanisms of the ion pump.

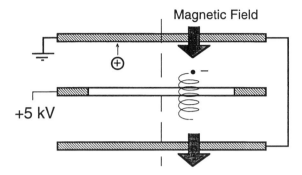

Figure 5. Configuration for the basic Penning ionization gauge. The magnetic field increases the electron path length in the discharge region. (After Tompkins[2])

In the ion pump, the geometry of the Penning gauge is changed somewhat to allow the discharge to be maintained down to pressures in the UHV region. In Fig. 6 we show an elemental ion-pump cell in which the anode ring of Fig. 5 has been extended as an anode tube. For most other types of pumps, greater capacity (pumping speed) is obtained by making larger pumps with larger parts. Since the ion pump depends on a magnetic field, this is not a practical way to obtain greater pumping capacity. Instead, we simply make a structure containing many of the elemental cells, as shown in Fig. 7. The enclosure is made of nonmagnetic material, typically type 304 stainless steel. The magnet is then placed around the pump, outside the vacuum envelope.

Another phenomenon occurs when an ion bombards the cathode plate, and this eventually accounts for the other pumping mechanism. When ions with energies of a few thousand electron volts strike a surface, some of the surface material is removed by sputtering and lands on nearby surfaces. The anode material is typically titanium, which is chemically very active, so nearby surfaces are coated with it. This is shown schematically in Fig. 8. Chemically active gas-phase molecules (either charged or neutral) which arrive at these titanium sites will form stable chemical compounds, thereby removing the molecules from the gas phase. This process of removal is called *gettering*. (Gettering is also used in other areas of vacuum technology, with solid getters or getter surfaces deposited by flash evaporation or sublimation.)

The gettering surface will pump chemically active molecules such as O_2, N_2, and CO very efficiently. More stable molecules such as CO_2, H_2O, CH_4, etc., can be pumped very efficiently if they can be broken into fragments. The Penning discharge provides this fragmentation, and hence the ion pump (but not an isolated getter pump) will also pump these molecules well. Noble gases such as He, Ne, Ar,

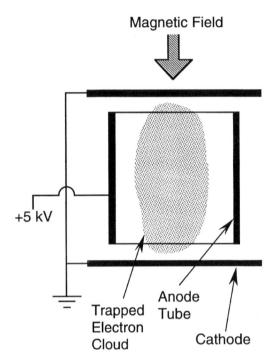

Figure 6. Cross section of a single-cell ion pump showing the Penning discharge. (After Tompkins[2])

Xe, etc., are pumped not by gettering but by burial, and the combination of these two mechanisms allows pumping of all types of molecules.

The selectivity of the gettering is due to chemical reactions (different for each type of gas molecule), and the effectiveness of the burial is due to ionization

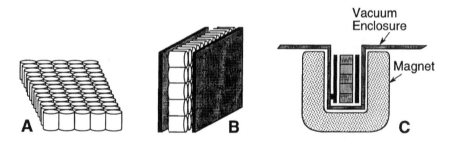

Figure 7. (A) The eggcrate anode ring assembly, (B) with cathode plates added, and (C) with vacuum enclosure and magnet added. (After Tompkins[2])

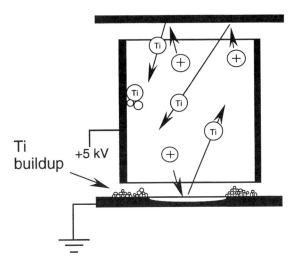

Figure 8. The sputtering process that leads to deposition of the gettering material. (After Tompkins[2])

efficiency. As a result, the ion pump will not pump all molecules equally. Typically, the pumping speed of H_2 is about twice that of O_2 and N_2, and the pumping speed of the noble gases is about one quarter that of the air molecules.

We shall not deal with the issue of noble-gas evolution, which occurs when the sputtering uncovers a previously buried noble-gas molecule. Manufacturers deal with this issue with creative designs.[2,4-6] Suffice it to say that these pumps will remove any type of residual gas molecule.

The ion pump works well when the pressure is below the 10^{-6} Torr range. Pumping speed curves, such as Fig. 9, show a nonzero pumping speed at pressures up to about 10^{-3} Torr. During the initial pumpdown after exposure to air, the discharge and the ions bombarding the anode surfaces may liberate adsorbed gases. Often when the pump is first started, the pressure goes up rather than down. It is helpful if an additional pump (of another type) is available to help deal with these desorption products. The ion pump is designed to pump only briefly between 10^{-6} and 10^{-3} Torr. It is expected, however, that the pressure will drop to lower ranges in a reasonably short time (typically minutes). Pumps which must operate above 10^{-6} Torr for extended periods must be water cooled. For most applications in surface analysis, the system pressure drops quickly below 10^{-6} Torr and no cooling is needed.

The discharge current is directly proportional to the pressure. Since the ion pump normally operates at very low pressures, very little power is required for operation. In addition, the failure mechanism is to sputter through the anode plate. If we suppose that the lifetime of a water-cooled pump with replaceable electrodes

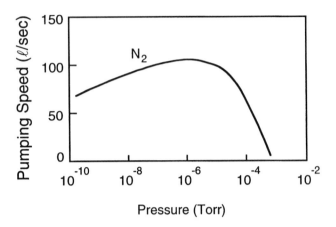

Figure 9. Typical pumping speed curve for an ion pump. (After Tompkins[2])

which is expected to operate in the 10^{-6} Torr range is two years, the lifetime of a pump which operates normally in the 10^{-9} Torr range would be 2000 years. Clearly these pumps will operate for very long times.

Ion pumps do not have particularly high throughputs and are not suited for applications where gas is purposely injected (e.g., sputter deposition systems) or where evolution of gas is expected to be high (e.g., vacuum ovens). Prior to the availability of other options, an application requiring the cleanliness of an ion pump would have a very large ion pump to deal with the high throughput. Currently, other options are available which deal with large gas loads and provide comparable cleanliness.

The ion pump clearly does not operate at atmospheric pressure, so the system must be rough-pumped by some other pump. Although an ion pump can be started (with difficulty) at a pressure as high as 10 mTorr, operation of the ion pump is enhanced if the pressure in the chamber can be reduced to below the 1-mTorr region. This requirement is significant in that it places requirements on the type of roughing system used. Note that the ion pump is a capture pump and, correspondingly, contains all of the gas which it has ever pumped. For radioactive gases, this may be an issue.

5.3. Cryopumps

For many years in the past, when cleanliness was paramount, the ion pump was the only choice. Currently, cryogenic pumps provide another alternative. Although vacuum technology of cryopumping has been known for a very long time, cryopumps were not used because economical refrigeration technology was not in place to provide the needed low temperatures.

Figure 10 shows a schematic of the basic cryopump. The refrigerator cold head is a closed-loop compression/expansion-type system which uses helium as the working fluid. The compressor is not shown. The refrigerator unit is capable of providing temperatures as low as 10 K, although the component "second stage 15 K" typically runs between 10 and 20 K.

Since the vacuum enclosure, at room temperature, radiates heat to the cold stage, it is advantageous to enclose the 15 K stage inside a "first stage" which is typically held at about 80 K. In order to exclude radiation but allow the molecules to pass through, the 80 K stage has a chevron panel, as indicated in Fig. 10.

From the vapor pressure curves in Fig. 11, we see that, at 80 K, CO_2 will be removed from the system until the pressure drops to about 10^{-7} Torr. The equilibrium pressure for H_2O (not shown) is significantly below 10^{-11} Torr; hence any H_2O molecule which strikes the chevron panel will remain there. N_2, O_2, and Ar (not shown) molecules which strike the chevron panel will not remain, but will go into the pump or back into the chamber.

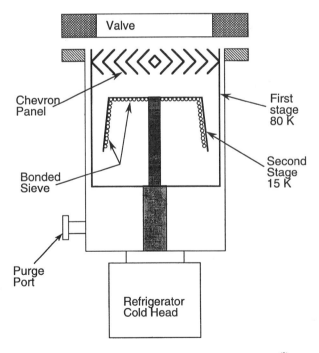

Figure 10. Schematic of a cryogenic pump. (After Tompkins[2])

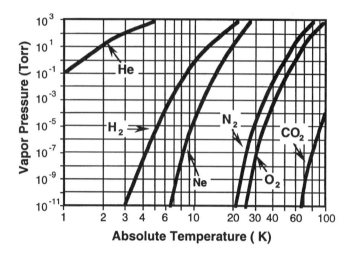

Figure 11. Vapor pressure curves. (From R. E. Honig and D. A. Kramer, *RCA Rev.* **30**, 285 (1969).)

When N_2, O_2, and Ar molecules which go into the pump strike the 15 K stage, they will remain (in Fig. 11 the equilibrium vapor pressure is below 10^{-11} Torr). At 15 K, the vapor pressure of Ne is to about 100 mTorr, that of H_2 reduced to about 50 Torr, and that of He not significantly reduced from atmospheric pressure. We see, then, that the chevron panel pumps H_2O and some CO_2 molecules, that the flat surface of the 15 K stage pumps the N_2, O_2, and Ar, and the Ne, H_2, and He are essentially not pumped by either of these surfaces.

A vacuum pump which does not pump Ne, H_2, and He would not be of much use. Correspondingly, it is necessary to add activated charcoal to the 15 K surface of the cryopump for these molecules. This material has minute cavities connected by passageways with radii of a few atomic dimensions. A gas molecule which may not condense on a flat surface at 15 K will "sorb" into these spaces which have extremely small radii of curvature. This activated charcoal is shown in Fig. 10 as "bonded sieve" and is bonded to the underside of the 15 K stage with epoxy. Although the use of epoxy is not recommended in UHV systems in general, outgassing is normally not a problem at 15 K.

The capacity of the chevron panel and the flat-surface part of the 15 K stage is limited only by closure of the passageway. The activated charcoal, or sieve material, however, will fill up much faster. Accordingly, the charcoal is placed in a location where several collisions with the flat 15 K surface are required before the molecule can arrive at the charcoal. In this way, the charcoal does not become filled with molecules which could have been pumped by the flat surfaces.

Cryopumps are capture pumps which handle a high throughput of most gases very well. As a capture pump, it must be regenerated periodically. For this process, they must be valved off from the chamber and allowed to warm up to room temperature, usually with a purge of heated dry nitrogen. The interval between purges depends on the amount of gas pumped. For sputter deposition systems where a large amount of argon is intentionally introduced, the pumps must be regenerated once every week or two. For secondary-ion mass spectrometers, the time interval between regenerations is several months.

Although it is possible to build a cryopump which will pump from atmospheric pressure, it is not economically feasible. The chamber (and the pump after regeneration) is typically rough-pumped from atmospheric pressure down to a few hundred mTorr with a rotary-vane oil-sealed mechanical pump. It is important to change over to the cryopump while the gas flow is still in the viscous range. Using a rotary-vane oil-sealed mechanical pump in the molecular flow regime may result in oil contaminants backstreaming into the chamber.

The ultimate pressure specification for cryopumps ranges from the low 10^{-8} to mid 10^{-10} Torr, depending on whether the pump is sealed with O-rings or copper gaskets. The relative pumping speed for various gases depends on which stage is used for pumping. If a gas molecule must interact with a surface before arriving at the pumping surface, the pumping speed will be lower, since the molecule may leave the surface going the wrong way. The number of collisions per unit time also influences the pumping speed. Table 1 gives the pumping speed for several gases relative to O_2 and N_2. As we would expect, H_2O has the greatest pumping speed since it is pumped at the first surface (the chevron panel). The reason why H_2 has a much higher pumping speed than might be expected (several wall interactions) is that it travels much faster than all of the other molecules. The relative capacity of a typical 12-inch pump is also shown in Table 1. As would be expected, the pump has the smallest capacity for the gases which must be pumped by the activated charcoal.

Table 1. Relative Pumping Speed and Typical Capacity of a Cryopump for Various Gases

Gas	Pumping speed (relative to air)	Capacity (Torr-L) (assuming a 12-in pump)
H_2O	2.6–4.0	1×10^6
H_2	1.1–1.4	5×10^3
N_2, O_2	1	2×10^6
Ar	0.7–0.9	2×10^6
He	0.3–0.6	8×10^2

5.4. Turbomolecular Pumps and Molecular Drag Pumps

Another type of pump capable of pumping into the UHV range is the turbomolecular pump. Like the cryopump, this pump also has a high throughput. It has the advantage over the classical diffusion pump in that there is no working fluid which might contaminate the vacuum chamber. It does have moving mechanical parts with bearings, however. Again, the vacuum technology of turbopumps has been known for a long time. The development of bearings and other materials which could withstand the rigorous requirements allowed the economical development of turbomolecular pumps in the 1970s.

The basic pumping mechanism involves a disk with a series of blades, as depicted in Fig. 12. The disk rotates at a very high rate, so the blades move at speeds which are comparable to the speed of the molecular species to be pumped. Typical rotation rates are 50,000 rpm. The molecules are driven in the desired direction, setting up a pressure gradient from one side of the disk to the other. (A more complete description of the mechanism of pumping is given in Ref. 2.) Several disks are mounted on a rotor which is mounted on the axis of a stator consisting of stationary blades, shown schematically in Fig. 13. The stationary blades are slanted

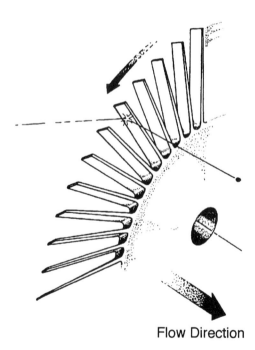

Flow Direction

Figure 12. Pumping mechanism of a turbomolecular pump. (After Tompkins[2])

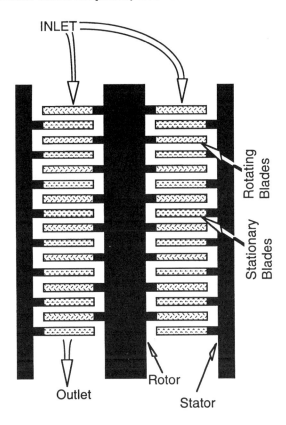

Figure 13. Rotor and stator of a turbomolecular pump, shown schematically. (After Tompkins[2])

at opposite angles to the rotor blades, as indicated in Fig. 14. At the inlet part of the pump, the openings between the blades are large, allowing molecules at a low pressure to pass through. At the outlet part of the pump, where the pressure is considerably higher, the openings are smaller and the blade angles sharper, to support a larger pressure gradient.

The optimum operation of the turbopump is when the mean free path of the molecules is at least 10 times the slot dimensions. This criterion means that the pump does not work at atmospheric pressure at the entrance of the pump; hence a roughing pump is required. In addition, the pump does not work when the pressure at the outlet of the pump is at atmospheric pressure; hence a forepump is required. (A forepump is often a rotary-vane oil-sealed mechanical pump attached to the outlet of the turbopump.) The need for a roughing pump and forepump reduces the cleanliness benefit slightly (although cleanliness is not affected when the pump is operated properly). Figure 15 shows a typical pumping speed curve. The curve for

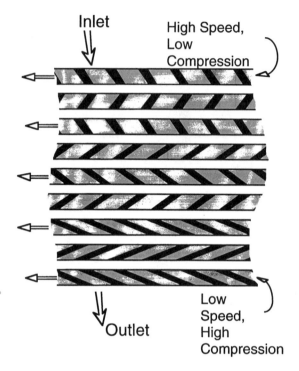

Figure 14. End view of the rotating disks and the stationary disks. (After Tompkins[2])

the lighter gases (H_2 shown, and He not shown) lies below the curves for most of the other gases, which do not significantly differ from each other.

A turbopump acts somewhat as a "pressure demagnifier" in that the ultimate pressure of a turbomolecular pump depends on the pressure in the foreline. A typical turbopump backed by a mechanical pump with an ultimate pressure of 100 mTorr might have an ultimate pressure of 10^{-7} Torr. The ultimate pressure of the same turbopump would be 10^{-8} Torr if the forepump could pump to 10 mTorr, and 10^{-9} Torr if the forepump could pump to 1 mTorr. We shall discuss this aspect more extensively later.

Early in the history of turbopumps, a pump was stopped by coasting it down under vacuum, which took about half an hour. Today's turbopumps are stopped rather quickly (a few tens of seconds) by exposing them to atmospheric pressure. Some systems are coasted down to three quarters of full speed, and then opened.

This situation leads to a rather interesting feature of turbopumps which is different from all other UHV pumps. They can, for a short time, be exposed to atmospheric pressure without damage. Thus, small chambers can be rough-pumped

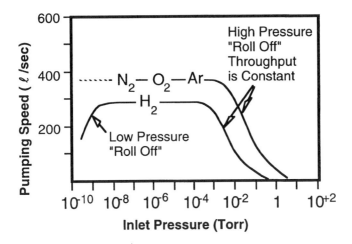

Figure 15. Pumping speed for several gases versus pressure for a typical turbomolecular pump. (After Tompkins[2])

through the turbopump. When the pressure is above about 100 mTorr, the turbopump is not doing any pumping (in fact, it is in the way), and it usually slows down during this period as gas moves into the roughing pump and the pressure is reduced. When the pressure is less than about 100 mTorr, the turbopump will increase its rotation rate to its normal operating value and begin pumping. The sequence from atmospheric pressure to high vacuum consists of only one operation rather than two (roughing and subsequent high vacuum pumping).

The ultimate pressure of the turbopump is a function of the pressure in the foreline. The ultimate pressure of a classic rotary-vane oil-sealed pump is no lower than 1×10^{-4} Torr, and this limits the ultimate pressure of a turbopump operated in this way to about 1×10^{-10} Torr. This can be improved by using the combination of a large turbopump backed by a small turbopump, which in turn is backed by a rotary-vane oil-sealed mechanical pump. This would result in significantly lower foreline pressures for the large turbopump and hence would allow a lower ultimate pressure for the system.

Another combination, which appeared on the market in the early 1990s, is a turbopump backed by a molecular drag pump. A molecular drag pump usually consists of a stator with spiral grooves like a helix, as shown schematically in Fig. 16. A cylindrical rotor rotates around the stator at speeds similar to those of the turbomolecular pump. As the molecules are passed back and forth from the rotating drum to the spiraling grooves, the molecules move from one end of the stator to the other. Most designs have stationary grooves facing both sides of the rotating drum (inside and outside).

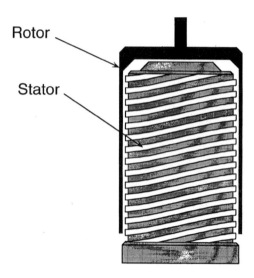

Figure 16. Basic design of a molecular drag pump. Cylindrical rotor around stator. (After Tompkins[2])

Molecular drag pumps have an ultimate pressure of about 10^{-6} Torr, but have the very useful feature that the outlet pressure can be as high as 10 Torr. This feature allows the drag pump to be backed by a diaphragm pump or other oil-less pump. The combination of turbopump, drag pump, and oil-less mechanical pump therefore totally eliminates any possibility of oil contamination. With the ultimate pressure of the drag pump at the outlet of the turbopump, an ultimate pressure of 10^{-11} Torr is possible. Since molecular drag pumps and turbomolecular pumps must operate at about the same rotation rate, some manufacturers have combined a turbopump with a drag pump.

One final comment concerning ultimate pressures is in order. The ion pump, cryopump, and turbopump (and the diffusion pump) are capable of pumping well into the ultrahigh vacuum range. Many applications can have ultimate pressures which are significantly short of ultrahigh vacuum. The reason for this situation is not that the pumps are inadequate, but that the fabrication and operating procedures are such that the gas load from desorbing species limits the ultimate pressure.

5.5. Roughing Pumps and Forepumps

Traditionally, roughing a chamber and backing a diffusion pump were done with a rotary-vane oil-sealed mechanical pump. The roughing function for diffusion pumps, turbopumps, and cryopumps requires that the pressure be reduced to a few hundred mTorr. At this pressure, the molecules are still moving in the viscous flow regime, flushing any oil contamination back into the mechanical pump. These mechanical pumps have ultimate pressures from 0.1 to 10 mTorr, which is in the molecular flow regime, where no flushing occurs. If a rotary-vane oil-sealed pump is used to rough a system to this latter pressure range, contamination of the main chamber could occur.

The system to be pumped by an ion pump must be rough-pumped to less than 1 mTorr. In the past, a cryosorption pump (not described here) was used for roughing purposes to avoid any possibility of oil contamination. Currently, ion-pumped systems often have introduction chambers which are pumped with turbopumps, and this turbo/backing pump is available for roughing the ion-pumped chamber. Because of this, cryosorption pumps are not used nearly as much as in the past.

To repeat a phrase used previously in connection with oil diffusion pumps, the rotary-vane oil-sealed mechanical pump is not inherently contaminating, but it is inherently unforgiving. Roughing with this type pump requires attention to detail to ensure that this pump is not attached to the vacuum chamber such that molecular flow from the mechanical pump can occur.

6. Pressure Measurement

It is important that we know the gas density in a surface analysis chamber, which we represent as "pressure." In the UHV range, the mean-free path of gas molecules is much larger than typical chamber dimensions, but the possible contamination of specimen surfaces by residual gases is of major concern. Accordingly, we require information on both the total amount and the identity of the residual gas species. These quantities are determined by measurement of the total gas pressure and by residual gas analysis.

Mechanical deflection and thermal properties of gases are used for pressure measurement in the range from atmospheric pressure down to about 1 mTorr. Below this range, the ionization of gases is used for quantitative measurement. Electrons are generated by a thermal source and are accelerated to ionize some of the gas molecules. Some of these ions are collected and the ratio of the ion current to the electron current is then a measure of the gas density and, hence, the pressure.

The most commonly used gauge is the Bayard–Alpert gauge, shown schematically in Fig. 17. The electrons are generated by the filament and attracted to the grid. The grid structure is such that electrons pass through the openings and oscillate back and forth before eventually striking the grid and being removed from the process. Ions which are created are attracted to the ion collector in the center.

With this somewhat inverted geometry (see Ref. 3 for description of the normal triode and the X-ray limit), the lower limit of detection is typically about 10^{-12} Torr.

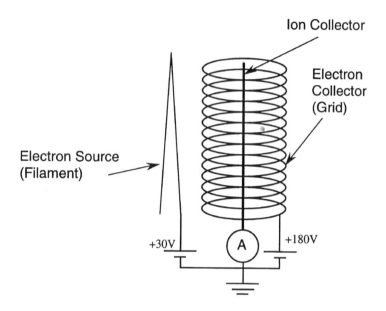

Figure 17. Bayard–Alpert ionization gauge, shown schematically. (After Tompkins[3])

Because of mean-free path considerations and oxidation phenomena, the upper limit for this type gauge is about 10^{-3} Torr. By using different geometries and different filament materials, the upper limit can be extended to as high as 1 Torr, with corresponding increase in the lower limit.

Different gas species are ionized with different efficiencies, and hence the ion current obtained for a given pressure depends on the gas species present. Calibration is normally done with N_2. The ionization efficiencies of most gases have been measured, and the calibration can be changed from N_2 to another gas such as Ar. This feature is useful when essentially all of the residual gas is one species. Normally, however, the residual gas is a mixture of species, and hence the N_2 calibration is commonly used for simplicity.

The total pressure is a useful quantity in following the pumpdown and subsequent processes. The amount of contamination, however, depends far more on the identity of the gas species than the quantity. A pressure of 10^{-5} Torr of argon may be completely benign, whereas the same pressure of O_2 might be disastrous to the process being conducted in the chamber. Accordingly, it is often necessary to determine the identity of the residual gas molecules. This type of measurement is called *residual gas analysis* or *partial pressure analysis* and the device for performing the measurement is called a residual gas analyzer (RGA).

Residual gas analysis is similar to total pressure analysis with the added feature that the gas species are separated and only one species with a given mass-to-charge ratio is measured at a given time. The intensities of ions with different mass-to-charge ratios are measured and presented as either an analog plot or a bar graph.

The quadrupole mass filter is used almost universally for mass separation in residual gas analysis. The structure consists of four rods (or poles) mounted as indicated in Fig. 18. The ions are injected down the axis of the structure. An rf field is superimposed on a dc field in such a way that the ions move in complex oscillatory paths. Most of the ions strike one or the other of the two sets of rods and are neutralized. Only the ions which have the chosen mass-to-charge ratio are able to traverse the entire length and be detected.

As a vacuum system is pumped down from atmospheric pressure, the N_2 and O_2 molecules are rapidly pumped, and for pressures below a mTorr, the main remaining species is water vapor. When the system is pumped into and below the 10^{-8} Torr range, the predominant species are H_2, CO, and CO_2. Figure 19 shows a typical residual gas spectrum for a clean ion-pumped system in the low 10^{-8} Torr range and the same system at a slightly higher pressure, when contaminated with water vapor and acetone. The mass peaks at masses 43 and 58 represent acetone. The mass peaks at 18 and 17 represent water vapor. Masses 28 and 44 are CO and CO_2, respectively. Masses 4, 20, and 40 are residual amounts of He, Ne, and Ar. The largest peak in the clean system is H_2 at mass 2.

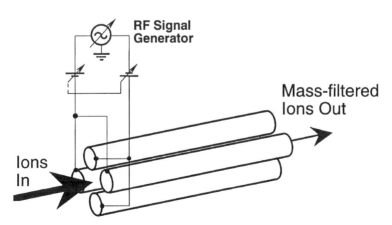

Figure 18. Schematic of a quadrupole mass separator for residual gas analysis. (After Tompkins[3])

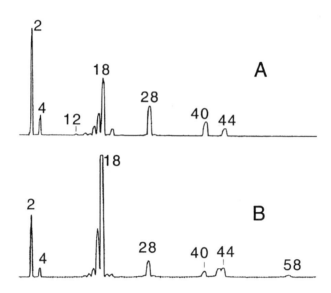

Figure 19. (A) Typical mass spectrum of an ion-pumped system in the low 10^{-8} Torr range. (B) Spectrum of ion-pumped system at a slightly higher pressure than that depicted in (A) which is contaminated with water vapor and acetone. (After Tompkins[3])

References

1. H. G. Tompkins, *The Fundamentals of Vacuum Technology*, 2nd ed., AVS Monograph Series, M-6, American Vacuum Society, New York (1991).
2. H. G. Tompkins, *Pumps Used in Vacuum Technology*, AVS Monograph Series, M-9, American Vacuum Society, New York (1991).
3. H. G. Tompkins, Vacuum Gauging and Control, AVS Monograph Series, M-12, (American Vacuum Society, New York (1994).
4. J. F. O'Hanlon, *A User's Guide To Vacuum Technology*, 2nd ed., Wiley, New York (1989).
5. A. Roth, Vacuum Technology, North-Holland, Amsterdam (1976).
6. G. L. Weissler and R. W. Carlson (eds.), *Vacuum Physics and Technology, Methods of Experimental Physics*, Vol. 14, Academic Press, New York (1979).

2

Cryogenic Sample Transfer for Preservation of Surface Chemistry

Leonard B. Hazell

1. Introduction

The chemistry at solid–liquid interfaces is extremely important in a wide range of industrial technologies ranging from cleaning and electroplating to corrosion and mineral ore processing. Preparation of analytical test pieces in solution is relatively straightforward, but there are no routine *in situ* techniques capable of providing molecular-layer-level analysis of the reaction products formed at the liquid–solid interface. To obtain the level of chemical analysis detail available from established techniques such as X-ray photoelectron spectroscopy (XPS) and secondary-ion mass spectrometry (SIMS), it is necessary to transfer the sample from the liquid environment to ultrahigh vacuum (UHV). This chapter describes a method based on cryogenic techniques which is designed to minimize the changes in surface chemistry that might occur during the transfer. The utility of the approach is illustrated with reference to examples in corrosion and dissolution.

1.1. The Problem of Sample Transfer from Liquids

Understanding the chemical reactions which occur at an iron surface in aqueous saline solution containing additives intended to inhibit corrosion is a key industrial problem that represents an archetypal system for study. Dedicated environmental

Leonard B. Hazell • CSMA (South) Ltd, Middlesex TW16 5DB, United Kingdom.

Specimen Handling, Preparation, and Treatments in Surface Characterization, edited by Czanderna *et al*. Kluwer Academic / Plenum Publishers, New York, 1998.

chambers have been attached to XPS spectrometers with the intention of transferring samples without chemical change from electrochemical cells.[1] In this process the sample must be extracted from the aqueous solution and dried, thereby losing volatile compounds from the surface. A washing step may be included to avoid drying stains formed from both dissolved solids and any "surfactant" film at the liquid–gas interface through which the sample must be extracted. It is, of course, inevitable that any applied electrical polarization will be disconnected, that the electrical double layer at the solid–liquid interface will be destroyed, and physisorbed species will be lost.

The problem of sample transfer from liquids may seem daunting, but there are some factors in our favor, particularly when dealing with adsorbed organic films. For example, successful corrosion-inhibiting films on iron tend to produce hydrophobic surfaces due to the orientation of chemisorbed molecules. The outermost atoms are usually hydrocarbon groups which repel polar water molecules. Since the interaction of this type of surface with water is minimal, the sample can be washed, at least briefly, without fear of destroying the adsorbed film. Also, the films formed are electrically insulating, so the galvanic corrosion circuit is already broken. There is, therefore, no intrinsic reason why UHV surface science techniques cannot be used for chemical analysis of these systems.

The analytical problem arises because commercial inhibitor formulations are often blends of impure organic compounds, and it is important to identify which is most active in any given situation and on what type of surface. Chemical changes, e.g., oxidation, which occur during sample transfer make this a much harder task.

2. Experimental Methods

The options for sample transfer involving highly reactive surfaces produced in solution are limited by the rate of reaction at ambient conditions. Glove box techniques are an attempt to slow the reaction rate by reducing the partial pressure of active gaseous species. An alternative approach is to cool the sample to cryogenic temperatures.

2.1. Glove Box Techniques

It is possible to use glove box techniques in an attempt to reduce the severity of oxidation, hydrolysis and/or contamination prior to evacuating the sample in the spectrometer air lock. However, the loss of volatile material from the sample is just as unavoidable in an inert-gas-filled glove box as it is in vacuum. Valuable results have been reported using glove box techniques to transfer activated (reduced) catalysts[2] and to prepare reference iron oxides.[3] However, extreme precautions are required to reduce the background level of oxygen in a glove box to much below

~1 ppm. This corresponds to a partial pressure ~10^{-3} Torr, easily sufficient to generate monolayer coverage in a millisecond on a reactive metal surface. Furthermore, sample preparation is restricted in a glove box, and special vessels are needed for transfers between it and the spectrometer, unless attached directly. These factors discourage routine use of glove boxes, unless imperative, because the benefit does not compensate for the amount of effort involved.

2.2. Cryogenic Sample Transfer Technique

Although on initial examination, cryogenic sample transfer may seem an unlikely method for transferring "wet" samples to UHV, careful selection of a suitable strategy can produce extremely successful results. The method, termed the SPLINT technique (sample preservation using liquid nitrogen transfer),[4] is highly compatible with UHV and can be used to obtain evidence of solid–liquid interface reaction chemistry confused relatively little by the effects of sample transfer. It would be desirable to transfer a reactive metal, say, with no oxidation at all, but a more practically realistic aim is to restrict surface oxidation during sample transfer to submonolayer levels.

The SPLINT approach is based on removing the sample from the liquid, after the surface chemistry has been established, and plunging it immediately into liquid nitrogen. The film of liquid on the sample provides a limited barrier against atmospheric oxidation during the brief transfer. To reduce the chances of drying stain formation, a rinse step in distilled water or organic solvent, for example, can be included prior to freezing, although this may not always be acceptable for the reactions under investigation. The speed at which the whole process is carried out is all important, but, once in the "frozen" state, submerged in a nonreactive medium at ~77 K, the sample may be "preserved" almost indefinitely. For samples constructed of more than a few micron-thick multilayers, there is the possibility of mechanical damage caused by sudden differential thermal contraction, but this is unlikely to be a problem for typical metal samples with submicron-thick oxides and/or monolayer coverages of organics. However, hydrophilic samples removed from aqueous solutions are likely to become encapsulated in a film of ice, e.g., from the frozen residual rinse solution.

The difficulties arise when the sample is removed from the liquid nitrogen to be placed in the spectrometer fast insertion lock and pumped down from atmosphere. An encapsulating ice film protects against ingress of reactive gases but must not be allowed to melt and restart reaction once in the UHV chamber. The advantage of the transfer method is lost if the whole process has to be via a glove box and the sample warms up before UHV conditions are reached. In practice, both problems can be easily avoided with an appropriate strategy.

While under the liquid nitrogen, the sample with any ice encapsulant is placed in an oversized (deep) boat and this is removed, filled with liquid nitrogen covering the sample, to the fast insertion lock. If the transfer is carried out rapidly, only a small amount of atmospheric moisture condenses as ice on the liquid nitrogen surface before evacuation begins. This condensation ice is blasted off by the volatilizing liquid nitrogen. Any ice film encapsulating the sample itself slowly sublimes, without melting, as the sample warms up to a temperature where the vapor pressure of water exceeds the ultimate pressure of the pumps. Initially, the latent heat of vaporization of the nitrogen contributes additional cooling of the sample, to such an extent that solid nitrogen can be formed if the pumping speed is too great. Pumping the evaporating liquid nitrogen has no detrimental effect on turbo or rotary pumps although, to avoid the possibility of adsorption of backstreaming rotary pump oil, use of an oil-free cryosorption pump is recommended.

The sample should be maintained at cryogenic temperatures for as long as possible, preferably until UHV conditions are achieved. This period can be increased by careful design of the sample mount such that it is in extremely poor thermal contact with the surroundings to reduce heat conduction. Sample warming by convection is avoided in the vacuum, so only radiation between bodies at a temperature differential of ~220 K is significant. Experience suggests that within a matter of minutes the sample can be transferred to the UHV conditions of the preparation or analysis chamber, but the sample can take hours to warm up to a temperature where it again becomes reactive. With suitable sample holder design, it should, if necessary, be possible to transfer to a cooled sample holder before this occurs.

3. Application to Corrosion of Mild Steel

The intention of this illustrative study was to define the adsorption mechanism of a corrosion inhibitor on an iron surface and to provide convincing evidence of the interaction chemistry for use in product development and molecular modeling of the solid–liquid interface. First, however, it was necessary to show the capability of the SPLINT technique and verify that the interpretation of the results was not likely to be significantly influenced by sample transfer to the spectrometer. Determining how much of the metallic state is present on the surface of acid-etched iron was thought to be both a realistic test of the method and a good system for developing the sample transfer strategy.

3.1. Sample Preparation

For this demonstration, a mild steel coupon was etched in 10% analar-purity hydrochloric acid for 1 min. It was then removed with stainless steel tweezers, rinsed for ~1 s by agitating in distilled water, removed and dropped immediately into liquid nitrogen before being transferred using the SPLINT technique. The whole process was carried out in the open laboratory, although both liquids were purged with nitrogen-gas bubblers, to remove dissolved oxygen, for several minutes before, and continuously during, the experiment. The sample was placed in the preparation chamber within approximately 5 min, pumped down to 5×10^{-7} Torr in 10 min, and left overnight at 5×10^{-9} Torr. In this experiment, the rough pumping speed was limited so that the nitrogen boiled off as liquid and a thin film of ice remaining on the sample could be observed to sublime in bursts, starting from the warmer outer edges and reaching the center after about 15 min. Analysis was carried out in 1 hr at 5×10^{-9} Torr in a VG Scientific ESCALAB using Al Kα radiation at 300 W and with a pass energy of 50 eV.

3.2. XPS Results

The XPS spectra in Figs. 1a–c are the Fe $2p$ doublet and the C $1s$ and O $1s$ peaks detectable on the acid-etched mild steel sample surface. The first column in Table 1 gives the quantified composition derived using Shirley background subtraction[5] with the quoted sensitivity factors applied to the accepted formula for a homogeneous sample.[6] The fact that the overwhelming majority of the iron can be assigned to the metallic state vindicates the SPLINT approach. There is barely a trace of oxide oxygen, but approximately half a monolayer of chlorine is present

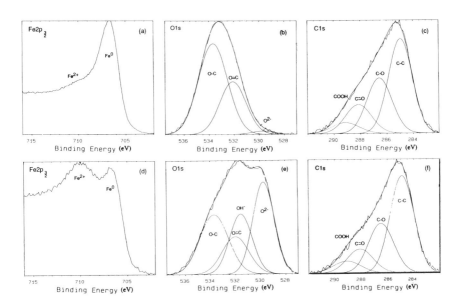

Figure 1. (a) Fe $2p$, (b) O $1s$, and (c) C $1s$ XPS spectra from acid-etched mild steel transferred to the spectrometer using the SPLINT technique; (d) Fe $2p$, (e) O $1s$, and (f) C $1s$ spectra obtained on reanalyzing the same sample after exposure to air for 1 min.

Table 1.   Cryogenic Sample Transfer Comparison Data

	Sample treatment			
	Acid-etched SPLINT transferred	Acid-etched 1-min air exposed	As received degreased	Argon-ion cleaned in UHV
Species	Sample composition from XPS (at%)			
Total C	36.3	35.9	30.4	9.5
Carbon fit index[a]	1.6	1.6	1.8	2.3
Degraded carbon	0.0	0.0	0.0	3.2
C–H or C–C	17.7	18.6	16.2	4.7
C–O	10.6	9.8	6.8	1.0
C=O or O=C–O	5.7	4.9	6.2	0.4
COOH	2.3	2.6	1.2	0.2
Total oxygen	17.0	30.4	44.9	11.1
Oxygen fit index[a]	1.2	1.3	0.8	1.0
Oxide	0.3	10.0	23.4	7.1
OH^-	0.0	6.8	9.0	2.1
O=C	5.7	5.0	4.5	1.3
O–C	11.0	8.6	8.0	0.6
Iron	42.8	30.0	24.2	76.2
Chlorine	4.2	3.9	0.2	0.2

[a]Fit index is unweighted rms% deviation between fit and data expressed as a percentage of the maximum peak height above the Shirley background.
[b]Sensitivity factors used: C1s (0.25), O1s (0.66), Fe 2p (1.1), Cl 2p (0.8).

as chloride. The small amount of Fe^{2+}/Fe^{3+} present is probably associated with the chloride, but it is difficult to distinguish credibly in the skewed tail of the metallic Fe $2p$ peak. There is clearly scope for more detailed examination of this type of sample using monochromated X-rays.

The O $1s$ peak has a trace of oxide component, possibly as oxychloride. The component at 531.5 ± 0.2 eV in Fig. 1c is characteristic of hydroxide, alcohol, or carbonyl (double bond) oxygen species, whereas that at 532.9 ± 0.2 eV is most likely organic ester or etheric (single bond) oxygen species. The C $1s$ peak indicates a high level of functionality, presumably associated with these organic oxygen components. The adsorbed material is thought most likely to be an organic con-taminant of the analar-purity-grade hydrochloric acid, but, since such a surface has not been encountered before, precise assignments are uncertain. Static SIMS analysis would complement the XPS results and better define the molecular species involved.

The amount of oxidation of the iron is barely detectable in comparison to either the posttreatment oxidation reaction experienced prior to using the SPLINT tech-nique of sample transfer or in earlier trial results where a majority of oxide species was obtained because the acid and rinse solutions were not properly deoxygenated during the experiment. These results indicate the great potential of the cryogenic method of sample transfer from solution to reveal differences in solid–liquid interface chemistry. It also highlights and warns of the potential for low levels of (possibly unsuspected) dissolved contaminants to become involved in reactions at interfaces.

To demonstrate the magnitude of the decrease in sample-transfer-induced oxidation, the spectra in Figs. 1d–f were obtained on taking the acid-etched sample after analysis into the fast insertion lock, leaking it up to atmosphere using nitrogen (from liquid nitrogen), and opening the door to air for 1 min before reevacuating and repeating the analysis. Growth of an oxide component is clear in both the Fe $2p$ and the O $1s$ peaks. The composition data in the second column of Table 1 shows that approximately 1.0 nm of oxide has grown and that this is probably a mixture of Fe_2O_3 and FeOOH. This amount of posttreatment reaction would be unacceptable and would adversely affect any interpretation of the results in terms of the mechanism of chemical reaction at the solid surface in aqueous solution. Clearly, transfer of the sample using the SPLINT technique prevents formation of this layer on the active surface.

3.3. Comparison with Argon-Ion-Cleaned Mild Steel

It is instructive to compare the results obtained by acid etching of iron with those from *in situ* cleaning in UHV using argon-ion bombardment (AIB). This is the usual method to prepare clean metal surfaces for experiments in the spectrometer under controlled conditions. The spectra of Figs. 2a–c are of a sample of the same mild steel as received, and those of Figs. 2d–f are after AIB cleaning in the spectrometer analyzer chamber. The quantified compositions are given in the third and forth columns of Table 1, respectively. A VG AG2 ion gun was used at 5 kV, 12 μA emission current, 45° incidence angle, with 5×10^{-4} Torr pressure of argon.

This sample shows a low level of mainly oxide component in the O 1s spectrum but no clear indication of oxide component in the Fe $2p$ spectrum. Much of the organic contamination has been removed, and the functionality of the residual carbonaceous material has become nonspecific due to the degradation introduced by the ion etching process. Residual inorganic oxide is present on the surface, almost certainly arising as a consequence of surface-roughness-induced shadowing effects during ion cleaning. In contrast, hydrochloric acid etching produces a surface where inorganic oxide is well removed, some chloride is formed, and organic contamination is preserved intact.

3.4. Corrosion Inhibitor Adsorption

Having established that cryogenic sample transfer preserves adsorbed organic molecules, this series of experiments was intended to demonstrate how the method

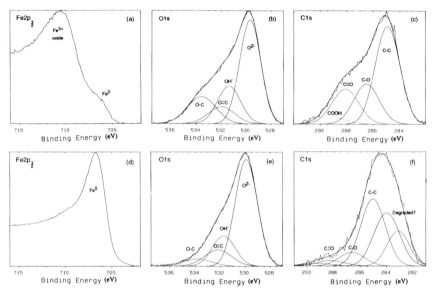

Figure 2. (a) Fe $2p$, (b) O $1s$, and (c) C $1s$ XPS spectra from "as-received" mild steel and (d) Fe $2p$, (e) O $1s$, and (f) C $1s$ spectra obtained after extensive argon-ion bombardment (AIB) cleaning *in situ*. Note that AIB degrades the organic carbon to produce a broad peak shape encompassing a variety of nonspecific reduced functionalities.

can be used to elucidate the mechanism by which an organic corrosion inhibitor functions on a mild steel surface in aqueous CO_2-saturated saline solution at pH 4. Corrosion of mild steel in aqueous saline solution is an oxidation process which is greatly exacerbated both in acidic conditions and in the presence of oxygen. The formation of iron oxides as "rust" is the well-known consequence.

The theoretical molecular structure of fresh oleic imidazoline inhibitor is shown in Fig. 3. The undiluted fresh commercial product is impure and is a liquid which becomes a waxy solid after long-term storage. The expected C_{Total}/N ratio is ~7, and the N $1s$ peak should have one component for the double-bond-ring nitrogen and another, of twice the intensity, for the ring and pendant amine functionalities. The quantitative composition determined by XPS analysis of the vacuum-dried undiluted inhibitor in both its fresh liquid and its aged solid condition are given in the fourth and fifth columns of Table 2, respectively. The N $1s$ spectrum of the fresh liquid shows evidence of the nitrogen ring structure, but after long-term storage the N $1s$ spectrum is characteristic of amine/amide, indicating that the imidazoline ring is sensitive to hydrolysis to the amide. There is also evidence for ethoxylated surfactant compounds present, either added deliberately or as an impurity.

Electrochemical inhibitor performance measurements indicate that adding the inhibitor at ~50 ppm concentration under deoxygenated conditions reduces the corrosion rate to ~100 μm per year, but in the simultaneous presence of oxygen the

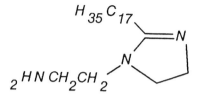

Figure 3. Structure of oleic imidazoline.

rate is as high as ~2.3 mm per year. Performance measurements where oxygen is admitted after the inhibitor has reacted are not normally carried out because it is not considered an important operational situation.

For the inhibitor adsorbed from aqueous solution, analysis by conventional XPS, without the precaution of cryogenic sample transfer, indicates only one amine/amide-like $1s$ component. This result suggests that the inhibitor undergoes hydrolysis in aqueous solution and behaves as an amide adsorbing onto oxidized iron.[7] There may be commercial implications in this result because imidazoline is

Table 2. Atomic Percentage Composition Of Mild Steel Samples Treated in Aqueous Saline Solution at pH 4 with 100 ppm Oleic Imidazoline as a Function of Oxygen Addition

Species	Mild steel treated with oleic imidazoline (OI) in aqueous NaCl at pH 4			Oleic imidazoline concentrate	
	O_2-free	O_2 after OI	O_2 with OI	Fresh (liquid)	Aged (solid)
Total carbon	**44.8**	52.7	**58.6**	**86.6**	**88.5**
C–H	37.5	41.9	47.0	65.6	71.2
C–N				17.5	12.8
C–O	3.9	6.5	6.8		
N–C=O	3.0	2.9	3.8	3.5	4.5
O=C–OH	0.4	1.4	1.0		
Total oxygen	**8.9**	12.7	**19.8**	**4.6**	**4.3**
Oxide	1.3	3.3	7.2		
OH–/O=C–N	5.6	7.7	9.9	3.6	4.3
O–C	2.0	1.7	2.7	1.0	
Total nitrogen	**2.2**	**3.4**	**5.1**	**8.8**	**7.2**
N–C=O	2.2	3.4	5.1	6.4	7.0
C–N=C				2.4	0.2
Total iron	**42.6**	30.2	**15.1**		
Fe^0	42.6	30.2	6.1*		
Fe^{3+}			9.0*		
Chlorine	1.7	1.1	1.4		
C_{Total}/N ratio	20.4	15.5	11.5	9.86	12.3

*Estimated

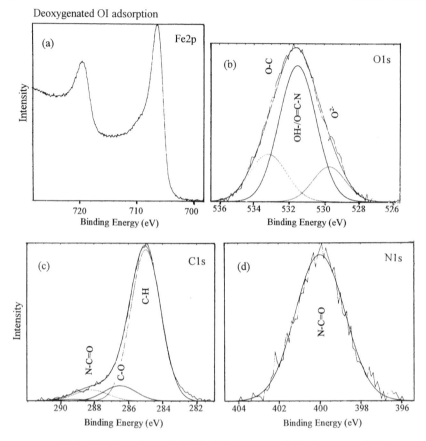

Figure 4. (a) Fe $2p$, (b) O $1s$, (c) C $1s$, and (d) N $1s$ XPS spectra corroded in deoxygenated, CO_2-saturated aqueous saline solution containing oleic imidazoline at pH 4 and transferred to the spectrometer using the SPLINT technique.

more expensive to manufacture than the amide. Furthermore, based on the acid-etched iron studies reported earlier, exposure of a metallic iron surface would be expected to occur at pH 4, but, due to air transfer oxidation, little evidence for this is present in the spectra acquired in the conventional way.

The Fe $2p$ spectrum in Fig. 4a was obtained from a mild steel coupon exposed to the continuously CO_2-purged deoxygenated saline solution for several minutes before the inhibitor was introduced at ~100 ppm concentration. The sample was then removed and transferred using the SPLINT technique. The first column of Table 2 gives the quantified surface composition derived from the XPS analysis. Interpretation of the result as an adsorbed organic layer on the mainly metallic iron surface is relatively straightforward. There is ~1.7 at% chloride and a small amount, ~1.3 at%, of oxide detectable in the O $1s$ spectrum (Fig. 4b), but a corresponding oxidized iron component is not resolvable in the Fe $2p$ peak shape. The exclusive presence of the amine/amide functionality is confirmed by a single N $1s$ peak (Fig. 4d), and the carbonyl/amide carbon functionality (Fig. 4c). Etheric carbon and oxygen species from the added surfactants or impurities are also adsorbed, leading to a much higher C_{Total}/N ratio than would be expected from either the theoretical stoichiometry or the value obtained from XPS analysis of the vacuum-dried raw products, also given in Table 2.

An almost identical Fe $2p$ spectrum (Fig. 5a) was obtained from a sample prepared similarly but with oxygen being deliberately added for 15 min after the inhibitor had reacted. The amount of oxide component in the O $1s$ spectrum (Fig.

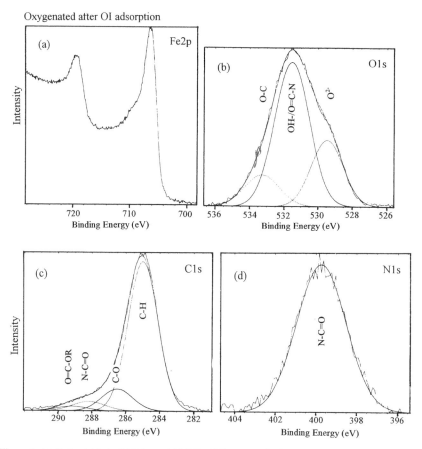

Figure 5. (a) Fe $2p$, (b) O $1s$, (c) C $1s$, and (d) N $1s$ XPS spectra of mild steel corroded in deoxygenated, CO_2-saturated aqueous saline solution containing oleic imidazoline at pH 4 and then exposed to oxygen in solution before transferring to the spectrometer using the SPLINT technique.

5b) has increased to ~3.3 at%, as indicated in the second column of Table 2, and there is a barely detectable corresponding oxidized iron component in the Fe 2p spectrum. This result provides direct evidence that the adsorbed film on the metallic iron surface substantially inhibits further reaction, even when solution corrosivity is severe in the presence of oxygen. To the author's knowledge, such an unequivocal demonstration of the effectiveness of a corrosion inhibitor mechanism in operation has not been reported previously.

In contrast, the Fe 2p spectrum in Fig. 6a was obtained from a sample prepared when a low level of oxygen was bled in with the CO_2 before the inhibitor was introduced. In this case, the electrochemically measured corrosion rate is greater, ~2.3 mm per year, and there is evidence in the spectrum of substantial oxidation having taken place. The fact that the levels of oxidation and inhibitor adsorption

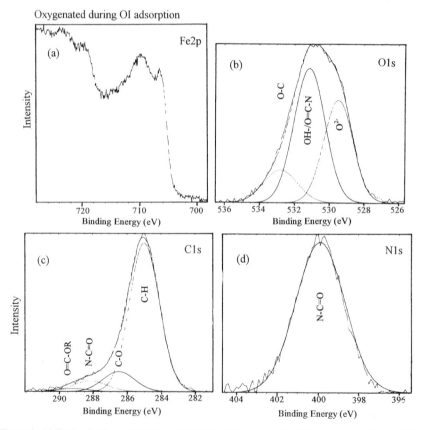

Figure 6. (a) Fe 2p, (b) O 1s, (c) C 1s, and (d) N 1s XPS spectra of mild steel corroded in partially oxygenated, CO_2-saturated aqueous saline solution containing oleic imidazoline at pH 4 and transferred to the spectrometer using the SPLINT technique.

indicated by XPS analysis are greater, as reflected in the nitrogen composition in the third column of Table 2, suggests that adsorption onto iron oxide "rust" is greater than on the metallic surface. However, the molecular conformation is a form which cannot be as effective as an inhibitor. This is an important result because it implies that corrosion rate reduction is not simply related to inhibitor adsorption. It is clear that full understanding of corrosion inhibition requires a detailed chemical knowledge of interfacial reactive adsorption processes.

4. Application to Corrosion of Aluminum

Aluminum is also industrially important but is an intrinsically more reactive metal than iron. However, compared to iron, objects manufactured from aluminum are better protected by the surface oxide formed in air. The aluminum oxide growth mechanism at elevated temperatures is known to be via oxygen anion diffusion into the metal, whereas for iron it is by outward diffusion of iron cations.[8] In aqueous acid solution, both aluminum and iron evolve hydrogen, and a soluble salt is formed. The SPLINT results on acid-etched iron indicate that a metallic iron–liquid interface exists in solution, but there is no published direct evidence for how the reaction proceeds on either iron or aluminum. The absence of data is probably a reflection of the lack of *in situ* techniques to detect and define the chemistry at the reacting solid–liquid interface. There is considerable commercial interest in surface modification of aluminum alloys, and, therefore, there is a need for analytical evidence which can be used to deduce fundamental mechanistic information on the dissolution behavior of aluminum in aqueous acidic solutions. This section describes the potential of UHV surface analytical techniques used in combination with cryogenic sample preparation and transfer methods for reactive materials.

The Al $2p$ spectrum in Fig. 7a is typical of a sample of aluminum etched in ~20% v/v hydrochloric acid, deoxygenated using nitrogen-gas bubblers, rinsed in deoxygenated isopropyl alcohol, and transferred to the spectrometer using the SPLINT technique. The result is not affected by substituting a rinse in deoxygenated distilled water. The oxide is significantly thicker than the ~2.5 nm that would be obtained on untreated aluminum cooking foil, for example. In contrast, the spectrum in Fig. 7b shows the Al $2p$ peak obtained from an aluminum sample which had been abraded and transferred under liquid nitrogen to the spectrometer. The level of oxide is much lower than on the aqueous-etched aluminum and represents a surface region that is more substantially metallic.

These results do not suggest that there is a problem with the cryogenic sample transfer technique but that there is an oxide–hydroxide formed on aluminum in aqueous acidic conditions that is genuinely thicker than the oxide that would be formed in air. It may be that dissolution of aluminum occurs through this interfacial oxide–hydroxide film, in contrast to the situation with iron, where the metal

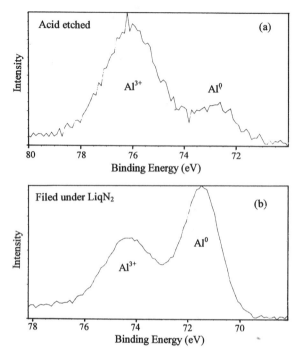

Figure 7. (a) Al 2p XPS spectrum of deoxygenated, ~20% v/v hydrochloric-acid-etched aluminum transferred using the SPLINT technique. (b) Al 2p XPS spectrum of aluminum sample abraded under liquid nitrogen and transferred to the spectrometer using the SPLINT technique.

dissolves directly into acid solution at the metal–solution interface. The implication is that the aqueous surface reaction mechanisms at room temperature parallel the known high-temperature oxidation behavior of these metals. Surface analysis combined with cryogenic sample transfer techniques has enabled a fundamental, unexpected, and as yet unconfirmed, discovery to be made about chemical reactivity at the solid–liquid interface.

5. Other Application Areas

In assessing the potential utility of cryogenic sample transfer techniques, we may address the problems of obtaining reliable standards or reference compounds in surface science. It is well known that most commercially available powders have surfaces that have been oxidized or hydrolyzed by atmospheric reaction during mechanical milling in production or in subsequent storage. These powders are inadequate as reference materials for metals in low oxidation states. To produce suitable reference compounds of metals in low oxidation states, a large single crystal could be crushed under liquid nitrogen and transferred to the spectrometer using the SPLINT technique.

In chemisorption experiments, there is a requirement for reference compounds that occur on surfaces as complexes but do not exist as pure stable compounds obtainable "in a bottle." Cryogenic sample transfer techniques offer a potential method to obtain monomolecular-layer reference samples of both chemisorbed organic compounds, such as carboxylic acid or amine salts, and inorganic compounds such as low-valence hydroxides, chlorides, phosphates, and sulfides.

The segregation of impurity elements in alloys has been widely studied using *in situ* liquid-nitrogen-cooled fracture stages in UHV.[9] On suitably designed test specimens, the fracture could equally well be carried out under liquid nitrogen in the open laboratory and the sample transferred to the spectrometer using the cryogenic technique. This procedure would avoid the need for brittle fracture stages and would facilitate batch sample analyses.

Finally, a potential application of cryogenic sample transfer techniques is in the collection of samples directly from the operating environment in an industrial plant, particularly in areas remote from analytical facilities. Once the practical problems of appropriate sample collection and transportation of cryogenic liquids have been overcome, samples could be sent for analysis without concern about chemical changes related to an uncertain sample history between collection and analysis. Clearly, not all samples will be amenable to this type of approach, but experience suggests that availability of this option would help bridge that "credibility gap" between the results from UHV surface analysis and what actually exists at the solid–liquid interface in solution.

6. Summary

This chapter has shown that the cryogenic technique of transferring samples from liquids to UHV surface analysis instruments is an extremely promising approach to overcoming the difficulty of providing credible interpretation of chemical reactions at solid–liquid interfaces. Evidence has been given to substantiate the accepted hypothesis that iron dissolves directly from the exposed metallic surface into aqueous acidic solution. Chemisorption of an organic (oleic imidazoline) corrosion inhibitor can occur directly on this surface, and the monomolecular layer film formed can be an effective barrier to further oxidation. However, it is the amide hydrolysis product that is adsorbed and not the original imidazoline. Where oxygen is present to compete with the inhibitor, iron oxy/hydroxides are formed and a higher level of inhibitor adsorption occurs, but the corrosion rate increases. This shows that inhibitor efficiency is not related directly to the level of adsorption, and it is important to understand fully the surface chemical interaction mechanism before attempting to predict the corrosion behavior of adsorbed films.

The SPLINT technique has been shown to be relevant to reactive materials such as aluminum. In this case, dissolution of the metal in aqueous acidic solution is proposed to be through an intermediate oxide–hydroxide layer that is thicker than would be found as an air-formed protective film.

The use of a suitable sample transfer strategy is paramount and manufacture of special sample holders is advised, but the results obtained allow a much more confident interpretation, so the effort is well rewarded. The SPLINT technique is worthy of consideration for preparation of reference materials, fracture surfaces, and samples which can only be produced under liquids. It also has potential in collecting samples from their operating environments and transferring them to remote laboratories for surface analysis.

Acknowledgments

Development of the SPLINT technique for cryogenic preservation of surface chemistry came about through application of lateral thinking to a perennial problem. Many of the author's colleagues at the BP Research and Engineering Centre contributed constructive ideas and practical ingenuity to make the approach successful and fun to develop. Special thanks are due to Richard A. Cayless, Carolyn J. Love, Dr. Kevin Harrison, Dr. Maurice Man, and, for the corrosion rate measurements, Dr. Simon Webster.

References

1. R. W. Revie, J. O'M. Bockris, and B. G. Baker, The Degraded Passive Film on Iron—An Application of Auger Electron Spectroscopy, *Surf. Sci.* **52**, 664–669 (1974).
2. E. S. Shapiro, V. I. Avaev, G. V. Antoshin, M. A. Ryashentseva, and K. M. Minachev, XPS Studies of the Rhenium State in Supported Rhenium Catalysts, *J. Catal.* **55**(3), 402–406 (1978).
3. C. R. Brundle, T. J. Chang, and K. Wandelt, Core and Valence Level Photoemission Studies on Iron Oxide Surfaces and the Oxidation of Iron, *Surf. Sci.* **68**, 459–468 (1977).
4. C. J. Love, R. A. Cayless, and L. B. Hazell, The SPLINT Technique for Sample Preservation using Liquid Nitrogen Transfer, *Surf. Interface Anal.* **20**, 565–568 (1993).
5. D. A. Shirley, High Resolution X-ray Photoemission Spectrum of the Valence Band of Gold, *Phys. Rev. B* **5**, 4709–4714 (1972).
6. L. E. Davis, N. C. MacDonald, P. W. Palmberg, G. E. Riach, and R. E. Weber, *Handbook of Auger Electron Spectroscopy*, 2nd ed., Physical Electronics Industries Inc., Eden Prairie, MN (1976); M. P. Seah, in: *Practical Surface Analysis by Auger and X-ray Photoelectron Spectroscopy* (D. Briggs and M. P. Seah, eds.), Wiley, Chichester (1990), pp. 181–216.
7. J. A. Martin and F. W. Valone, The Existence of Imidazoline Corrosion Inhibitors, *Corrosion-NACE* **41**(5), 281–287 (1985).
8. O. Kubaschewski and B. E. Hopkins, *The Oxidation of Metals and Alloys*, 2nd ed., Butterworths, London (1967).
9. J. P. Coad, J. C. Riviere, M. Guttmann, and P. R. Krahe, Reversibility of Temper Embrittlement Studied by Combined XPS and AES, *Acta Met.* **25**, 161–171 (1977).

3

Specimen Handling: Cleaning and Processing

Paul A. Lindfors

1. Introduction

When surface-sensitive analytical techniques are used, proper handling of specimens will improve efficiency and interpretation of data. Whenever possible, minimize handling of a specimen, and especially do not touch the surface area that will be analyzed. In addition to general rules for specimen handling, many specialized procedures have been developed that allow analyses which would otherwise be difficult or impossible. The list of specialized procedures is constantly expanding.

The methods of specimen handling are similar for all surface analysis techniques, but there are occasional differences, determined by either the analytical technique or the type of analytical system. When a method of handling is especially well suited, or ill suited, to a technique or system, this will be noted in the text. The analytical techniques to which this chapter is applicable are Auger electron spectroscopy (Auger), X-ray photoelectron spectroscopy (XPS or ESCA), secondary-ion mass spectrometry (SIMS), time-of-flight SIMS (TOF SIMS), ion scattering spectrometry (ISS), and other similar techniques.

Paul A. Lindfors • Department of Electrical Engineering and Electronic Engineering Technology, Mankato State University, Mankato, MN 56002-8400.

Specimen Handling, Preparation, and Treatments in Surface Characterization, edited by Czanderna *et al.* Kluwer Academic / Plenum Publishers, New York, 1998.

2. General Considerations

The degree of cleanliness required in surface analysis is often much greater than that required for other analytical techniques, because surface analysis techniques are sensitive to surface layers that range from only the outermost layer of atoms to layers that are a few nanometers thick. Such thin layers are subject to severe alteration if specimens are not handled properly.

Whenever possible, the surface to be analyzed should not be handled. When a specimen is contacted, even away from the area that will be analyzed, it should be done with care and preferably with clean tools. The tools should be of materials that will not transfer to the specimens and cleaned in high-purity solvents prior to use. Previous solvents of choice were methanol and Freon, but environmental problems with chlorofluorocarbons have prompted exploration of other less-damaging solvents, and alternatives have been identified and tested.[1-3]

Although gloves are sometimes used to handle specimens, it is likely that they will result in some contamination. In addition to transfer of bulk material, talc, silicone compounds, and other contaminants are often found on gloves.

Contact with specimens can often be minimized when the specimen mount is convenient for the specimen. Modification of normal mounts, or use of special fixtures, should be considered whenever normal specimen mounting is cumbersome. Special fixturing and alteration of mounts need not require exotic machining. Simple holes, slots, and blocks combined with some imagination are often all that is required.

Surface analysis may only be part of the characterization of a specimen. When other techniques are used, it is normally best to perform surface-sensitive analyses before applying them, because other techniques often contaminate the surfaces. For example, the environment in most scanning electron microscopes, together with the energetic electron beam, will cause a carbonaceous film to be deposited on the surface of a specimen.

There is also a preferred hierarchy among surface analysis techniques. The incident ion beams used in SIMS and ISS will definitely alter a specimen. The incident electron beam used in Auger is normally much more intense than the incident X-ray flux used in XPS. Therefore Auger is more likely to alter a specimen than XPS. XPS should be done first, Auger next, and SIMS or ISS last. Static SIMS analysis, where the density of the flux of incident ions is low enough that only a small percentage of surface atoms are sputtered, may represent an exception to the preferred order of analysis.

The analytical systems utilize ultrahigh vacuum conditions (total pressure < 1.33×10^{-6} Pa, or 1×10^{-8} Torr) in the analytical chamber. In addition, the analytical techniques will bombard the specimen surface with energetic electrons, or X-rays, and/or ions. Either the vacuum or the bombardment may cause specimens to deteriorate and contaminate the vacuum chamber and/or modify the surface of the

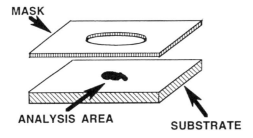

Figure 1. Masking a specimen to avoid charging adjacent to the area of analysis.

specimen. Organic and especially biological materials are usually the most sensitive to vacuum environments and bombardment. However, the vapor pressure of some elements can be a problem. Species that should be of concern if they are present in elemental form are Hg, P, S, Na, K, Cs, F, Cl, Br, I, Se, Zn, As, Cd, Rt, Po, and Te.[4] These elements are not usually encountered in elemental form except as decomposition products; and in compounds, these same species are often stable.

To eliminate the effects of possible charging of areas adjacent to the analysis area, one can mount the specimen under a mask of conducting material with a window exposing the area of analysis as shown in Fig. 1. (See Chapters 1 and 2 in Vol. 5 of this series for additional discussion of charging effects in Auger and XPS.) The ideal mask will provide uniform contact along all points at the edge of the mask window. Commercially available masks made of material having spring tension are excellent, but clean aluminum foil can also be used as a mask. If a specimen is wrapped in an envelope of aluminum foil, pockets of gases may be enclosed and may interfere with subsequent analysis. Great care must be exercised when using masks for SIMS analysis because fringe area effects of the incident ion beam and/or analyzer optics may be significant in the data acquired.

The American Society for Testing and Materials (ASTM)[5] has available recommended practices for handling specimens when using surface-sensitive analytical techniques. Another excellent source for information regarding specimen handling in surface analysis is the Applied Surface Science Division of the American Vacuum Society.[6]

3. Visual Inspection

Visual examinations should be done before and after analysis. Depending upon the dimensions of the specimen, a light microscope may be required. The visual examination prior to analysis is important to efficiently locate the area to be analyzed later when the specimen is in the analytical chamber. Following analysis, one should make another visual examination, noting especially possible effects of sputtering, electron beam radiation, X-ray radiation, heating, and vacuum exposure. The results of the postanalysis examination can be important for proper interpretation of data.

Not all features that are visually apparent will be detected in secondary-electron and/or absorbed current images. For this reason, it might not be possible to determine positively the desired location for analysis once the specimen is in the analytical chamber. In such cases, it may be necessary to scribe the surface adjacent to locations that are to be analyzed or to identify the locations by windows in a mask.

4. Removal of Particulate Debris

Compressed gases can be used to blow particles from the surface of a specimen. However, such gases must be considered as a source of possible contamination. Use only high-purity gases and pay particular attention to the cleanliness of any lines conveying the gas. Occasionally, a gas stream will produce static charge in specimens, and this could result in attraction of more particulate debris. Use of an ionizing nozzle on the gas stream should eliminate this problem.

A somewhat recent option for cleaning with aerosols utilizes high-velocity carbon dioxide "snow."[7-10]

5. Vacuum Conditions

Interaction between a specimen and the ultrahigh vacuum environment of the analytical chamber should be considered. A poor vacuum in the analytical chamber may result in contamination of the surface to be analyzed. Conversely, a specimen with high-vapor-pressure components could result in contamination of the analytical chamber.

The chemical activity of surface species is an important factor controlling the rate of possible contamination. A specimen that has previously been exposed to the atmosphere is not likely to be chemically active in the analytical chamber. On the other hand, specimens that are sputtered, fractured, cleaved, or scribed in the analytical chamber are often very active chemically. Sputtered gold is not particularly active. In contrast, sputtered specimens containing Mg, Al, Ti, Zr, etc., are very active. In addition to elemental species chemical compounds can also be chemically active on surfaces, but the list of such compounds is too long to include here. General knowledge of the reactivity of compounds is an indication of their reactivity on surfaces.

A second factor controlling the possible contamination rate is the composition of residual gases in the analytical chamber. There are always some chemically active gases remaining in the analytical chamber. The gases most likely to be present are hydrogen, methane, oxygen, water vapor, carbon monoxide, and carbon dioxide. For Auger and XPS, an increase in the signals for carbon and oxygen is a good indication that contamination is occurring. For SIMS, one might also detect an increase in the signal for hydrogen. Of course, an increase in signals from contamination on the surface would be accompanied by a decrease in signals from species being covered by the contaminants. Whenever a chemically active surface is present, strict attention must be paid to the vacuum conditions in the analytical chamber, and the time required for analysis is a critical factor.

A plot of vacuum chamber pressure versus the time for surface contamination is shown in Fig. 2. The diagonal band indicates a worst-case situation, wherein every gas particle that strikes a surface sticks to the surface or reacts with surface species. This condition is approximated by a combination of a chemically active surface and an active residual gas. At a pressure of 1.33×10^{-4} Pa (1×10^{-6} Torr), only 1 s would be required for a monolayer of contamination to accumulate on the

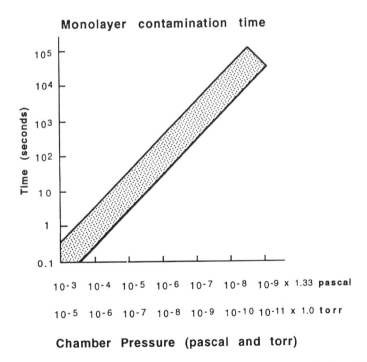

Figure 2. Approximate time to obtain one monolayer of contamination as a function of chamber pressure, assuming every molecule striking the surface sticks or reacts.

surface of a specimen. Reduction in the pressure of an active gas species will result in a corresponding reduction in the maximum possible contamination rate due to that species.

A specimen containing volatile materials may contaminate the analytical chamber. For analytical systems with rapid specimen-insertion probes, one can quickly check for outgassing by introducing the specimen and monitoring the pressure in the analytical chamber. If a rapid-insertion probe is not available, then pretesting may need to be done in a separate vacuum chamber. It is possible to make a specimen compatible with the required vacuum conditions in an analytical chamber by prepumping the specimen in a separate vacuum chamber for periods ranging from a few hours to several days. For most specimens that are prepumped, a brief exposure to air during transfer to the analytical chamber will not normally result in recontamination and continued outgassing. When present, the introduction chamber that is part of rapid-insertion probes can serve as the prepumping chamber. The incident electron, ion, and photon beams used in the analytical techniques may influence the interaction of the specimen and the environment of the analytical chamber (see Chapters 1–3, Volume 5 in this series). These incident beams may stimulate reactions between the surface of a specimen and the residual gases in the analytical chamber[11,12] or degrade the specimen, resulting in contamination of the analytical chamber. If there are questions regarding the effects of the incident beams, it is obviously prudent to begin any analysis by using incident beams of very low intensity. One can test for possible interaction effects between specimens, residual gases, and/or the incident beams by setting the analysis system as though doing a sputter-depth profile, but do not turn on the ion gun. The data obtained will be signal changes as a result of time of exposure to the vacuum environment and/or the incident beam.

Contamination by surface diffusion can sometimes be a problem, especially with silicone compounds.[13] It is possible to have excellent vacuum conditions in the analytical chamber and still have contamination by surface diffusion.

6. Analysis of Interfaces

Often the surface or interface of interest lies beneath a layer of contaminants or other constituents. The problem is then to remove the overlayer without perturbing the surface of interest. Exposing an interface may require more than one method for maximum efficiency. For instance, mechanical or chemical methods may be used to thin an overlayer to a depth that is close to the interface. Then sputtering may be used to probe to the interface.

High-purity solvents can be used to remove soluble overlayers. To aid the removal process, solvents are often used in conjunction with ultrasonic agitation or Soxhlett extractors. When using solvents, an analyst should be alert to the possible introduction of artifacts. Poorer grades of solvents and solvents with more complex molecules are more likely to leave residues. Solvents sometimes dissolve contaminants from a part of a specimen that will not be analyzed and then deposit them as residue on a surface that will be analyzed. A critical inspection of the entire specimen should be done prior to placing the specimen in a solvent. Solvents can also be applied via tissues or other similar materials which are then used to wipe a specimen. Care should be exercised in such cases because contaminants can leach from the tissues and be deposited on the specimen.

It has been reported that bombardment with low-energy hydrogen ions can effectively remove organic overlayer contaminants.[14] The removal process is thought to result from formation of volatile organic species. This approach can sometimes remove organic material with minimal alteration of the interface of interest.

Interfaces can be exposed via cross sections of the specimen in a plane that is orthogonal to the surface. The spatial resolution of modern Auger systems will allow analysis of an interface shown in a cross-sectional specimen and reveal gradation of intermixing at, and adjacent to, the interface plane.[15] Such analyses often reveal that the interface is not planar on an atomic scale.

Cross sections can be formed by one of the following techniques:

- Breaking the specimen if it is brittle enough. The geometry of such a break is often jagged and requires a search to find an adequate location for analysis.
- Breaking, polishing, and, possibly, chemically etching the section as is often done in the preparation of metallurgical specimens.
- Ion milling. With modern ion milling systems it is possible to etch rapidly enough to make this a reasonable approach.[16–20] The advantage of a cross section prepared by ion milling is its precise location. This precision can be critical in analysis of integrated circuits and other specimens of very small and known geometry.

Sputtering (ion etching) is often used to expose interfaces. The effects of sputtering in surface analysis can be quite complex and are discussed in Chapter 3 of Volume 5 in this series. If the overlayer is thick (greater than 1 micron), then it may be difficult to expose and analyze an interface uniformly via sputtering. Sputter-ion craters produced at high sputtering rates and without simultaneous analysis have been combined with high-spatial-resolution Auger electron spectroscopy to investigate interfaces. This technique is called *crater edge profiling*.[21–23]

To minimize the effect of a thick overlayer, mechanical methods of angle lapping[24] or ball cratering[25] may be applied. Angle lapping is illustrated in Fig. 3 and ball cratering in Fig. 4. The angle should be very shallow, often 1–5 degrees. For ball cratering, the radius of the ball must be very large compared to the thickness of the films being analyzed. This criterion will result in a geometry at the interface that is very similar to shallow angle lapping. A standard metallurgical apparatus

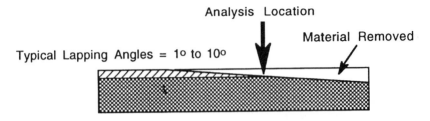

Figure 3. Angle lapping to expose a deep interface.

and polishing compounds can be used. The specimen should be cleaned with solvents before it is mounted and inserted into the analysis chamber. A glass wheel can be used in the final stages of polishing for greater precision.[26] The process of polishing many types of specimens, including integrated circuit chips, can be automated to a great extent by using a simple jig which maintains the specimen at a constant angle relative to the surface of the polishing disk and uses gravity to hold the specimen against the polishing disk.[27]

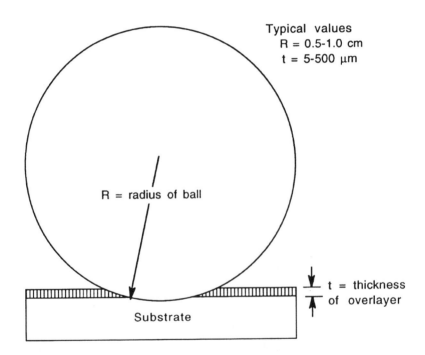

Figure 4. Ball cratering to expose a deep interface.

Sometimes it may be possible to separate layers mechanically and expose the surface of interest in this manner. This procedure is often very useful if delamination at the interface of interest has already begun, and it can be extended by additional mechanical force just prior to insertion of the specimen into the analytical chamber. Sticky tape may be used to delaminate structures. Except for possible reactions with the atmosphere, a surface exposed in this way is generally suitable for analysis. The inside surfaces of blisterlike structures are often investigated in this way. Sputter-depth profiling is often not a good method to use on blisterlike structures. At the point when the outer skin is penetrated by the ion beam, the data acquisition becomes dominated by artifacts.

In some specimens, it may be easier to approach the interface of interest by removing the substrate rather than the overlayer. This situation could arise when the composition of the substrate is known and the composition of the overlayer material is unknown. It may not be necessary to completely remove all of the substrate layer. One could etch to a point that is close to the interface of interest and continue with sputtering from that point.

Overlayers can sometimes be removed by vacuum pumping, reactions with gaseous species, exposure to ultraviolet light, or a combination of these methods.[28-35] These methods are most applicable to removal of organic materials. The time necessary to remove an overlayer may be reduced if the specimen is also heated. Heating would be most applicable with substrate materials that can withstand high temperatures, such as refractory metals and ceramics. The addition of reactive gases such as oxygen, hydrogen, etc., may result in the transformation of overlayers to volatile species. Ozone has also been used extensively to remove organic overlayers.

7. Wires, Fibers, Particles, Powders, and Pellets

Wires, fibers, particles, powders, and pellets can present interesting challenges concerning suitable methods for mounting. When preparing such specimens of such geometries, the spatial resolution of the analytical technique and the electrical conductivity of the specimen are important considerations. Specimen charging is generally more of a problem when analytical systems with high spatial and/or energy resolution are employed.

For a given system the type of analysis will also determine the extent to which specimen charging is a problem. As an example, it is often possible to obtain an Auger spectrum while rastering the incident electron beam over a relatively large area on the surface of many powder particles. The rastered beam allows various regions of the specimen to discharge before the beam touches them again. The resulting spectrum appears quite normal. However, attempts to map the locations of the elements found in the raster survey spectrum can be frustrated when the longer residence time of the incident beam in the mapping mode results in significant charging and modification of the Auger signals. This problem has occurred in the analysis of catalyst species on insulating powder substrates. Attempts to reduce charging by varying the angle of incidence of the incident beam are of little value because powder particles already include a wide range of possible angles of incidence.

Specimen charging is generally not a problem for XPS systems that do not have a monochromator and whose signal comes as an average from an area of millimeter dimensions. Under such circumstances specimen mounting only requires that the powders and fibers be uniform over an area somewhat larger than the area of analysis. Thus, sticky tape can be used for mounting. The best type of tape for this purpose consists of a metal substrate covered with a thin layer of adhesive and a polymer strip overlayer which is removed immediately before use. This type of tape generally has the best compatibility with the vacuum requirements of the analytical chamber. Powders are sometimes ground with a mortar and pestle to expose fresh material immediately before mounting. The powder or fibers can be sprinkled onto the tape, or the tape can be dipped into a reservoir of loose material.

Pellets can be formed from powders or fibers, but the pellets are often more difficult to mount than either fibers or powders. Special fixtures may be needed to hold the pellets securely. The additional handling required to form pellets and to mount them may also introduce artifacts. Forming pellets from powders and fibers is usually not a good approach for Auger unless the material is a conductor. If the material is an insulator, then a high level of charging will often arise during analysis of pellet-making, which is difficult at best.

Some specimens to be analyzed may be of such small size that it is not possible for either the probing beam or the analysis area to remain only on the specimen, and signals from the substrate may be a source of interference. If the analyzer used has a plane of maximum sensitivity, then it may be possible to hold the specimen so that it protrudes beyond the position of the normal mounting surface. This would position the mount so far out of the maximum sensitivity plane that signals from the mount do not interfere appreciably with signals from the specimen. An example for a cylindrical mirror analyzer is shown in Fig. 5. Not all analyzers have a plane of maximum sensitivity.

Powders, fibers, and particles may be easier to analyze if they are mounted on a conducting substrate. Indium foil has been used because it is soft at room temperatures and the powders, fibers, and particles can be partly embedded in the foil. When embedding powder or fiber pieces in foil, use pressure against an overlayer of the same type of foil. This overlayer would be removed before the foil holding the specimen material is inserted in the analytical chamber. A problem with indium foil is that it splatters easily when sputtered. Foils of copper and aluminum and sheets of graphite have been used successfully to mount powders and fibers. When using these materials that are hard at room temperature, relative to indium

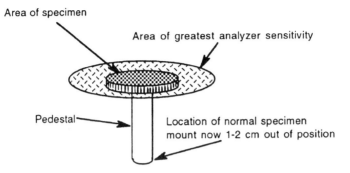

Figure 5. Mounting a specimen to minimize signals from mount.

foil, one can simply sprinkle the pieces on the foil. Some pieces will generally stick to the surface of the foil, and the excess can be removed by gently tapping the substrate. Mechanical jarring should be avoided when inserting the specimen into the analytical chamber. Some caution should be observed if there is nonuniformity in the specimen material, because it is possible that only a subset of all powders and fibers will adhere in such cases.

Moderately large particles can be transferred on the end of a very sharp point to a substrate suitable for insertion in the analytical chamber. Such a transfer must often be done under a microscope because of the small size of the particles. This procedure requires a steady hand and considerable patience. Occasionally, one can float insoluble particles on, or suspend particles in, a liquid and pick them up on a conducting mesh that is raised from within the liquid. This last method has been successfully applied with asbestos specimens using methanol and a silver grid. Particles suspended in a liquid can also be moved to a substrate and then the liquid can be evaporated. A pure, smooth, and conducting substrate such as a polished silicon wafer is recommended.

8. Storage, Transfer, and Shipment

The longer a specimen is in storage the more care must be taken to ensure that the surface to be analyzed will not be contaminated. Containers used for storage should not transfer contaminants to the specimen via particles, liquids, gases, or surface diffusion. The surface to be analyzed should not contact the container. Glove boxes, vacuum chambers, and desiccators may be excellent choices for storage of specimens.

Volatile species may leave a specimen in storage and will do so more quickly at elevated temperatures. Storage at low temperatures is often advantageous, although care must be taken to avoid condensation of water on the surface when it is subsequently exposed to the atmosphere.

Special chambers that allow transfer of specimens from a controlled environment to an analytical chamber have been reported in the literature.[36–38] Controlled environments could be other vacuum chambers, glove boxes (dry boxes), glove bags, reaction chambers, etc. Glove bags can be temporarily attached to an analytical chamber with transfer of a specimen achieved by removal and then replacement of a flange on the analytical chamber inside the glove bag.

Coatings can sometimes be applied to specimens, allowing transfer in the atmosphere. The coating is then removed by heating and/or vacuum pumping in either the analytical chamber or its introduction chamber. This concept has been successfully applied for the transfer of GaAs[39] and Si.[40]

Shipping specimens requires the same considerations as storing them. However, shipping will normally imply less control of temperature and time conditions. In addition, shipping requires the need to hold specimens securely.

Specimens that have been charged with hydrogen are often required for *in situ* fracture analysis. They will usually lose the hydrogen if they are allowed to remain at room temperature for a relatively short time period. They can be shipped imbedded in dry ice via overnight express and/or stored in liquid nitrogen for many days without serious degradation of the hydrogen charge.

9. Organic and Biological Samples

Perhaps the most rapidly developing area of specimen handling concerns organic and biological materials. The effects of incident beam alteration can be serious, and specimen charging is almost always a problem. Cooled stages can often be used to advantage.

Viscous liquids can sometimes be analyzed in an ordinary XPS spectrometer by placing a very thin layer on a smooth substrate material. Such a layer can be made by starting with a thick layer and then wiping away most of the liquid. The remaining layer of the viscous liquid is usually of sufficient thickness that no signal from the substrate is detected, yet the vacuum requirements of the analytical chamber are met. Further reduction of outgassing is possible through cooling.

A problem of considerable importance in analysis of biological materials is preservation of the cellular structure of the specimen. Unfortunately the specimens must almost always be chilled before insertion into the vacuum environment of the analysis chamber. A very fast chilling technique has been developed by electron microscopists that will preserve the cellular structure.[41–43] The chilling is so rapid that water freezes in noncrystalline form, and this is often required to preserve the cellular structure.

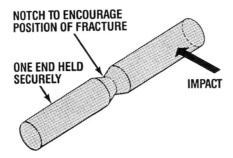

Figure 6. Typical specimen for fracture.

10. Fracture Specimens of Unusual Geometries

Most fracture analysis is done by impact, and many fracture devices allow fracture of several specimens without breaking vacuum. In addition, impact-fracture systems generally have provisions for cooling the specimens prior to fracture. The impact-fracture devices generally have a preferred geometry for the specimen to be fractured. Notching the specimen is usually done to control the location of the fracture. A typical specimen for impact fracture is shown in Fig. 6.

Specimens of smaller size than the ideal fracture specimen can still be fractured in an impact device by attaching additional pieces to allow the nonideal shape to approximate the ideal form. Such a technique is indicated in Fig. 7. The key points to consider are the method of securing the specimen in the fracture device and the geometry of the impact hammer. The specimen may also be notched at the preferred point of fracture, or the break may occur within the sandwich structure and make analysis very difficult. To avoid premature fracturing or a fracture in the wrong location it may be advisable to wrap the end of the specimen held in the fixturing with a soft foil such as aluminum or indium, as shown in Fig. 8.

Wires of small diameter can be successfully fractured using the modification of a standard fracture specimen as shown in Fig. 9. The standard fracture specimen

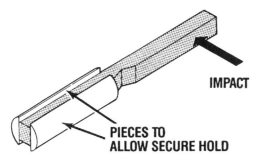

Figure 7. Holding fracture specimens of nonstandard geometry.

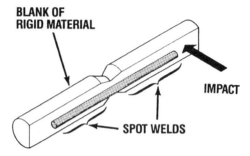

Figure 8. Modification of a standard fracture specimen to support fracture of wires of small diameter.

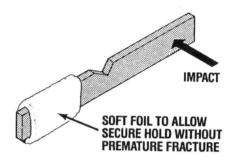

Figure 9. Wrapping a nonstandard fracture specimen to minimize premature fracture.

should be made of a material that is known to fracture sharply and will provide secure spot welds. The fracture device hammer should hit on the wire side.

When fibers of very small diameter are to be fractured, it may be possible to use the scheme illustrated in Fig. 10. Trial and error will probably be necessary to determine if set screws or solder will hold the fibers securely enough to result in a

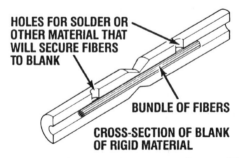

Figure 10. Modification of a standard fracture specimen to support fracture of fibers of small diameter. In application, the standard specimen would have a hole drilled along its axis and would not be cut as a cross section as shown.

fracture. The electrical conductivity of the combination of fractured fibers and the mount must also be considered.

11. Miscellaneous Procedures

11.1. Growth and Analysis of an Interface

Occasionally an interface between overlayer and a substrate cannot be suitably exposed for analysis. In such instances it is sometimes possible to obtain information regarding that interface by growing the overlayer very slowly (monoatomic scale layers) on the substrate and applying surface analysis techniques until the analyses show a transition from purely substrate material to purely overlayer material. The composition at the interface measured in this way, however, may not always be identical with that for a thicker-overlayer film.

11.2. Dissolved Materials

Material dissolved in a liquid, including trace concentrations of an impurity, can sometimes be analyzed. To prepare a specimen, some of the liquid is placed on a suitable substrate and evaporated, possibly leaving a residue of interest for analysis.[44] The evaporation may be done in several stages to obtain a sufficient concentration of the residue. A pure, smooth, conducting substrate such as a polished silicon wafer is recommended.

11.3. Preservation of Electrochemical Salt Films

Some salt films can be formed by electrochemical means, but will be present only when a potential is applied. The salt film will dissolve once the potential is removed. Such films may be withdrawn from solution with the potential still applied by placing a layer of benzene over the electrolyte solution and turning off the potential when the film is within the benzene layer where it will not dissolve.[45] Specimens obtained in this way can be inserted directly into the analytical chamber without further treatment.

11.4. Flat Mounting of Foils

A foil specimen often needs to be mounted flat, particularly for angle-resolved measurements. However, many foil-like specimens have a tendency to curl and twist. A border ring of a rigid material to push the foil against a flat substrate can be useful in such cases.[46] The masks with spring tension discussed in Sec. 2 would serve nicely for this purpose.

11.5. Using an Overlayer to Stabilize SIMS Signals

When starting SIMS analysis on a surface where concentrations and/or oxidation states are different from the bulk, there can be a transition zone wherein signals vary greatly due to rapid changes in matrix effects. As a result, interpretation of data, including detection of trace elements, may be difficult in the transition zone. By applying an overlayer of composition similar to the bulk, one can achieve stable SIMS signals at the position of the original surface. This technique only provides an advantage if a noble-ion species is used for bombardment of the specimen because bombardment by an active ion species will create its own overlayer. Using an overlayer of 30–100 nm of SiO_2 on a silicon wafer, this procedure has resulted in detection of trace contaminants on the silicon surfaces.[47]

11.6. Replicating

Replicating can be used in surface analysis under special circumstances.[48] The key elements for success are a replicating compound that is an electrical conductor and a specimen where the critical film or particles will transfer from the sample to the replicating compound. This approach may be considered when the sample to be analyzed is large relative to the insertion probe and/or the analytical chamber.

12. Acknowledgments

The author would like to thank Kenton D. Childs, Peter E. Sobol, John. F. Moulder, and Lawrence. E. Davis from the Analytical Laboratory, Physical Electronics, Inc., Eden Prairie, MN, for their contributions and many helpful suggestions.

References

1. Y. Tito Sasaki, A survey of vacuum material cleaning procedures: A subcommittee report of the American Vacuum Society Recommended Practices Committee, *J. Vac. Sci. Technol. A* **9**(3), 2025–2035 (1991).
2. J. D. Herbert, A. E. Groome, and R. J. Reid, Study of cleaning agents for stainless steel for ultrahigh vacuum use, *J. Vac. Sci. Technol. A* **12**(4), 1767–1771 (1994).
3. E. A. Hill and K. D. Carter, Jr., Using ethyl lactate for precision cleaning of metal surfaces, *Microcontamination* **11**, 27–31 (1993).

4. R. W. Hoenig and D. A. Kramer, Vapor pressure data for the solid and liquid element, *RCA Rev.*, **30**(2), 285–305 (1968).

5. ASTM Standard E1078–90, Guide for Specimen Handling in Auger Electron Spectroscopy, X-ray Photoelectron Spectroscopy, and Secondary Ion Mass Spectrometry, *1994 Annual Book of ASTM Standards*, American Society for Testing and Materials, Philadelphia, PA, 3 (1994), p. 776.

6. American Vacuum Society, Applied Surface Science Division (AVS/ASSD), New York.

7. R. Sherman, D. Hirt, and R. Vane, Surface cleaning with the carbon dioxide snow jet, *J. Vac. Sci. Technol. A* **12**(4), 1876–1881 (1994).

8. R. Sherman, J. Grob, and W. Whitlock, Dry surface cleaning using CO_2 snow, *J. Vac. Sci. Technol. A* **B**(9), 1970–1977 (1991).

9. L. Layden and D. Wadlow, High velocity carbon dioxide snow for cleaning vacuum surfaces, *J. Vac. Sci. Technol A* **8**(5),1881–1883 (1990).

10. R. Sherman and W. Whitlock, The removal of hydrocarbons and silicone grease stains from silicon wafers, *J. Vac. Sci. Technol. A* **8**(3), 563–567 (1990).

11. J. P. Coad, H. E. Bishop, and J. C. Riviere, Electron-beam assisted absorption on the Si (111) Surface, *Surface Sci.* **21**, 253–264 (1970).

12. H. H. Madden and G. Ertl, Decomposition of carbon monoxide on a (110) nickel surface, *Surface Sci.* **35**, 211–226 (1973).

13. N. M. Kitchen and C. A. Russell, Silicone oils on electrical contacts—effects, sources and countermeasures, *IEEE Trans. Parts, Hybrids, Packaging* **PHP-12**, 24–28 (1973).

14. R. Bouwman, J. B. van Meachelen, and A. A. Holscher, Surface cleaning by low-temperature bombardment with hydrogen particles: an AES investigation on copper and Fe-Cr-Ni steel surfaces, *Appl. Surface Sci.* **15**, 224–237 (1983).

15. J. F. Moulder and D. F. Paul, Characterization of metal matrix composites, Physical Electronics Laboratories Application Note L9401, Eden Prairie, MN (1994).

16. J. Orloff, Focused ion beams, *Sci. Am.*, p. 96 (Oct. 1991).

17. G. R. Matusiewocz, S. J Kirch, and V. J. Seeley (eds.), The role of focused ion beams in physical failure analysis, *Proc. 29th Annual IEEE Intl. Reliability Physics Symposium,* IEEE Cat. No. 91CH2974–4 (1991).

18. J. Melngailis, Critical review: focused ion beam technology and applications, *J. Vac. Sci. Technol. B*, **5**, 469–495 (1987).

19. K. Nikawa, K. Nasu, M. Murase, T. Taito, T. Adachi, and S. Inoue, New applications of focused ion beam technique to failure analysis and process monitoring of VLSI, *Proc. 27th Annual IEEE Intl. Reliability Physics Symposium*, IEEE Cat. No. 91CH2974-4 (1989), pp. 43–52.

20. R. G. Lee and J. C. Morgan, Integration of a focused ion beam system in a failure analysis environment, *ISTFA '91: The 17th Intl. Symposium for Testing and Failure Analysis*, ISTFA (1991).

21. A. Zalar and S. Hoffman, Crater edge profiling of Ni/Cr multilayer thin films by scanning Auger microscopy (SAM), *Surf. Interface Anal.* **2**(5), 183–186 (1980).

22. N. J. Taylor, J. S. Johannessen, and W. E. Spicer, Crater-edge profiling in interface analysis employing ion-beam etching and AES, *Appl. Phys. Lett.* **29**(8), 497–498 (1976).

23. Y. X. Wang, Y. D. Cui, Z. G. Chen, E. Lambers, and P. H. Holloway, Auger crater-edge profiling of multilayer thin films by scanning Auger spectroscopy, *J. Vac. Sci. Technol. A* **8**(3), 2241–2245 (1990).

24. M. L. Tarng and D. G. Fisher, Auger depth profiling of thick insulating films by angle lapping, *J. Vac. Sci. Technol.* **15**(1), 50–53 (1978).

25. I. K. Brown, D. D. Hall, and J. M. Walls, The depth resolution of composition-depth profiles obtained by ball-cratering and Auger electron spectroscopy, *Vacuum* **31**(10–12), 625–629 (1981).

26. L. L. Levenson, Sectioning thick coatings for scanning Auger electron spectroscopy, *Test Meas. World*, 84–92 (Nov. 1985).

27. K. L. Black, Carborundum, Inc., Phoenix, AZ, private communication.
28. M. Tabe, UV ozone cleaning of silicon substrates in silicon molecular beam epitaxy, *Appl. Phys. Lett.* **45**,1073–1075 (1984).
29. J. R. Vig, UV/ozone cleaning of surfaces:, *J. Vac. Sci. Technol. A* **3**(3), 1027–1034 (1985).
30. J. M. Lenssinck, A. J. Hoeven, E. J. van Loenen, and K. Kijkkamp, Carbon removal from as-received samples in ultra-high vacuum using ultraviolet light and ozone beam, *J. Vac. Sci. Technol. B* **9**(4),1963–1967 (1991).
31. R. R. Sowell, R. E. Cuthrell, D. M. Mattox, and R. D. Bland, Surface cleaning by ultraviolet radiation, *J. Vac. Sci. Technol* **11**(1), 474–475 (1974).
32. P. H. Holloway and D. W. Blushmire, Detection by Auger electron spectroscopy and removal by ozonization of photoresist residues, *Proc.12th Annual IEEE Reliability Physics Symp.* IEEE Cat. No. 74CH0839-1 (1974), pp. 12–16.
33. B. S. Krusor, D. K. Biegelsen, R. D. Yingling, and J. R. Abelson, Ultraviolet-ozone cleaning of silicon surfaces studied by Auger spectroscopy, *J. Vac. Sci. Technol. B* **7**(1), 129–130 (1989).
34. T. Momose, Y. Maeda, K. Asano, and H. Ishimaru, Surface analysis of carbon on ozone treated metals, *J. Vac. Sci. Technol. A* **13**(3), 515–519 (1995).
35. C. G. Worley and R. W. Linton, Removing sulfur from gold using ultraviolet/ozone cleaning, *J. Vac. Sci. Technol. A* **13**(4), 2281–2284 (1995).
36. J. P. Hobson and E. V. Kornelsen, A target transfer system at ultra-high vacuum, *Proc. 7th Intl. Vac. Congress and 3rd Intl Conf. on Solid Surfaces*, Vienna, Austria (1977), pp. 12–16.
37. J. P. Hobson, First intercontinental test of UHV transfer device, *J. Vac. Sci. Technol.* **15**, 1609–1611 (1978).
38. T. Fleisch, A. J. Shephard, T. Y. Ridley, W. E. Vaughn, N. Winograd, W. E. Baitinger, G. L. Ott, and W. N. Delgass, A system for transferring samples between chambers in UHV, *J. Vac. Sci. Technol.* **15**,1756–1760 (1978).
39. G. L. Price, Preservation and Regeneration of an MBE Grown Surface, *Collected Papers of 2nd Int. Symp. MBE and Related Clean Surface Technology*, Tokyo, Japan, Jpn. Soc. Appl. Phys. (1982), pp. 259–262.
40. R. Lieberman and D. L. Klein, Temporary protection of silicon surfaces by iodine films, *J. Electrochem. Soc.* **113**(9), 956–958 (1966).
41. J. G. Linner, S. A. Livesley, D. S. Harrison, and A. L. Steiner, A new technique for removal of amorphous phase tissue water without ice crystal damage: a preparative method for ultrastructural analysis and immunoelectron microscopy, *J. Histochem. Cytochem.* **34**(9), 1123–1135 (1986).
42. S. A. Livesey, A. Del Campo, A. W. McDawa, and J. T. Stasny, Cryofixation and ultra-low-temperature freeze-drying as a preparative technique for TEM, *J. Microsc.* **161**(2), 205–215 (1990).
43. S. A. Livesey and J. G. Linner, Cyrofixation methods for electron microscopy, in: *Low Temperature Biochemistry: Emerging Applications and Engineering Contributions*, BED Vol. 10, HTD Vol. 98 (J. J. McGrath and K. R. Diller, eds.), Am. Soc. Mech. Engrs., New York (1988).
44. J. B. Bindell and P. F. Schmidt, The detection of trace impurities in liquids by Auger electron spectroscopy, Bell Laboratories memo, Murray Hill, NJ (1976).
45. W. H. Holt and C. M. Blackmon, Simple techique for the flat smooth mounting of thin foils, *Rev. Sci. Instrum.* **51**(5), 671–673 (1980).
46. A. J. Bevolo, B.J. Beaudry, and K. A. Gschneidner Jr., Auger analysis of the passivation of gadolinium by electropolishing, *J. Electrochem. Soc.* **127**, 2556–2557 (1980).
47. G. J. Slusser, IBM-Essex Junction, VT, private communication.
48. P. B. DeGroot and R. H. Scott, Extending replication methods to Auger electron spectroscopy by using conductive replicas, *Microbeam Analysis–1979*, San Francisco Press, San Francisco, CA (1979).

4

Atomically Clean Surfaces of Elemental Solids

R. G. Musket, W. McLean, C. A. Colemenares, and W. J. Siekhaus

Glossary

AES	Auger electron spectroscopy
bcc	Body-centered cubic crystal structure
cub	Cubic crystal structure
cub-dia	Cubic-diamond crystal structure
fcc	Face-centered cubic crystal structure
EELS	Electron energy loss spectroscopy
ELS	Energy loss spectroscopy
ESCA	Electron spectroscopy for chemical analysis (i.e., XPS)
FEM	Field electron microscopy
FIM	Field ion microscopy
hcp	Hexagonal close-packed crystal structure
hex	Hexagonal crystal structure
ISS	Ion scattering spectroscopy
L	Langmuir (1 L = 0.13 mPa · s)
LEED	Low-energy electron diffraction
mon	Monoclinic crystal structure

R. G. Musket, W. McLean, C. A. Colemenares, and **W. J. Siekhaus** • Lawrence Livermore National Laboratory, Livermore, CA 94551-9900.

Specimen Handling, Preparation, and Treatments in Surface Characterization, edited by Czanderna *et al.* Kluwer Academic / Plenum Publishing, New York, 1998.

orth   Orthorhombic crystal structure
rhdr   Rhombohedral crystal structure
RHEED   Reflection high-energy electron diffraction
SEM   Scanning electron microscopy
SIMS   Secondary-ion mass spectroscopy
STM   Scanning tunneling microscopy
SXAPS   Soft X-ray appearance potential spectroscopy
tetr   Tetragonal crystal structure
UHV   Ultrahigh vacuum
UV   Ultraviolet
XPS   X-ray photoelectron spectroscopy (i.e., ESCA)
XTEM   Cross-sectional transmission electron microscopy

1. Introduction

Preparation of atomically clean, elemental surfaces is required prior to experimental studies of their physical and/or chemical properties. The investigator in search of a cleaning recipe often must track the origin of any cleaning procedure back through several references only to locate a procedure that has not been verified by element-specific surface characterization techniques. Furthermore, that procedure may be neither the only one nor the best one. The investigator's alternative is the tedious process of developing his own procedure. We have been in similar situations and have concluded that a review that documents the cleaning procedures and recommends one "best" procedure for each elemental surface would be quite valuable. The procedures discussed here are, with the exception of the wet chemistry used to form hydrogen-terminated silicon surfaces, those used after the sample is in an ultrahigh vacuum environment (i.e., after mechanical and/or electrochemical polishing and residue removal). The maintenance of the cleanliness for time scales consistent with the experimental study is a matter of applying the appropriate vacuum technologies and is beyond the scope of this review. Usually, subsequent recleaning can be accomplished by a shorter-time repeat of the specified procedure.

In 1963, Roberts published a paper on the generation of clean surfaces in high vacuum [1]. He described the advantages and disadvantages of six general methods for preparing clean surfaces and gave specific procedures for a few elements. However, techniques for surface analysis were not available at that time to verify the effectiveness of the procedures. We published a previous version of this document [2], and Grunze *et al.* [3] provided an update of our compilation for cleaning procedures of 18 metal surfaces using oxidation–reduction cycles. We have now made updates to our original version to reflect key advances and new data. In this document, we have reviewed the literature for the specific cleaning techniques applicable to 75 elements with vapor pressures of less than 1.3×10^{-7}

Pa at room temperature. Wherever possible only those procedures verified by element-specific surface-analytical techniques (e.g., AES, XPS, SIMS, ISS) are given. Details are discussed for polycrystalline and various single-crystalline surfaces. Procedures for which no element-specific analyses were reported have been included with the appropriate caveats. From the review and assessment of the various methods, recommended procedures for each element have been combined in a table.

For the purposes of this chapter a clean surface is defined to be an annealed surface (except where noted) at ambient temperature with a total surface contamination level of less than a few percent of a monolayer. Annealing is required to ensure that any studies performed will be reproducible and characteristic of a nearly defect-free surface. Estimates of the structural perfection for annealed surfaces of single crystals have been obtained from LEED intensities. A total contamination level of less than a few percent of a monolayer was specified as acceptably clean because (1) the two most widely used techniques for surface analysis (AES and XPS) have detection limits of about 1% of a monolayer for individual elements, and (2) such limited contamination in the first few atomic layers should not markedly affect the results of most experimental studies.

Cleaning procedures considered are (1) heating to a high temperature in UHV or in a partial pressure of a reactive gas, (2) ion sputtering with the sample at ambient or elevated temperatures, (3) *in situ* fracturing or cleaving, and (4) *in situ* machining or scraping. For the sputtering processes, we have indicated the ion current density whenever it was given in the referenced paper. The less useful total ion current is presented in other cases; nevertheless, because typical sputtered areas are about 0.2 cm^2, a rough estimate of the current density can be made in these cases. Since many methods call for repetition of processing steps until the surface is clean, the need for *in situ* element-specific analysis of the surface is not eliminated by following the recipes given here.

Specifically excluded are preparation of clean surfaces of elements by *in situ* deposition of films, by laser irradiation, by UV–ozone exposure, and by field desorption methods. For some elements, *in situ* deposition of a film may be the only known method of preparing a large, clean surface (i.e., ~1 cm^2); this fact is noted in the table of recommended surface cleaning procedures. Laser irradiation and UV–ozone processing are not generally available and have limited applications. Field desorption cleaning techniques are not reviewed because the sample size, which is limited to ≤500 nm diameter by the requirements of FEM and FIM, is not large enough to be generally useful. The sources reviewed and quoted are generally in English from easily accessible books and journals. These sources were found by searching for published papers using both key-worded computerized systems and journal indices.

2. Review of Surface Cleaning Procedures

The elements are discussed in alphabetical order with details and references for individual crystallographic planes presented wherever appropriate. Both the crystalline structure type and the melting point T_m are listed parenthetically after the elemental name and abbreviation. Recommended cleaning procedures for all of the reviewed elements have been summarized in a table at the end of Section 3: Discussion and Recommendations.

Actinium (Ac): cub, T_m = 1323 K

We found no surface studies on bulk or thin-film actinium.

Aluminum (Al): fcc, T_m = 933 K

Polycrystalline and single-crystalline rods with orientations (100), (110), and (111) are sold commercially with 99.9995% purity; thus, the major cleaning task was the removal of the thin oxide layer that formed on the surface during exposure to air and during electropolishing.

Polycrystalline Surface [Al-1 to Al-15]

Polycrystalline surfaces of aluminum have been investigated ever since surface science tools became available. Two cleaning procedures have been used: (1) heating the sample to 873 K, and (2) repeated cycles of ion bombardment and annealing. For the first procedure, AES peaks of oxidized aluminum faded away and eventually disappeared after several hours of annealing at 873 K under UHV conditions, such that the ratio of the oxygen peak (510 eV) to the Al peak (68 eV) was less than 0.001 [Al-12]. In the second procedure, the sample was bombarded with noble-gas ions at room temperature (5 keV, 78 μA/cm^2) [Al-10], and thereafter annealed at various temperatures (e.g., at 550 and >723 K). The majority of studies included the ion bombardment and annealing cycle. The sputtering time necessary to remove the initial oxide layer was approximately 10 h; for recleaning, sputtering times of about 30 min were required.

(100) *Surface* [Al-15 to Al-24, Al-26 to Al-29]

Many studies [Al-16, Al-18, Al-19, Al-21, Al-23] dealt exclusively with the (100) surface, and the other reports cited above treated the (100) surface in conjunction with other low-index planes. Essentially, the same procedures used for cleaning the polycrystalline surface were shown to be successful for cleaning the (100) surface. Some authors used heating to 873 K without sputtering as a cleaning step [Al-26, Al-28]; most researchers used argon-ion bombardment and annealing cycles, but other inert-gas ions (neon [Al-20], xenon [Al-19]) have also been used. Most investigators performed the ion bombardments at room temperature and annealed at 723 K, but some raised the target temperature during ion bombardment (473 K [Al-29], 673 K [Al-23]), and others preferred an annealing temperature of 805 K [Al-16, Al-19].

(110) *Surface* [Al-15, Al-17, Al-20, Al-22 to Al-30]

One research report [Al-30] dealt exclusively with the (110) surface; the others treated the (110) surface in conjunction with the other low-index planes. The (110) surface was subject to faceting and produced, in general, less satisfactory LEED patterns than either the (100) or the (111) surface [Al-27, Al-30]. The cleaning procedures used were the same as those described for the (100) surface.

(111) *Surface* [Al-15, Al-17, Al-20, Al-22, Al-24 to Al-34]

Several groups studied the (111) surface exclusively [Al-25, Al-31 to Al-34]. Other researchers dealt with the (111) surface in conjunction with other low-index planes [Al-15, Al-17 to Al-24, Al-28, Al-29]. The UHV cleaning procedures were the same as reported for the (100) and (110) surfaces.

(421) *Surface* [Al-35]

A clean surface was obtained by a series of argon-ion bombardment and annealing (823 K and then 893 K) cycles.

Americium (Am): hex, T_m = 1267 K

We found no surface studies on bulk or thin-film americium.

Antimony (Sb): rhdr, T_m = 903 K

No surface studies on polycrystalline antimony were found, but cleaning procedures for four single-crystal planes consisted of cleaving and/or sputter–anneal treatments.

(0001) *Surface* [Sb-1, Sb-2]

Antimony has been cleaved to yield (0001) surfaces in air and dry nitrogen. These surfaces were immediately transferred into UHV systems and produced a well-defined LEED pattern [Sb-1, Sb-2]. The small oxygen contamination was removed by argon-ion sputtering (200–300 eV, few μA/cm^2, 1 min). This surface plane is relatively inert to residual gases since no oxygen or carbon contamination was observed over the 6–10 h period required for XPS measurements [Sb-1].

Anneals for a few hours at ~520 K have been shown to improve surface structure as indicated by improved LEED patterns.

$(01\bar{1}2)$ *and* $(11\bar{2}0)$ *Surfaces* [Sb-2]

In LEED-only studies the $(01\bar{1}2)$ and $(11\bar{2}0)$ surfaces were given the same sputter–anneal treatment used for the (0001) plane to improve the LEED pattern; however, inclined facets in the $(01\bar{1}2)$ plane caused maverick LEED beams, whereas this surface had a well-defined LEED pattern before any treatment. Only after the treatment did the $(11\bar{2}0)$ surface yield a LEED pattern indicative of a reconstructed surface.

(111) *Surface* [Sb-3]

Cleaving 99.999 +% pure antimony using a razor blade at 233 K in a vacuum of 8×10^{-9} Pa should have produced a clean (111) surface.

Arsenic (As): rhdr, $T_m = 1090$ K

Arsenic is not a UHV-compatible material because of its high vapor pressure (10^{-2} Pa at 483 K). However, this high vapor pressure proved useful for preparation of an atomically clean (0001) crystal plane for LEED, AES, and XPS investigations. Impurities of concern were carbon and oxygen. An air-cleaved arsenic crystal was heated to 493 K in vacuum to evaporate the oxide and all surface impurities. The sample was surrounded by cold surfaces to condense the arsenic vapor and prevent contamination of the UHV system. The amount of carbon and oxygen on cleaned arsenic surfaces did not increase with time of exposure in the UHV system [As-1, As-2]. Clean As(111) surfaces were probably created by cleaving 99.999 +% pure arsenic (233 K, 8×10^{-9} Pa) using a razor blade [As-3].

Barium (Ba): bcc, $T_m = 998$ K

We found no surface studies on bulk barium; however, high-quality XPS and UPS spectra were obtained for freshly evaporated films [Ba-1 to Ba-3]. One author [Ba-2] reported obtaining XPS data of similar quality from bulk samples that had been sputtered with argon ions and annealed (no additional parameters were listed); however, the spectra presented in this work were of films.

Beryllium (Be): hcp, $T_m = 1551$ K

Removal of silicon is the most difficult part of the cleaning procedure; the difficulty increases with increasing concentration of silicon in the bulk [Be-1].

Cleaning was usually accomplished by repeated cycles of argon-ion sputtering and annealing.

Polycrystalline Surface [Be-1, Be-2]

For beryllium containing less than 1 ppm silicon, several cycles of argon-ion sputtering and annealing at less than 900 K were sufficient to produce a clean surface; however, long-term annealing at 1240 K resulted in resegregation of silicon to this surface [Be-1]. Beryllium with 60 and 600 ppm silicon was similarly cleaned by a sputter–anneal procedure, but lower temperatures were sufficient to bring silicon to the surface. A temperature of 1070 K was high enough to create a silicon-contaminated surface for a bulk impurity level of 600 ppm silicon. Evidence indicates that the segregation of silicon results in a layer on top of the beryllium [Be-2].

(0001) *Surface* [Be-3, Be-4]

Cycles of argon-ion bombardment (0.6 or 3 keV, few $\mu A/cm^2$, 1–2 h, 300 or 1070 K) and annealing (1020–1070 K ~1 h) have produced clean surfaces.

Bismuth (Bi): rhdr, T_m = 544 K

A clean (0001) crystal face was prepared for LEED, AES, and ELS studies by removing carbon, oxygen, and chlorine with argon-ion bombardment (150–175 eV, 5 $\mu A/cm^2$, 6 h) and a UHV anneal (510 K) [Bi-1, Bi-2]. Sharp LEED patterns of the $(01\bar{1}2)$ and $(11\bar{2}0)$ planes were produced by a similar argon-ion bombardment (300 eV, 1–2 $\mu A/cm^2$) and UHV annealing (523 K) procedure; however, surface purity was not verified by an element-specific analytical technique [Bi-3].

Boron (B): α rhdr, β rhdr > 1470 K, T_m = 2573 K

Boron is commercially available in both polycrystalline and single-crystalline forms. Although the main crystal structure is rhombohedral, other crystal structures have been reported to exist. The exact temperatures for the various phase transformations are not well known. For the few surface studies reviewed, carbon and nitrogen impurities were the most difficult to remove.

Polycrystalline Surface [B-1]

A polycrystalline sample produced by fusing 1-μm particles in a solar furnace showed, after various alternate cycles of heat treatment and ion bombardment, an AES spectrum free of chlorine, sulfur, and phosphorus but containing small oxygen and carbon peaks.

(001) *Surface and Surfaces Parallel to the c Axis* [B-2]

Crystals were heated indirectly by radiation from a hot tungsten filament mounted behind the sample or by a quartz–iodine lamp. During the heat treatment (923–1023 K, 5 min) the vacuum was $< 1 \times 10^{-6}$ Pa. Traces of carbon, oxygen, and nitrogen were still visible in the AES spectrum even after a heat treatment at 1123 K. No LEED pattern could be seen.

(111) *Surface* [B-3]

Heat treatment at high temperature was used to obtain a clean ordered surface. After heating to 1473 K, large amounts of oxygen and carbon still remained at the surface, and the LEED pattern showed a completely disordered surface structure. Oxygen completely disappeared after heating for some minutes at 1573 K. To remove carbon and to obtain a good LEED pattern, temperatures higher than 1673 K were found to be necessary. Heating to 1823 K did not seem to improve the LEED pattern or the surface purity; the standard treatment was to heat at 1723 K for 1 min. Nitrogen could be removed by argon-ion bombardment but reappeared after the high-temperature anneal necessary to obtain a regular surface structure. The minimum ratio between nitrogen and boron AES peaks was about 1:50.

Cadmium (Cd): hcp, T_m = 594 K

No bakeout of the vacuum system containing cadmium is possible due to its high vapor pressure (10^{-9} Pa at 293 K) and low melting point (594 K). In common with other soft metals, cadmium was easily cleaned.

Polycrystalline Surface [Cd-1 to Cd-4]

Specimens cut from 99.999% pure cadmium ingots were mechanically scraped with an oxide-free tungsten carbide blade in a preparation chamber at about 10^{-6} Pa and transferred to the measuring chamber without breaking the vacuum. Photoelectron spectra of scraped specimens showed no trace of the oxygen $1s$ XPS line and only slight traces of the carbon $1s$ XPS line [Cd-1]. Sputter cleaning with argon ions at 3 keV has also produced surfaces free of contaminants [Cd-2]. Scraping or micromilling with a rotating diamond is preferred whenever it is important to keep the vacuum system free of cadmium contamination [Cd-3, Cd-4].

(0001) *Surface* [Cd-5]

After ion bombardment (700 eV, 3 μA/cm^2, 295 K, 24 h) the AES spectrum showed no traces of sulfur, nitrogen, or oxygen. A small carbon peak, if present, could have been concealed by the metal peak at 285 eV. The clean surface LEED pattern was well ordered and had a high background intensity, as expected from a metal with a low melting temperature.

Calcium (Ca): fcc, T_m = 1112 K

We found no surface studies on bulk calcium. In one study [Ca-1], XPS was used to verify the cleanliness resulting from scraping calcium surfaces *in situ*, but the form of the calcium was not given.

Carbon (C)

Bulk carbon exists in five forms: amorphous, glassy, graphitic, diamond, and fullerene crystals. Because we are not aware of any surface cleaning procedures for fullerene crystals, the review is limited to the first four forms.

Amorphous Carbon

Amorphous carbon surfaces have been produced by evaporating high-purity carbon under UHV conditions or by ion bombardment of graphitic carbon surfaces.

Glassy Carbon [C-1]

A contamination-free surface of glassy carbon has been produced by fracturing a disk-shaped ingot under dry nitrogen in a glove bag and inserting it directly into an UHV spectrometer without exposure to the atmosphere.

Graphitic Carbon (hex, T_m = 3820 K)

Graphite surfaces are very sensitive to ion bombardment. A dose of 3×10^{15} Ar/cm^2 at 300 eV resulted in a completely amorphous surface [C-2]. A temperature of 1773 K was necessary to anneal the radiation damage.

Graphite (0001) Surface [C-2 to C-10]   Many AES and LEED experiments have shown that this basal plane is an inert surface. At least one group [C-10] has shown by LEED that there is no chemical adsorption of H_2O, CO, oxygen, iodine, or bromine on the (0001) surface of graphite in UHV at 300 K. An air-cleaved (0001) surface inserted quickly into a vacuum system and annealed (723 K, 5 h) produced a clean surface as judged by AES [C-4]. However, neither AES, nor LEED, nor the growth pattern of inert gases condensed on the surface was established as the most sensitive test of surface cleanliness. Use of the gold decoration techniques under UHV conditions [C-9] demonstrated that surfaces, indistinguishable by these conventional techniques of surface analysis had different reaction site densities. Furthermore, the reaction site density depended on the number of charged particles (from ion gauges, ion pumps, etc.) that impacted onto the surface. The gold decoration technique does confirm the observation made earlier by AES: an air-cleaved (0001) surface is slightly contaminated, and the contamination can be removed by annealing (723 K, 5 h). This low annealing temperature reflects the weakness of the adsorbate–substrate interaction. In contrast, a UHV-cleaved surface exposed for 24 h to a UHV vacuum system having sources of free charges must be annealed at 1273 K; unfortunately, the resulting surface was still not perfectly clean. One must conclude that a clean (0001) graphite surface can be maintained only in a charge-free UHV system. If the sample is cleaved under a nitrogen atmosphere and placed within seconds under vacuum, contamination of the cleaved surface is avoided almost completely.

Graphite ($10\bar{1}0$) *and* ($11\bar{2}0$) *Surfaces* [C-11]   The ($10\bar{1}0$) and ($11\bar{2}0$) surfaces fractured in air and analyzed within minutes by XPS showed, in contrast to the (0001) planes, considerable oxygen contamination.

Diamond (*cub-dia*, $T_m \geq 3550$ K) [C-12 to C-15]

The (100), (110), and (111) surfaces of diamond have been studied. Producing a crystalline surface for this allotropic form of carbon was more difficult than cleaning it. All authors agreed that ion sputtering was useless because the damage generated could not be annealed out without graphitizing the surface. Therefore, all authors used an annealing procedure for cleaning. For high-purity diamond, annealing in UHV ($<10^{-8}$ Pa) between 1173 and 1573 K produced clean surfaces.

Cerium (Ce): fcc, $T_m = 1072$ K

Polycrystalline samples of 99.9% pure cerium have been cleaned by cycles of argon-ion sputtering (1 keV, 10 μA/cm^2) and low-temperature annealing (temperature not specified). The main residual contaminants on the surface were determined by SXAPS. After four cleaning cycles, all the impurities which were observed between excitation potentials of 50 and 1500 V, except oxygen and zirconium, had been reduced below the sensitivity of the instrument (estimated to be a few percent of a monolayer). Additional cycles did not further reduce the presence of these two impurities [Ce-1]. A more recent study also made use of argon-ion sputtering (900 eV, 11 μA) to produce clean surfaces on cerium (99.9% pure) as verified by XPS [Ce-2].

Chromium (Cr): bcc, $T_m = 2130$ K

The major impurities found to segregate onto chromium surfaces (~99.99% pure) during *in situ* cleaning processes were sulfur, carbon, nitrogen, oxygen, and chlorine. Oxygen was somewhat difficult to detect by AES because the O KLL peak at ~512 eV is located between two major chromium peaks with almost equal sensitivity factors (i.e., the LMM peak a ~484 eV and the LMV peak at ~529 eV). However, the Cr MMM transitions at ~37 and ~41 eV were very sensitive to minute traces of contamination [Cr-1]. In general, cleaning of chromium was accomplished by a combination of argon-ion bombardment and annealing as detailed below.

Polycrystalline Surface

The only studies we located were on evaporated films.

(100) *Surface* [Cr-1 to Cr-4]

Iterations of argon-ion sputtering (~500 eV, 3–5 μA, 1000 K [Cr-1] or 300 K [Cr-2]) followed by annealing in UHV (900 K [Cr-1] to 1170 K [Cr-2]) led in one case to a clean, unreconstructed surface with a (1 × 1) LEED pattern [Cr-1] and in another case to a clean, reconstructed surface exhibiting a $c(2 \times 2)$ LEED pattern [Cr-2]. A $c(2 \times 2)$ reconstructed surface was also observed when the following cleaning routine was used [Cr-3]: Outgas sample (670 K, several h), sputter with argon ions (1.5 keV), heat in UHV (≤1170 K), cool, and repeat the sputtering until the near-surface region is depleted of sulfur. To deplete nitrogen in the near-surface region, heat in UHV (~970 K, several days) and then perform a final anneal (670 K, 15 min). An argon-ion bombardment (450 eV, 4 μA/cm^2) and annealing cycle performed for an unspecified number of times produced a (1 × 1) LEED pattern that appears to have been stabilized by sulfur [Cr-4]. Temperatures between 870 and 1070 K drove carbon and sulfur from the bulk to the surface.

(110) *Surface* [Cr-1, Cr-5, Cr-6]

The (110) surface was cleaned by argon-ion bombardment (600 eV, 3–5 μA/cm^2, 1000 K) and annealing (900 K) while removing the argon gas used for cleaning [Cr-1]. Some authors used variations on the basic sputter (500 eV, 25 μA, 300 K, 30 min [Cr-5]; 200 to 500 eV [Cr-6]) and annealing (670 K, 15 min [Cr-5]; 870 K [Cr-6]) cycle. Clean (110) surfaces with (1 × 1) LEED patterns were produced in all cases.

(111) *Surface* [Cr-7]

Clean, unreconstructed (111) surfaces have been produced by removing the initial oxide coating using argon-ion bombardment (starting at 8 keV, 60 μA/cm^2 and gradually lowering the voltage and beam current density to 2 keV, 5 μA/cm^2). This treatment was followed by alternately heating the sample in UHV (1170 K) and argon-ion sputtering (2 keV, 5 μA/cm^2) until heating (770 K, 15 min) did not result in further impurity segregation to the surface.

Cobalt (Co): hcp; fcc > 693 K, T_m = 1768 K

Cobalt is difficult to clean because it undergoes a hcp-to-fcc phase transition at 693 K. Therefore, cleaning must be done below this temperature for single-crystalline specimens. However, several authors report obtaining sharp LEED patterns after brief anneals above the transition temperature [Co-1 to Co-5]. The most common impurities found on the surfaces of cobalt samples (99.999% pure) were sulfur, oxygen, chlorine, carbon, and nitrogen. Carbon was the most persistent and difficult to remove. Sputtering with low-energy argon ions was reported to be effective in removing oxygen and chlorine, but not carbon [Co-1, Co-4]. Similar sputtering with neon ions effectively removed carbon and oxygen [Co-1, Co-2, Co-4]. The same result was achieved with a 1-keV krypton ion beam [Co-3].

Polycrystalline Surface [Co-1, Co-2, Co-6]

Polycrystalline specimens could be cleaned by heating in oxygen (900 K, 1.3 $\times$ 10^{-4} Pa), bombarding with neon ions (530 eV, 1 μA/cm^2, 600 K), and then annealing (1000 K, 15 h) [Co-1]. LEED studies indicated that 90% of a surface cleaned in this way had recrystallized into ~1 mm crystals with surfaces oriented parallel to the (0001) plane, while the remainder had (10$\bar{1}$2) surface orientations [Co-1, Co-2]. Sputtering with argon ions at 3 keV also produced a clean surface [Co-6].

(0001) *Surface* [Co-1 to Co-3, Co-7 to Co-9]

Sputtering with neon ions (530 eV, 1 μA/cm^2) at elevated temperatures (600 K) followed by cyclical annealing between 650 and 800 K produced clean (0001) surfaces that yielded sharp LEED patterns [Co-1, Co-2]. Krypton bombardment (1 keV, 300 μA/cm^2, 300 K, 15 h) with intermittent annealing at 900 K in UHV also proved to be effective in producing a clean, well-ordered surface [Co-3]. Clean (0001) surfaces have also been obtained by many iterations of inert-gas-ion bombardment (500 eV, ~200 μA/cm^2) and annealing (620 K) for a total of 15 h of ion bombardment and 25 h of annealing; this treatment was followed by 20 h of annealing in UHV at 620 K to improve the ordering of the clean surface [Co-7, Co-8]. One author [Co-9] used a multistep procedure that included (a) removal of oxygen with argon-ion bombardment or heating in hydrogen, and (b) removal of carbon by heating in ~3 $\times$ 10^{-6} Pa oxygen at 590 K. At pressures greater than 2 $\times$ 10^{-3} Pa, oxygen was found to remain on the surface [Co-9].

(10$\bar{1}$0) *Surface* [Co-3]

Sputtering with krypton (1 keV, 300 μA/cm^2, 300 K, ~15 h) coupled with intermittent anneals at 900 K was an effective method for cleaning this crystal face.

($10\bar{1}2$) *Surface* [Co-4]

A cycle of neon-ion bombardment (300 eV, 600 K) followed by annealing at 700 K was continued over a period of several weeks to eventually produce a carbon-free surface.

($11\bar{2}0$) *Surface* [Co-10]

Initial contamination was removed from this surface by argon-ion bombardment (140–300 eV, 3 $\mu A/cm^2$, 0.5–2 h). This treatment was followed by more than 100 cycles of sputtering and annealing at 650 K. Carbon residue (resulting from CO created on the ion gun filament) was removed by a few minutes of sputtering with a neon–oxygen mixture in a 1000:1 ratio with the sample at 550 K followed by a short sputtering period with pure neon.

(100) *Surface* [Co-11, Co-12]

One set of authors reported on the cleaning of a cobalt sample in the fcc form at room temperature; therefore, heating above the transition temperature was not a concern. However, it was quite difficult to remove carbon from the surface. The reaction of surface carbon with up to about 10^{-4} Pa oxygen at temperatures up to 1020 K was so slow that carbon concentration on the surface actually increased (presumably due to segregation from the bulk). It was found that argon-ion bombardment (150–180 eV) and annealing at 420 K produces a clean, well-ordered surface.

Copper (Cu): fcc, T_m = 1356 K

The impurities most commonly observed on polycrystalline [Cu-1] and single-crystalline [Cu-2, Cu-3] copper surfaces were carbon, oxygen, and sulfur. Chlorine and nitrogen were also detected on the (111) crystal plane [Cu-3]. Most investigators used repeated cycles of low-energy (≤600 eV) argon-ion bombardment and annealing in UHV (600–1000 K) to clean single-crystal copper surfaces.

Polycrystalline Surface [Cu-1]

Low-temperature hydrogen-ion bombardment was used to prepare a clean polycrystalline surface without a high-temperature annealing step. Omission of the annealing step prevented segregation of the bulk impurities to the surface. The chemical reactivity of the hydrogen ions enhanced the cleaning process, and the low mass of the ions resulted in minimal surface damage from ion bombardment. AES revealed that all carbon and sulfur had been removed after only 1 h of ion bombardment (800 eV, 10 $\mu A/cm^2$). A trace of oxygen remained on the surface even after 16 h of ion bombardment.

(100) Surface [Cu-2, Cu-4 to Cu-9]

Clean (100) crystal planes were prepared by several cycles of low-energy argon-ion bombardment (500–550 eV, 1 μA/cm^2) and annealing in UHV at 700 to 850 K [Cu-4 to Cu-6]. Clean (100), (110), and (111) crystal planes were prepared by cycles of 3-keV argon-ion bombardment and 1000 K UHV anneals [Cu-7]. Other investigators also reported preparing clean (100) crystal planes with argon-ion bombardment and UHV annealing [Cu-8, Cu-9]. Another cleaning procedure for the (100) surface consisted of heating the crystal in oxygen (800 K, 1.3×10^2 Pa) to remove carbon and sulfur followed by heating in atomic hydrogen to remove the surface oxide [Cu-2].

(110) Surface [Cu-7, Cu-10 to Cu-13]

Cycles of argon-ion bombardment (600 eV, 5 μA/cm^2) and UHV annealing at 723 K were used to prepare clean (110) surfaces [Cu-10]. Other investigators also reported using cycles of argon-ion bombardment and annealing (650–1023 K) but failed to give any other details of their procedures [Cu-11 to Cu-13].

(111) Surface [Cu-3, Cu-4, Cu-7, Cu-9]

Sulfur, chlorine, carbon, nitrogen, and oxygen impurities were detected with AES on a (111) crystal surface [Cu-3]. These impurities could be removed by argon-ion bombardment (100–300 eV, 100 μA/cm^2, 20 min) and annealing (573–723 K) in UHV. However, the carbon and sulfur contamination on the surface increased by diffusion from the bulk when the sample was heated between 773 and 1023 K in UHV. The amounts of carbon and sulfur on the surface decreased when the sample was heated above 1023 K.

(311) Surface [Cu-14]

A (311) plane was cleaned with cycles of xenon-ion bombardment (300 eV, 1 μA/cm^2, 2 h) and annealing (650 K, UHV, 1 h). A total of 16 h of ion bombardment was required to remove all traces of carbon and oxygen from the surface.

Dysprosium (Dy): hex, T_m = 1685 K

Polycrystalline surfaces were cleaned by repeated cycles of argon-ion bombardment (5 keV, 20 μA) and annealing at 500 K. Significant segregation of carbon and oxygen to the surface was observed in the early stages of the cleaning process [Dy-1].

Erbium (Er): hcp, T_m = 1802 K

We found no surface studies on bulk erbium. Erbium has been evaporated to form films.

Europium (Eu): bcc, T_m = 1095 K

We found no surface studies on bulk europium. Europium has been evaporated to form films.

Gadolinium (Gd): hcp, T_m = 1586 K

Polycrystalline samples of gadolinium were found by AES to contain "small amounts" of carbon and oxygen after cleaning by repeated cycles of argon-ion bombardment and low-temperature annealing (conditions not specified) [Gd-1].

Gallium (Ga): orth, T_m = 303 K

Gallium is a liquid at slightly above room temperature; however, it is a UHV-compatible material with a vapor pressure of 1.3×10^{-7} Pa at 840 K. The only surface study on gallium found in the literature was one using AES with imaging capability to study the surface of a liquid gallium drop held in a pyrolytic carbon cup [Ga-1]. When the gallium was heated to 600 K, the surface became almost free of carbon and oxygen. Surface precipitates (mainly carbon, gallium oxide, and occasionally sulfur) were first observed when the gallium was cooled to 390 K. These precipitates slid down the drop and accumulated near the edge of the carbon cup. The thin oxide and carbon overlayer on the liquid gallium could be cleaned by ion bombardment (3 keV, 20 μA/cm^2). Unexpectedly, impurity precipitates originating outside the ion impact area migrated into the impact region. Thus, an area larger than the actual ion impact areas was cleaned by ion bombardment. The impurities collected near the carbon cup were redistributed by heating to 600 K and then cooling to 300 K. The entire liquid gallium surface was cleaned by a "few" sequences of ion bombardment followed by temperature cycling.

Germanium (Ge): cub-dia, T_m = 1210 K

Single-crystal surfaces have been prepared using one of the following procedures: (a) chemical pretreatment and heating in UHV, (b) ion bombardment and annealing, (c) cleaving in UHV, or (d) growth and evaporation of a sulfide layer. Ion bombardment and annealing cycles may result in a surface structure that is not the true equilibrium termination of the bulk lattice. However, the use of chemical pretreatment to form a very thin contamination-free passivative oxide film that can be removed in UHV by low-temperature heat treatment alone leaves an equilibrium surface structure.

Polycrystalline Surface

No surface studies were found on polycrystalline germanium.

Amorphous Surface [Ge-1, Ge-2]

Amorphous layers 3.5 nm thick resulted from evaporation of 99.9% pure germanium from a tungsten boat onto carbon films over 30 min at 1.5×10^{-6} Pa [Ge-1]. An amorphous layer was also created by argon-ion sputtering of a clean Ge(111) surface to achieve a disordered surface [Ge-2].

(100) Surface [Ge-3 to Ge-8]

Neon-ion bombardment and a short anneal at 1273 K produced clean surfaces [Ge-3]. Heating at high temperatures resulted in pitting and, consequently, exposure of new surfaces; the LEED diffraction pattern exhibited spots that originated from the initial (100) surface directly beside spots that originated from the newly developed (111) facets [Ge-4]. One other procedure utilized chemical pretreatment: (a) the Ge wafers were cleaned in deionized, running water and etched in HF; (b) a thin oxide layer was prepared by dipping in a mixture of H_2O_2 and H_2O for a few seconds, and the oxide layer was removed by dipping in HF (this cycle was repeated several times to ensure the removal of several atomic layers of Ge); and (c) the final oxide layer was removed by ultrahigh vacuum annealing between 573 and 773 K. The resulting carbon and oxygen levels were undetectable by AES or XPS, and no structural defects were seen using XTEM [Ge-5]. Another chemical procedure involved ozone treatment and UHV annealing at >663 K for >30 min; neither carbon nor oxygen were detected by XPS, AES, or EELS [Ge-6]. One author reported that several cycles of sputtering with 1-keV argon ions and annealing at 823 K for 30 min and cooling to room temperature at a rate of less than 1 K/s produces surfaces having good LEED patterns and negligible impurities, as shown by both AES and helium-ion backscattering [Ge-7]. In another study, AES showed only a small residual carbon contamination after cycles of sputtering (800-eV Ar ions, 1 μA/cm^2, 20 min, angle of incidence 45°) and annealing (850 K, 60 min). During sputtering the crystal was heated to ~700 K for about 15 min; this simultaneous ion bombardment and annealing procedure was followed by slow cooling [Ge-8].

(110) Surface [Ge-9 to Ge-11]

A clean surface was prepared by evaporating a previously grown sulfide film in the UHV system at 623 K. However, there was no documentation for the purity of this surface with an element-specific technique. This method allowed preparation of a clean surface at a lower temperature than was possible by argon-ion etching and annealing [Ge-9, Ge-10]. Ion bombardment plus UHV annealing produced clean surfaces [Ge-11].

(111) *Surface* [Ge-2, Ge-11 to Ge-15]

Samples (*n* type) were submitted to repetitive argon-ion sputtering and heating cycles to achieve a clean surface [Ge-2]; annealing has been carried out at temperatures up to 1173 K (about 40 K below the melting point) for 1 h [Ge-12]. Clean surfaces have been prepared by cleaving in vacuum [Ge-13, Ge-14]. Haneman [Ge-11] documents that a wide range of surface structures can be observed on the (111) surface depending on the preparation technique.

Gold (Au): fcc, T_m = 1337 K

High-purity gold (99.999% pure) is commercially available in polycrystalline and single-crystalline forms. The major contaminants important in preparing clean surfaces are carbon, sulfur, and calcium.

Polycrystalline Surface [Au-1 to Au-5]

An effective cleaning cycle for calcium-free materials consisted of a series of heatings at 1000 K: 6 h in vacuum of less than 6×10^{-8} Pa, then 24 h in 6×10^{-3} Pa hydrogen [Au-2]. If the metal contained calcium, sputter cleaning was essential to obtain a clean surface. One recommendation was that the sample be subjected to alternate cycles of heating, exposure to oxygen, and ion bombardment in various combinations over a long period of time [Au-3]. The purpose of this treatment was to purify the gold sample by continuously segregating calcium to the surface where it could be removed by ion bombardment. The temperature of the target during exposure to oxygen at 5×10^{-5} Pa was 873 K. Cycles of heating under ultrahigh vacuum ($T > 1073$ K), ion bombardment, and annealing at 873 K have also been effective [Au-4].

(100) *Surface* [Au-5 to Au-10]

All investigators produced clean surfaces by repeated cycles of argon- or xenon-ion sputtering [Au-5, Au-6] and annealing (500–1000 K). Annealing temperatures close to the melting point produced irreversible high-temperature surface structures [Au-6]. Heating in oxygen has also been included in the cleaning process [Au-10].

(110) *Surface* [Au-11 to Au-15]

Clean surfaces were produced by repeated cycles of ion bombardment and annealing at temperatures from 523 K [Au-11] to 800 K [Au-12]. A two-stage process has been used to minimize the annealing temperature and calcium segregation [Au-15]. In the first stage, the crystal was bombarded by argon ions (0.3–3 keV) at a variety of angles and annealed at temperatures up to 1073 K. After the contamination was reduced, low-energy neon ions (~300 eV) were used in the second stage to bombard the sample at oblique angles of incidence; this treatment was followed by annealing at less than 573 K, which proved to be sufficient to produce a sharp (1 × 2) LEED pattern.

(111) *Surface and Its Vicinals* [Au-15 to Au-18]

Carbon was successfully removed by exposing the surface to 10^{-5} Pa oxygen at 973 K for 10 h; all other impurities were eliminated by argon-ion sputtering at 773 K [Au-17] or by successive cycles of argon-ion bombardment at room temperature and annealing at 973 K [Au-16].

Hafnium (Hf): hcp, T_m = 2500 K

The major contaminant found in hafnium foil samples was zirconium (~3%) that was not effectively removed in the refining process [Hf-1]. Hafnium is found in zirconium-bearing minerals. Because of their nearly identical chemical nature, these two elements are extremely difficult to separate [Hf-2]. Zirconium tended to segregate to the surface at 1070 K and became the major surface species in ≤30 s. Inert-ion bombardment and annealing at 2300 K produced surfaces containing approximately 10% zirconium [Hf-3]. Other bulk impurities that migrated to the surface during the heating operation were sulfur, carbon, chlorine, and oxygen [Hf-4]. Sulfur segregation increased at 1370 K; the sulfur content of the surface region decreased but did not disappear at 1570 K [Hf-5]. Most of the contaminants listed above could be removed by argon-ion bombardment, [Hf-1, Hf-4 to Hf-6]. This result suggests that the most effective cleaning method would consist of a prolonged period of sputtering at 1070–1370 K [Hf-5, Hf-6].

Holmium (Ho): hcp, T_m = 1747 K

Polycrystalline samples of high-purity holmium have been cleaned by argon-ion sputtering and annealed by heating with an electron beam. SXAPS measurements showed a weak oxygen peak that probably resulted from the reaction of the metal with residual gases [Ho-1].

Indium (In): tetr, T_m = 430 K

Only polycrystalline indium has been subjected to surface studies. Argon-ion bombardment (0.5–5 keV) quickly produced a clean surface in either the molten or solid state. This result does not imply that the material is free of impurities or that the surface will stay clean. On solidification from the molten state, sulfur, carbon, and oxygen reappeared on the surface. Since the melting point of indium is low, it is reasonable to assume that these impurities will diffuse at room temperature from the bulk to the surface. It is therefore recommended that a clean surface on solid indium be prepared by many repeated cycles of sputtering at a temperature above the melting point and then cool to solidify [In-1 to In-5]. *In situ* scraping has also been used to create clean surfaces [In-6, In-7].

Iridium (Ir): fcc, T_m = 2683 K

Cleaning iridium was routinely accomplished by heating in oxygen. The impurities most often encountered on bulk iridium surfaces during the cleaning process were carbon, oxygen, nitrogen, sulfur, and phosphorus. Studies on clean single crystals were complicated by reconstruction in the (100) and (110) planes.

Polycrystalline Surface [Ir-1 to Ir-6]

A polycrystalline foil was cleaned by repeated heating (1500 K) in oxygen (1.3 × 10^{-4} Pa, several h) and then heating in UHV (1700 K, ~20 min) until no impurities could be detected by AES [Ir-1, Ir-2]. Clean Ir films have been grown by the technique of metal–metal epitaxy on silicon (MMES) [Ir-3] and by the decomposition of metal organic Ir compounds in a H$_2$ carrier gas by pyrolysis (MOCVD) [Ir-4] at substrate temperatures of 673 to 1023 K. Iridium films have been prepared on clean Mo(110) [Ir-5] and on W(110) [Ir-6]. Films on W(110) were cleaned with a slight variation of the method described above.

(100) *Surface*

A. *(5 × 1) Reconstructed Surface [Ir-7 to Ir-18].* A (5 × 1) LEED pattern corresponding to a distorted hexagonal overlayer on the normal (100) surface resulted when the (100) surface was subjected to alternating heating cycles in oxygen (1400–1500 K, 7 × 10^{-6} to 1.3 × 10^{-5} Pa, 5–10 min) and in UHV (1500 K, ~1 min) until no impurities were detectable by AES [Ir-7 to Ir-13, Ir-17]. In most cases, the initial contamination on the surface was removed by argon-ion bombardment prior to heating in oxygen [Ir-7 to Ir-12]. Variations of this basic procedure included lower-temperature (1200–1300 K) oxidation cycles [Ir-14, Ir-15, Ir-17, Ir-18] and higher-temperature anneals (2100 K, >1 h) [Ir-15]. One author obtained a clean (5 × 1) surface by repetition of the following steps: (a) heating in UHV

(~1670 K), (b) argon-ion bombardment (800 eV, ~100 μA/cm^2), and (c) annealing (~670 K) in UHV [Ir-16].

B. *(1 × 1) Unreconstructed Surface* [Ir-7 to Ir-11, Ir-18]   Clean (100) surfaces that reflect the underlying bulk structure can be generated by chemical means. The treatment involved adding 10–20 L of oxygen at ~7 × 10^{-6} Pa to the previously prepared (5 × 1) surface held at 475 K [Ir-7] to 670–870 K [Ir-8 to Ir-11, Ir-18]. It was important that the sample be heated to ~660 K at some point during [Ir-8 to Ir-11] or after [Ir-7] the addition of oxygen. The foregoing operation converted the clean (5 × 1) surface to an oxide. At this point, the oxygen was removed by the addition of hydrogen (400–700 K, ~10^{-5} Pa [Ir-7]; ~30 L, ≤400 K [Ir-7]) or CO (~300 L, ≤400 K) [Ir-8 to Ir-11]. Gradual heating to less than 800 K removed adsorbed gases and yielded a clean, unreconstructed (1 × 1) surface [Ir-8 to Ir-11, Ir-18]. At temperatures higher than 800 K, the (1 × 1) structure converted irreversibly to the (5 × 1) structure [Ir-7].

(110) *Surface* [Ir-19 to Ir-35]

The (110) surface is also subject to reconstruction, as evidenced by the (2 × 1) LEED pattern observed for this crystal face. The general procedure for cleaning this surface was as follows: bombard with argon ions to remove the initial surface contaminants, heat in oxygen (~800 K, 7 × 10^{-6} to 1.3 × 10^{-5} Pa, several min), and anneal (~1600 K) in UHV; repeat the heating and annealing cycle until the surface is atomically clean [Ir-28 to Ir-35]. Variations on this procedure were to hold the sample at 1470 K for several days to aid in calcium removal [Ir-26] or to anneal at lower temperatures (970 K) [Ir-27]. In another study, this surface was cleaned by sputtering with magnetically separated 2-keV neon ions and annealing between 873 and 1173 K [Ir-31].

(111) *Surface* [Ir-14, Ir-23, Ir-36 to Ir-51]

No reconstruction has been observed on this face of iridium. The most commonly used cleaning method was to remove initial surface contaminants by argon-ion bombardment, then heat in oxygen (800–1100 K, 7 × 10^{-6} to 1.3 × 10^{-4} Pa), and anneal in UHV (1200–1600 K) [Ir-14, Ir-23, Ir-36 to Ir-42, Ir-46, Ir-49, Ir-51]. One variation on this procedure was to conduct the initial cleaning by first heating the sample in oxygen (1100–1200 K, ~10^{-5}–10^{-6} Pa), and then bombarding it with xenon or argon ions (600 eV, ~2 μA/cm^2, 10–30 min) before proceeding with the heating cycle described above [Ir-43, Ir-44, Ir-46, Ir-50]. It was possible to obtain a clean surface by cycles of argon-ion bombardment (500 eV, ~200 μA/cm^2, 1 h) with intermittent flashing to 1770 K and annealing [Ir-45]. In one case the sample was cleaned by heating in 6.6 × 10^{-6} Pa oxygen at 1100 K and then heating to 1600–1900 K to desorb the oxygen [Ir-47, Ir-48].

(755) *Surface* [Ir-40, Ir-41]

This surface has 6-atom-wide (111) terraces separated by 1-atom high (100) steps; this configuration is stable at temperatures up to 1470 K. It has been prepared by heating in oxygen (770–870 K, 7×10^{-5} Pa, several minutes) followed by flashing in UHV to 1470 K [Ir-40, Ir-41].

Iron (Fe): bcc; fcc > 1183 K; bcc > 1663 K, T_m = 1808 K

Iron is difficult to clean for two reasons: (1) the bcc-to-fcc phase transition at ~1183 K necessitates that all cleaning be done below this temperature, and (2) most commercially available samples have high concentrations of nonmetallic impurities, particularly carbon and sulfur. The nominal purity of single-crystal samples ranged from 99.92% to 99.995% pure, while most polycrystalline samples were nominally 99.999% pure. The use of some form of pretreatment in hydrogen to remove those impurities that form volatile hydrides increased as a function of the openness of the iron crystal face under consideration (i.e., increased use in going from (100) to (111) to (110) to polycrystalline Fe). All investigators pretreated polycrystalline Fe with hydrogen. The major contaminants found on single crystals from a variety of sources were carbon, sulfur, oxygen, nitrogen, phosphorus, and chlorine (ordered in decreasing amounts). In general, the cleaning of iron surfaces was accomplished by removing gross contamination and oxides by gentle sputtering with argon ions and then depleting carbon, sulfur, and phosphorous in the bulk by repeated cycles of argon-ion sputtering and annealing in UHV. Recleaning between exposures to gases was accomplished by argon-ion sputtering and a flash anneal.

Polycrystalline Surface [Fe-1 to Fe-3]

Polycrystalline iron samples have been cleaned by argon-ion sputter–anneal cycles [Fe-1, Fe-2] and by the combination of sputter–anneal cycles with oxidation–reduction cycles [Fe-3]. Few details were listed for these processes, but they can be inferred from the discussions below.

(100) *Surface* [Fe-4 to Fe-16]

The most frequently used method for cleaning this crystal face consisted of cycles of argon-ion sputtering (~500 eV, ~20 μA/cm^2, 15–30 min) and annealing in UHV (700–800 K, 1–60 min) [Fe-4 to Fe-15]. Sputtering was carried out both at elevated temperatures (600–700 K) [Fe-4 to Fe-7] and at ambient temperature [Fe-8 to Fe-15]. This basic cycle has been combined with additional chemical cleaning steps that involved heating the sample to ~770 K and exposing it to oxygen (0.5–10 L), followed by oxygen "titration" with acetylene [Fe-13, Fe-14] or by heating in either oxygen or water vapor (770–870 K, $\sim10^{-4}$ Pa) [Fe-15]. Another method included cycles of argon-ion sputtering at 1070 K and annealing at 670 K for 12-h periods for one week and, finally, flashing to 970 K [Fe-16]. Sulfur has been observed to segregate to the Fe(100) face between 500 K [Fe-6, Fe-12, Fe-13] and 700 K [Fe-4].

(110) *Surface* [Fe-4, Fe-11, Fe-17 to Fe-26]

The preferred method for cleaning the (110) surface consisted of the sputter–anneal cycle outlined for the (100) plane [Fe-4, Fe-17 to Fe-21], although higher ion energies (1–4 keV) were sometimes used for the (110) place [Fe-11, Fe-22, Fe-23]. In some cases, sputtering was carried out at elevated temperatures (920 K, ~1 h) [Fe-4, Fe-20, Fe-22]. Variations included combining the basic method outlined above with a hydrogen treatment (650 K, 10^{-5} Pa) [Fe-24], an oxygen treatment (970 K, 7×10^{-6} Pa, several h) [Fe-25], or oxidation–reduction treatments in oxygen (1070 K, $\sim10^{-3}$ Pa, long periods of time) and hydrogen (1070 K, $\sim10^{-2}$ Pa, long periods of time) [Fe-26]. Oxygen and sulfur segregated to the surface at 820 K and above 870 K, respectively [Fe-17, Fe-19, Fe-20].

(111) *Surface* [Fe-11, Fe-20, Fe-27 to Fe-30]

The (111) surface has been cleaned by argon-ion sputtering (250–500 eV, 1–10 μA/cm^2) and annealing (900–1120 K) cycles [Fe-11, Fe-20, Fe-27 to Fe-30]. The sputtering was done at 700 K [Fe-20] or at ambient temperature [Fe-11, Fe-27, Fe-30]. Sulfur migration to the surfaces began at 670 K and produced a monolayer of sulfur at 970 K [Fe-27, Fe-29]. Oxygen, carbon, and nitrogen were observed to dissolve into the bulk metal at 720 K [Fe-27].

Lanthanum (La): hcp, $T_m = 1194$ K

Lanthanum foil (99.9% pure) was sputtered in static argon (900 eV, 11 μA, ~6 $\times 10^{-3}$ Pa), but the surface still contained oxygen [La-1].

Lead (Pb): fcc, $T_m = 601$ K

Lead is commercially available in polycrystalline and single-crystalline forms with nominal purity of 99.999%. The major nonmetallic impurities are carbon, oxygen, nitrogen, and hydrogen. All investigators succeeded in reducing the surface concentrations of oxygen and carbon to about 1% of a monolayer.

Polycrystalline Surface [Pb-1 to Pb-3]

Polycrystalline lead foils were cleaned by prolonged argon-ion bombardment (700 eV, 5–10 μA, ~2.7 $\times 10^{-3}$ Pa, several h) until the oxygen 1s XPS signal fell below the detection level (~0.1% of Pb 4f intensity) and the carbon 1s signal had ceased to diminish. Bombardment of the surface with microwave-discharged oxygen (~10^{-3} Pa, 2450 MHz, 80 W) followed by further ion sputtering reduced this residual carbon to a negligible level (~0.1% of lead 4f intensity) [Pb-1]. Polycrystalline surfaces have also been cleaned successfully by *in situ* scraping with a tungsten carbide blade [Pb-2] or micromilling with a rotating diamond edge [Pb-3].

Single-Crystal Surface [Pb-4]

Single-crystal surfaces (planes not specified) were cleaned by repeated argon-ion bombardments (900 eV, 10 μA, 15 min). All impurity XPS lines were eliminated by this procedure.

(100) Surface [Pb-5]

Low-energy argon-ion bombardment (420 eV, angle of incidence = 45°, total dose = 1.4×10^{18} Ar/cm^2) at temperatures near the melting point (570 K) produced a clean surface for this crystal face.

(110) Surface [Pb-5 to Pb-9]

This surface has also been cleaned *in situ* by argon-ion bombardment and annealing. Treatment at 430 K and otherwise unspecified bombardment conditions reduced the impurity levels to the detection limits for oxygen and carbon; the O 1s: Pb 4f and C 1s: Pb 3f XPS intensity ratios were <0.001 [Pb-5]. Other investigators reported success with simultaneous sputtering and annealing near the melting point [Pb-6]. A combination of argon-ion sputtering (700 eV, angle of incidence = 45°, ion dose = 2×10^{15} Ar/cm^2) followed by annealing in UHV at 590 K for 1 h was also successful [Pb-7]. A similar treatment (Ar$^+$ at 700 eV, normal incidence, 1 μA/cm^2 for 10 h cycles at 300 K and 530–580 K for a total of 100 h of treatment) also produced the desired result [Pb-8]. One set of authors found that several cycles of sputtering with xenon ions at 1 keV followed by flashing to 580 K in UHV resulted in sharper LEED patterns, but reported that oxygen contamination often appeared upon cooling to ambient temperature [Pb-9].

(111) Surface [Pb-9]

These authors reported that several cycles of sputtering with xenon ions at 1 keV followed by flashing to 580 K in UHV resulted in high-quality LEED patterns but reported that oxygen contamination often appeared upon cooling to ambient temperature [Pb-9].

Lithium (Li): bcc, T_m = 454 K

Bulk samples of lithium (99.97% pure) have been cleaned by abrading in UHV with a file [Li-1] and by argon-ion bombardment (1–3 keV, ~5 µA/cm^2, 300 K, ~2.7 h) [Li-2, Li-3]. Sodium segregation to the lithium surface at room temperature was found to be a major source of contamination [Li-2]. To our knowledge, no work has been reported on single crystals of lithium.

Lutetium (Lu): hcp, T_m = 1936 K

Attempts to clean the (100) surface by ion bombardment and annealing were unsuccessful because oxygen and carbon segregated at the surface upon cooling [Lu-1]. Clean surfaces have been prepared by evaporation procedures.

Magnesium (Mg): hcp, T_m = 922 K

The contaminants of concern for magnesium are carbon, nitrogen, oxygen, and phosphorus.

Polycrystalline Surface [Mg-1, Mg-2]

In both studies, a polycrystalline magnesium surface was cleaned by argon-ion bombardment (3 keV) [Mg-1] and annealing in vacuum.

Monocrystalline Surface [Mg-3]

A 99.99% pure magnesium single crystal (face not specified) was cleaned by repeated argon-ion bombardment (~25 µA/cm^2) and annealing at 473 K. All impurities, except phosphorous, were below the detection limit of AES. The phosphorus signal was approximately three times the noise level.

(100) *Surface* [Mg-4]

Since annealing at temperatures above approximately 435 K caused significant magnesium sublimation and resulted in a macroscopically rough surface, argon-ion bombardment (400 eV) and annealing at less than 435 K were combined to produce an atomically clean surface and, yet, preserve the mirrorlike appearance of the freshly polished surface.

(0001) *Surface* [Mg-5 to Mg-7]

This plane has been cleaned by argon-ion bombardment (500 eV, ~10 µA/cm^2, ≤24 h) followed by annealing at 473–573 K. Although these surfaces were clean and yielded sharp LEED patterns, they were macroscopically rough, as shown by

SEM and loss of specular reflectance. The increased surface roughness was apparently caused by the combination of prolonged sputtering and excessive sublimation [Mg-5, Mg-6]. In another study by the same laboratory, clean surfaces were produced by successive cycles of neon-ion bombardment and annealing at 390 K [Mg-7].

Manganese (Mn): cub, T_m = 1517 K

Polycrystalline manganese (99.99% pure) was cleaned by iterations of 3-keV Ar^+ bombardment for periods of hours followed by heating to 673 K, presumably in UHV [Mn-1].

Molybdenum (Mo): bcc, T_m = 2883 K

Sulfur and carbon are the contaminants that must be purged from molybdenum to achieve a clean surface. Heating to approximately 800 K for a short time led to sulfur segregation to the (100), (110), and (112) surfaces; it also led to carbon segregation to the (110) and (112) surfaces, but not appreciably to the (100) surface [Mo-1]. Although AES can be used to uniquely monitor the presence of carbon, detection of small amounts of sulfur is difficult because of the overlap of the S KLL and the Mo MNN AES peaks at approximately 150 eV. However, the sum of the amplitudes of the sulfur and molybdenum peaks (S + Mo) at 150 eV relative to the molybdenum peak near 163 eV has been shown to be practically constant for incident electron energies over 400 eV when sulfur contamination was quite low [Mo-2]. For a fixed incident energy, the ratio of AES peak amplitudes was monitored as a function of cleaning steps or cycles; the minimum value of this ratio was taken to be that corresponding to a clean surface [Mo-3]. The intensity ratio of the S + Mo peak at ~150 eV to the molybdenum peak at ~188 eV is less than 0.10 for normally clean surfaces [Mo-3 to Mo-5]. Quantifying the level of sulfur contamination present requires careful spectrum modeling and stripping and is not normally undertaken as part of a cleaning process. The best established cleaning procedure is reaction with oxygen to remove carbon (as CO gas) and possibly sulfur, followed by flashing to high temperature in UHV to desorb the molybdenum oxides formed.

Polycrystalline Surface [Mo-6 to Mo-8]

When only sulfur segregated to the surface, either argon-ion bombardment at elevated temperatures (~1200 K), heating at 1100–1500 K in oxygen, or flashing at ~2100 K in UHV has been used to clean the surface. When carbon was present, heating at 1500 K in 1.3×10^{-4} Pa oxygen followed by flashing to 1900–2100 K in UHV yielded a surface free of carbon, sulfur, and oxygen.

(100) *Surface* [Mo-3, Mo-4, Mo-9 to Mo-14]

Carbon, oxygen, and sulfur appeared on the surface after heating above 1300 K. All three contaminants were removed by repeated application of the oxygen-anneal treatment detailed for the polycrystalline surface. A lack of flatness was found for surfaces that had been subjected to oxygen-anneal treatments with flashes limited to 1900 K; however, prolonged annealing at 2200–2300 K led to a flattening of the surface [Mo-12]. Carbon and oxygen have also been removed by multiple cycles of argon-ion bombardment (at 45° from the surface normal) and annealing at 1700 K; these cycles were preceded by an initial oxygen-anneal treatment (heating at 1700 K in ~7 × 10^{-4} Pa oxygen followed by annealing at 1700 K in UHV) [Mo-10].

(110) *Surface* [Mo-4, Mo-5, Mo-15, Mo-16]

This surface plane has usually been cleaned by repeated heating at 1200–1800 K in ~10^{-4} Pa oxygen and subsequent flashing at 1900–2200 K in UHV. In one case, the oxygen treatment was performed at 1800 K, and the oxide was removed by heating in 10^{-2} Pa hydrogen followed by flashing to 2400 K [Mo-16]. The usefulness of this hydrogen reduction step appears negligible because other investigators needed only the high-temperature flash after the oxygen treatment to obtain a clean surface.

(111) *Surface* [Mo-4, Mo-17]

The oxygen-anneal treatment detailed for the polycrystalline surface has been used successfully to clean this plane. However, prolonged heating at temperatures above 1700 K in ~7 × 10^{-6} Pa oxygen may lead to irreversible faceting to (112) planes [Mo-17].

(112) *Surface* [Mo-18]

Heating to 1300 K resulted in sulfur segregation to the surface, but the sulfur disappeared at temperatures above 1500 K. Carbon was removed by the oxygen-anneal treatment detailed for the polycrystalline surface.

Neodymium (Nd): hex; bcc > 1135 K, T_m = 1294 K

A neodymium crystal (orientation not specified) was "cleaned" by heating in UHV at 1000 K, but SXAPS measurements disclosed that carbon and oxygen contamination remained [Nd-1]. It is doubtful that this treatment alone would produce atomically clean surfaces. A useful procedure should be cycles of argon-ion bombardment and annealing in UHV.

Neptunium (Np): orth; tetr > 551 K; cub > 773 K, T_m = 903 K

We found no surface studies on bulk or thin-film neptunium.

Nickel (Ni): fcc, T_m = 1726 K

Nickel was the first element to be subjected to surface analysis [Ni-1] and has been one of the most thoroughly studied elements. For a compilation of work done prior to the advent of techniques to verify the state of surface cleanliness, see the useful bibliography in Ref. Ni-2. Surfaces prepared from nickel samples of 99.99% to 99.9995% nominal purity have been found to contain carbon, sulfur, and oxygen contaminants. These can usually be eliminated by a number of sputter–anneal cycles. However, many workers found carbon to be very persistent and resorted to high-temperature treatments in oxygen for cleaning. The latter treatment often left an oxygen residue on the surface which could be removed in most cases by a high-temperature treatment in hydrogen and/or additional sputter–anneal cycles. The successful utilization of these chemical treatments varied with the crystal face.

Polycrystalline Surface [Ni-3 to Ni-8]

Argon-ion bombardment (0.5–1 keV) with periodic annealing to 1000 to 1300 K either during bombardment [Ni-3] or in UHV [Ni-4] has been found to produce clean nickel surfaces. This basic treatment has also been combined with oxygen exposures (900 K, ~10^{-5} Pa, ~1000 s) and hydrogen reduction (900 K, ~10^{-5} Pa, ~1000 s) to remove residual carbon [Ni-5 to Ni-7]. Removal of sulfur has been effected by sputtering heated nickel with hydrogen ions (~2 keV, 10–30 μA/cm^2, 920 K) [Ni-8].

(100) Surface [Ni-5, Ni-9 to Ni-47]

Many authors found cycles of argon-ion sputtering (100–500 eV, 0.3–10 μA/cm^2) and annealing (625–1170 K) sufficient to obtain clean nickel surfaces [Ni-9 to Ni-23]. The number of cycles reported varied from 15 [Ni-21] to 100 [Ni-10]. Although most researchers sputtered the sample at room temperature, some reported satisfactory results with the sample at ~620 K [Ni-13, Ni-23]. One author reported the generation of a clean nickel surface by cycles of sputtering with neon ions at elevated temperatures (670–970 K) and annealing (670 K) in UHV [Ni-24]. Surfaces cleaned in the manner described above were frequently found to contain ~0.05 monolayer of carbon which resisted all attempts at removal. In those cases, a chemical treatment in oxygen was necessary [Ni-5, Ni-25 to Ni-39]. Most groups carried out this operation at pressures from 7×10^{-7} to 7×10^{-4} Pa and at elevated temperatures (550–1200 K) [Ni-5, Ni-25, Ni-27 to Ni-39]; however, exposure to oxygen at temperatures above 300 K has been reported to result in solution of oxygen into the bulk [Ni-26]. One interesting variation on this procedure consisted of simultaneously bombarding the sample with potassium ions in a background of oxygen gas [Ni-40]. The use of oxygen to aid in carbon removal occasionally left an oxygen residue on the cleaned surface [Ni-41 to Ni-47]. This residue was removed by high-temperature treatments in hydrogen (950–1400 K, 10^{-5} to 10^{-2} Pa).

(110) *Surface* [Ni-3, Ni-10, Ni-18, Ni-25, Ni-38, Ni-41, Ni-48 to Ni-78]

The (110) surface was usually cleaned by cycles of argon-ion sputtering (250–500 eV, 1–40 µA/cm^2) and annealing (470–1170 K) [Ni-3, Ni-10, Ni-18, Ni-48, Ni-49 to Ni-70] for up to 100 cycles [Ni-10]. However, some groups did use higher ion beam energies (2.8–3 keV) [Ni-56 to Ni-58]. Sulfur was observed to segregate to the surface between 870 and 1120 K [Ni-50, Ni-52]. Generally, repetitions of the sputter–anneal cycle outlined above depleted the sulfur from the near-surface region. However, many authors reported that carbon contamination persisted after the sputter–anneal cycles. This residual carbon contamination was removed by heat treatments in oxygen gas. This step was done at either elevated sample temperature with subsequent heating to (820–1070 K) [Ni-71, Ni-25] or at ambient temperature with subsequent heating to 670–870 K [Ni-72 to Ni-74]. Occasionally, this treatment left residual oxygen on the surface which was removed by reduction in hydrogen (650–1300 K, ~10^{-4} Pa) [Ni-41, Ni-44, Ni-45, Ni-75 to Ni-78]. Carbon and oxygen were also removed by electron-stimulated desorption (1500 eV) at 1070 K and 1270 K, respectively [Ni-78].

(111) *Surface* [Ni-5, Ni-10, Ni-18, Ni-25, Ni-26, Ni-31, Ni-44, Ni-45, Ni-79 to Ni-95]

Cycles of argon-ion sputtering (400–600 eV, 1–10 µA/cm^2) and annealing (473–1200 K) were reported to produce clean (111) surfaces in some cases [Ni-10, Ni-18, Ni-79 to Ni-84]. Carbon remaining on the surface could be removed by oxygen treatments (800–1100 K, 10^{-4} Pa) [Ni-5, Ni-25, Ni-31, Ni-85 to Ni-91]. In some studies, the decarburization process consisted of a lengthy treatment in oxygen (1170 K, 10^{-4} Pa) followed by repolishing of the crystal [Ni-85 to Ni-87]. Oxygen adsorption at ~300 K followed by flash heating to 850–1200 K was also effective in carbon removal [Ni-26, Ni-41, Ni-90, Ni-91]. Removal of residual oxygen was effected by treatment in hydrogen (820–1300 K, ~10^{-3} Pa) [Ni-41, Ni-44, Ni-45, Ni-76, Ni-92 to Ni-94]. Cleaning by heating alone has been reported (~100 cycles, 1450 K in UHV for 45 s) [Ni-95].

Niobium (Nb): bcc, T_m = 2741 K

Removal of carbon was the most difficult aspect of cleaning niobium surfaces. Many investigators found that mere heating for prolonged times above 2300 K was sufficient to clean the surface [Nb-1 to Nb-4]. An oxygen-anneal treatment was found to be effective by one group [Nb-5, Nb-6]. Multiple cycles of argon-ion sputtering followed by an anneal also produced clean surfaces [Nb-7, Nb-8].

Polycrystalline Surface [Nb-1, Nb-2, Nb-5]

Prolonged heating above 2300 K in UHV has been demonstrated to produce clean surfaces [Nb-1, Nb-2]. Decarburization at 2000 K in 1.3×10^{-4} Pa oxygen followed by heating in UHV at 2300 K also purified the surface [Nb-5].

(100) *Surface* [Nb-1, Nb-3, Nb-4, Nb-7, Nb-8]

Heating for several hours at 2300 K in UHV was shown to be quite effective in cleaning this plane [Nb-1, Nb-3, Nb-4]. An oxygen-anneal treatment was attempted but found unsatisfactory [Nb-3]. Ion bombardment with anneals above 2300 K has also yielded clean surfaces [Nb-7]. Numerous cycles of argon-ion sputtering (0.3–3 keV, ≤ 10 $\mu A/cm^2$) and anneals at 1000 K purged the sulfur and carbon from the sample but left a small, unspecified amount of oxygen on the surface of a crystal having approximately 0.1 at% oxygen in the bulk [Nb-8]. However, this low-temperature procedure may prove effective for higher-purity niobium.

(110), (111), *and* (750) *Surfaces* [Nb-6 to Nb-8]

Heating at 2300 K with a flash to 2700 K [Nb-7] and oxygen-anneal treatments (similar to those used to clean the polycrystalline surfaces) [Nb-6] have been found effective in cleaning the (110) planes. The only documented procedure for the (111) surface is to heat it at 2300 K in UHV followed by flashing to 2700 K [Nb-7]. A low-temperature bombardment-anneal may prove quite effective for cleaning the (750) surface [Nb-8].

Osmium (Os): hcp, T_m = 2973 K

The major contaminants in annealed osmium surfaces are carbon and sulfur. Heating osmium single crystals to high temperatures in a background of oxygen followed by heating in UHV or heating in UHV alone generally produced clean surfaces.

Polycrystalline Surfaces [Os-1]

Atomically clean surfaces of polycrystalline osmium have been prepared by repeated cycles of flashing to 1670 K in UHV followed by bombardment with 2-keV argon ions; holding the sample at 1670 K in UHV for periods of 10 min has been reported as a method for repeatably forming carbonized surfaces with approximately 0.4 monolayer of carbon.

(0001) *Surface* [Os-2, Os-3]

These crystal surfaces have been cleaned by cycles of heating to 1075 K in 1.3 $\times$ 10^{-5} Pa O$_2$ to reduce carbon, followed by heating to 2275 K in UHV to remove residual oxygen and sulfur [Os-2]. A similar treatment cycle of heating in 10^{-5} Pa O$_2$ at ~1000 K and heating in UHV to ~2000 K was also successful [Os-3]. These authors also found that heating in UHV to ~2000 K for several hours was sufficient to produce a clean surface [Os-3].

Palladium (Pd): fcc, T_m = 1825 K

The main contaminants of bulk Pd are sulfur and carbon which segregate to the surface during mild heating in UHV. Sulfur concentrations on the surface could be estimated directly from AES, but AES detection of less than 0.25 monolayer of carbon is difficult due to overlap of the C KLL peak at approximately 270 eV and the large Pd MNN peak at 279 eV [Pd-1]. Lower levels of carbon contamination were detected indirectly by CO and CO$_2$ desorption from a palladium surface following adsorption at room temperature [Pd-2]. Only O$_2$ was desorbed from a clean surface after oxygen adsorption [Pd-2, Pd-3]. The three most effective cleaning techniques were (1) reaction with oxygen to remove carbon as CO and CO$_2$ followed by flashing in UHV to desorb the oxygen as O$_2$, (2) alternate noble-ion sputtering and annealing, and (3) combinations of (1) and (2).

Polycrystalline Surface [Pd-4 to Pd-8]

Heating a palladium ribbon to 1500 K for a few minutes in UHV was reported to be sufficient to obtain a clean surface [Pd-6]. However, most investigators used more complicated cleaning procedures. The most effective oxygen-anneal treatment was to heat the material at 1000 K in 1.3 $\times$ 10^{-4} Pa oxygen and follow with flashing at 1200 K in UHV [Pd-4]. A similar oxygen anneal and flashing treatment with each of the two temperatures reduced by 200–300 K resulted in a sample that exhibited sulfur segregation after subsequent heating to temperatures above 600 K in UHV [Pd-7]. Repeated oxygen exposure and argon-ion bombardment at approximately 1200 K were found to generate a clean surface [Pd-5]. Cycles of argon-ion bombardment (10 keV, 1 μA/cm^2) and annealing at 900 K were also useful in obtaining a clean surface [Pd-8].

(100) Surface [Pd-1, Pd-9, Pd-10]

Sulfur and carbon were completely removed using argon-ion sputtering (500 eV, ~5 μA/cm^2, 900 K) for about 300 h per millimeter of sample thickness followed by annealing at 1100–1300 K. The rate of impurity removal appeared to be limited by the rate of segregation of sulfur and carbon to the surface [Pd-1]. In one case [Pd-1], elimination of the last traces of carbon required additional heating in oxygen (1000 K, ~7 × 10^{-5} Pa) followed by brief argon sputtering to remove the oxygen.

(110) Surface [Pd-3, Pd-11]

Initial cleaning has been accomplished by alternating cycles of inert-gas-ion sputtering and annealing [Pd-3, Pd-11]. Remaining traces of carbon were removed by treatment in ~3 × 10^{-5} Pa oxygen at 800 K followed by flashing at 1300 K to desorb oxygen [Pd-3].

(111) Surface [Pd-2, Pd-9, Pd-11 to Pd-13]

Most investigators found that repeated cycles of argon-ion sputtering and annealing were sufficient to remove sulfur, but complete carbon removal required an additional oxygen-anneal treatment (similar to that given for the polycrystalline surface).

(210) Surface [Pd-9, Pd-11, Pd-14]

Cleaning procedures were essentially the same as for the (111) surface.

(311) Surface [Pd-11]

Argon sputtering and annealing left some residual carbon.

Phosphorus (P): cub, T_m = 317 K

We found no surface studies on bulk or thin-film phosphorus.

Platinum (Pt): fcc, T_m = 2045 K

The major impurities detected on high-purity platinum single-crystal surfaces were carbon, calcium, and phosphorus. Other impurities commonly observed on platinum were sulfur, silicon, chlorine, and oxygen. Carbon presented a special problem because it dissolved into the bulk above 1423 K and precipitated onto the surface below 1348 K [Pt-1]. Carbon has been removed by three different procedures: (1) heating in a partial pressure of oxygen [Pt-2] (2) extended argon-ion bombardment [Pt-3], and (3) argon-ion bombardment at elevated temperatures (950–1150 K) [Pt-4]. Calcium segregated to the surface during heating [Pt-5, Pt-6]. Calcium and the other impurities could be easily removed by ion bombardment or by heating in UHV [Pt-5, Pt-7].

Polycrystalline Surface [Pt-8]

A high-purity platinum foil was cleaned by heating to 1400 K in UHV, sputtering with argon ions, heating in 1.3×10^{-4} Pa oxygen for 10 min at 1300 K, and, finally, annealing in UHV (1300 K, 10–20 min).

(100) *Surface* [Pt-2, Pt-7, Pt-9 to Pt-16]

Several investigators prepared clean surfaces by one or more cycles of heating in oxygen (1000–1300 K, 10^{-3}–10^{-5} Pa) and annealing in UHV (973–1700 K) [Pt-2, Pt-9 to Pt-11]. Others added an argon-ion bombardment step (300–500 eV) to the treatment cycle [Pt-7, Pt-12 to Pt-16]. One of the more lengthy cleaning procedures consisted of argon-ion bombardment (300 eV, 2 h) followed by anneals in oxygen (1173 K, 7×10^{-6} Pa, 15 h) and in UHV (1373 K, 7×10^{-8} Pa, 30 h) [Pt-12]. Two different surface structures have been observed by LEED on the cleaned surfaces after use of similar cleaning procedures: (5×1) [Pt-2, Pt-9, Pt-12 to Pt-14] and (5×20) [Pt-7, Pt-10, Pt-15, Pt-16]. The reasons for this apparent anomaly are not clear at this time.

(110) *Surface* [Pt-5, Pt-17, Pt-18]

A clean surface was prepared by one or more cycles of heating in oxygen (973–1300 K, 10^{-4}–10^{-5} Pa) and argon-ion bombardment. The ion bombardment was required to remove calcium [Pt-5, Pt-18]. Even after annealing the surface in oxygen ($>10^{-4}$ Pa), no oxygen was detected by AES on the reconstructed surface, but the surface structure was reported to be a function of the cleaning procedure used [Pt-18]. Argon-ion bombardment (500 eV) resulted in an unstable (1×1) structure that changed to a (1×2) structure when heated to 773 K. Heating above 1073 K in 10^{-4} Pa oxygen resulted in a (1×3) surface structure [Pt-18].

(111) *Surface* [Pt-1 to Pt-4, Pt-6, Pt-7, Pt-9, Pt-14, Pt-15, Pt-17]

Calcium segregated to the surface above 1400 K [Pt-6]; carbon precipitated on the surface when the sample was cooled below 1348 K [Pt-1]. Several cleaning procedures have been used to generate clean surfaces: (1) a sequence of one or more argon-ion bombardments (~500 eV), followed by heating in oxygen (673–1273 K, 7×10^{-4} Pa) and annealing in UHV at 1373 [Pt-7, Pt-14]; (2) argon-ion bombardment and heating in oxygen (673 K, 3×10^{-5} Pa) [Pt-15]; (3) oxygen treatment and

annealing without argon-ion bombardment [Pt-2, Pt-6, Pt-9]; (4) cycles of prolonged argon-ion bombardment (600 eV) and heating to 1400 K in UHV [Pt-3];
and (5) argon-ion bombardment for several hours at 950–1150 K followed by a
final flash in UHV to 1300 K [Pt-4].

(112), (113), (133), (122), and (012) Surfaces [Pt-19]

Carbon, phosphorus, and sulfur were removed by oxidation in 7×10^{-6} Pa
oxygen at 1075 K. The oxygen was then removed by heating to 1375 K in vacuum.
Removal of calcium required extensive oxidation heat treatments (24–48 h in 10^{-3}
Pa oxygen at 1500 K). This fixes the calcium on the surface in the form of a stable
oxide, which decomposes with Ca vaporization from the surface during a brief
heating to 1900 K. A small calcium impurity may also be removed using cycles of
argon-ion bombardment and 1100 K anneals.

Plutonium (Pu): mon; bc mon > 388 K, fc orth > 458 K, fcc > 583 K; bc tetr > 725 K; bcc > 753 K, T_m = 913 K

The cleaning of plutonium surfaces is complicated by the radiation hazard
resulting from alpha particle emission and by the low-temperature α-monoclinic to
β-body-centered-monoclinic phase transition, which is accompanied by a large
change in density. Interstitial impurities (sulfur, chlorine, phosphorus, oxygen, and
carbon) segregated to the surface during cleaning with carbon and oxygen were the
most difficult to remove. An oxygen-free surface could not be obtained by in situ
scraping with a titanium carbide blade [Pu-1]. AES and XPS measurements on a
plutonium sample showed that removal of oxygen, carbon, chlorine, and sulfur was
effected by the use of argon-ion-sputtering (1 keV, ~130 μA/cm^2); however, a large
amount of chlorine segregated to the surface upon heating (773 K, 30 min) and
plutonium carbide was formed from residual gases. After 25 sputter–heating cycles,
the surface was estimated to have 10 at% oxygen as oxides and 5 at% carbon as
carbides [Pu-2]. Because of the great affinity of plutonium for oxygen and carbon,
it is doubtful that a surface free of these two contaminants can be produced using
the purest metal currently available. The following recommended cleaning procedure has been formulated based on our experience with thorium and uranium [Th-2
to Th-6, U-8, and U-9]. Remove the gross contamination with argon-ion bombardment (0.5–4 keV, 10–50 μA/cm^2, 300 K) and follow with bombardment of the
sample beginning at a temperature near 676 K and continue with gradual cooling
to room temperature; this procedure may have to be repeated many times before a
clean surface is produced. Then anneal at a temperature below 676 K for several
minutes in a vacuum of less then 10^{-8} Pa. This procedure is applicable to polycrystalline specimens only, and the sample must be mounted to accommodate changes

in volume caused by the phase changes. For single-crystalline α-plutonium, the heating temperature cannot exceed 388 K.

Polonium (Po): mon, T_m = 527 K

We found no surface studies on bulk or thin-film polonium.

Praseodymiun (Pr): hex, T_m = 1204 K

We found no surface studies on bulk praseodymium. Praseodymium has been evaporated to form films.

Promethium (Pm): hex, T_m = 1350 K

We found no surface studies on bulk or thin-film promethium.

Protactinium (Pa): tetr, T_m = <1873 K

We found no surface studies on bulk or thin-film protactinium.

Radium (Ra): bcc, T_m = 973 K

We found no surface studies on bulk or thin-film radium.

Rhenium (Re): hcp, T_m = 3453 K

The majority of the authors surveyed found that rhenium is effectively cleaned by heating in oxygen ($2000 \leq T \leq 2500$ K, $\sim 10^{-4}$ Pa) and then annealing in UHV. The polycrystalline and single-crystalline samples obtained from various suppliers ranged from 99.95% to 99.995% pure. The contaminants most frequently encountered on rhenium surfaces were sulfur, chlorine, carbon, and nitrogen.

Polycrystalline Surface [Re-1 to Re-9]

Heating in oxygen (2000–2500 K, $\sim 10^{-4}$ Pa, 2–24 h) and then flash annealing in UHV ($\sim$2200 K) was the most cited method for producing clean rhenium surfaces [Re-1 to Re-6]. Samples subjected to this treatment produced crystallites that were oriented parallel to the (0001) plane [Re-1 to Re-5]. An alternative method that was also successful consisted of the following sequence: (a) conduct preliminary decontamination by argon-ion bombardment (2 keV); (b) heat in oxygen (1470 K, 2.7×10^{-6} Pa); (c) flash anneal in UHV (1970 K); (d) repeat (b) and (c) until clean [Re-7]. Another variation consisted of repeating argon-ion sputtering (2 keV) and annealing (1970 K) cycles until a contaminant-free surface was obtained [Re-8, Re-9].

(0001) *Surface* [Re-1, Re-10 to Re-14]

Unreconstructed (0001) surfaces were obtained by heating in oxygen (2200–2770 K, 10^{-5} to 10^{-4} Pa, ~2 min) followed by several flash anneals in UHV (2200–2770 K, ~20 s) [Re-1, Re-10, Re-11]. Argon-ion bombardment has also been utilized to produce clean (0001) surfaces by repeated cycles of sputtering (390 eV, ~10 $\mu A/cm^2$, 300 K) [Re-12] and annealing in UHV (1300 K or 1770 K, 30 s) [Re-13] until all impurities were removed. In one case [Re-13], this procedure was followed by a long anneal (1500 K, several h) without segregation of any impurity to the surface. One author reported obtaining a clean surface after heating the sample to 1770 K in UHV several times [Re-14].

Rhodium (Rh): fcc, $T_m = 2239$ K

Bulk rhodium samples of 99.9–99.999% purity have been reported to contain the following impurities: carbon, sulfur, boron, oxygen, silicon, and magnesium. Of these, boron was often the most troublesome because it migrated to the surface at ~1300 K. However, most workers reported methods for boron depletion that involved either argon-ion sputtering [Rh-1, Rh-2, Rh-5, Rh-6] or heat treatments in oxygen [Rh-1, Rh-5, Rh-6]. Heating in hydrogen (~600 K, ~10^{-5} Pa) was found to be a good way to remove traces of adsorbed oxygen in the final cleaning stage [Rh-1, Rh-5].

(100) *Surface* [Rh-1 to Rh-3]

In the initial cleaning stage, various surface contaminants were removed by argon-ion sputtering (500–2000 eV, ~5 μA, 300 K, ~10 min), and carbon was removed by either annealing in UHV or in oxygen (10^{-4} Pa) at 1000–1270 K [Rh-1, Rh-2]. This treatment was followed by repetition of the following steps: (a) argon-ion sputtering (as above), (b) heating in oxygen (~1270 K, ~10^{-5} to 10^{-4} Pa), (c) annealing in UHV at ~900 K for a few minutes, and (d) heating in hydrogen (~600 K, ~10^{-5} Pa). This treatment sequence eventually depleted boron in the samples [Rh-1, Rh-2]. In another study [Rh-3], carbon and sulfur were removed by repeated argon-ion bombardment (400 eV, 5 $\mu A/cm^2$, 30 min). A clean surface was achieved after subsequent annealing in oxygen (1.3×10^{-5} Pa, 1300 K, 20 min) and hydrogen (1.3×10^{-5} Pa, 600 K, 20 min) followed by a final flash to 1400 K.

(110) *Surface* [Rh-4, Rh-5]

Pretreatment by annealing the sample in flowing hydrogen at 1270 K for 90 h was reported to alleviate the boron problem [Rh-4]. This surface has been successfully cleaned by a combination of argon-ion sputtering (~300 eV, 10 $\mu A/cm^2$,

300–970 K), heating in oxygen (670–1300 K, 7×10^{-6} to 10^{-4} Pa), and annealing in UHV (1270–1300 K) [Rh-4, Rh-5].

(111) *Surface* [Rh-2, Rh-6, Rh-7 to Rh-9]

Argon-ion bombardment (1–2 keV, 1–10 μA, 300–1000 K, ~10 min) followed by "high temperature" treatment in oxygen (~10^{-5} Pa) and flash annealing to 1250–1300 K have been reported to produce clean (111) surfaces [Rh-2, Rh-6, Rh-7]. In another study [Rh-8], heating to 1300 K and cooling to room temperature removed the surface carbon but left sulfur and boron. The sulfur ceased segregating to the surface after an argon-ion bombardment treatment (500 eV, 4 μA/cm^2, several h), but the boron remained. The boron was removed by heating in oxygen (3×10^{-5} Pa, 1300 K, 20 h), and the oxygen was eliminated by a short anneal in vacuum at 1400 K. Others reported that carbon, not boron, was the troublesome impurity [Rh-9]. In this latter study, oxygen was easily removed by high-temperature flashing to >1100 K, but many cycles of high-temperature annealing (<950 K, several h) followed by heating in oxygen (1.3×10^{-5} Pa, 920 K, 10 s) served to segregate the carbon to the surface and to oxidize and volatize the carbon from the surface. Periodic flashes to >1200 K enhanced the carbon mobility; however, longer times at temperatures of >950 K caused the carbon to dissolve back into the bulk.

(210) *Surface* [Rh-10]

Initial cleaning was done by repeated cycles of argon-ion bombardment (500 eV, 1.7 μA), first at low temperature, then at increasingly higher temperatures (<1200 K). This treatment was followed by combined reactive gas and sputter cleaning. Carbon was removed by thermally cycling between 800 and 1200 K in 3×10^{-5} Pa oxygen. Excess oxygen was reacted off in hydrogen gas (750 K, 5×10^{-4} Pa, 30 min), but smaller amounts were desorbed by repeated heatings at 1400 K.

(331) *Surface* [Rh-9, Rh-11]

In one study [Rh-9], oxygen was easily removed by high-temperature flashing to >1100 K, but many cycles of high-temperature annealing (<950 K, several h) followed by heating in oxygen (1.3×10^{-5} Pa, 920 K, 10 s) served to segregate the carbon to the surface and to oxidize and volatize the carbon from the surface. Periodic flashes to >1200 K enhanced the carbon mobility; however, longer times at temperatures of >950 K caused the carbon to dissolve back into the bulk. In another study [Rh-11], carbon was removed by heating to 1270 K in UHV, sulfur was removed by argon-ion etching (2 keV, 300 K), and boron was eliminated by a few cycles of argon-ion bombardment (500 eV, 300 K) and annealing at 1070 K.

(332) *Surface* [Rh-12]

The major contaminant, carbon, was effectively removed by one or two routine cleaning cycles consisting of argon-ion bombardment followed by high-temperature catalytic reaction with oxygen. The resulting surface oxygen was eliminated by an anneal in hydrogen (3×10^{-5} Pa, 1020 K).

(775) *Surface* [Rh-11]

Carbon was removed by heating to 1270 K in UHV, sulfur was removed by argon-ion etching (2 keV, 300 K), and boron was eliminated by a few cycles of argon-ion bombardment (500 eV, 300 K) and annealing at 1070 K.

Ruthenium (Ru): hcp, T_m = 2583 K

The impurities most often found in ruthenium single crystals were carbon, sulfur, and oxygen. The absence of carbon on cleaned ruthenium surfaces was difficult to confirm by AES because the Ru MNN AES peak at 274 eV interferes with the C KLL AES peak at ~272 eV. One method for establishing the state of cleanliness has been to obtain a minimum in the intensity ratio $R = (Ru_{274} + C_{272})/Ru_{235}$, because the Ru AES peak at 235 eV is free of interfering effects from carbon [Ru-1, Ru-2]. The lowest R value found in a survey of published AES spectra was 1.57 for the (110) surface [Ru-1]; values between 1.85 and 2.05 were more typical [Ru-2 to Ru-4]. The absence of CO in the thermal desorption spectra from ruthenium exposed to several Langmuirs of oxygen has also been used to deduce the absence of carbon [Ru-5]. A series of weak peaks whose intensities varied with temperature was found between 300 and 500 eV in the AES spectrum of almost all of the crystal faces of ruthenium [Ru-1 to Ru-4, Ru-6, Ru-7]. These peaks could not be removed by up to 200 h of argon-ion sputtering [Ru-6]; they have been interpreted to be diffraction peaks characteristic of clean surfaces [Ru-7], although some ambiguity remains [Ru-6]. The low-index planes of ruthenium could be cleaned by a combination of heating in gas (usually oxygen at low pressures) and a brief anneal at a higher temperature.

(001) *Surface* [Ru-3 to Ru-16]

Most authors [Ru-3 to Ru-14] cleaned the (001) plane by repeatedly heating in oxygen (1300–1500 K, 7×10^{-6}–1.3×10^{-5} Pa, 10–15 s) and annealing (1500–1600 K, UHV) to remove adsorbed oxygen. An alternative method [Ru-15, Ru-16] was to heat the sample in hydrogen (1450–1500 K, ~7×10^{-5} Pa). This latter procedure was reported to remove adsorbed oxygen completely [Ru-15]. Some authors also carried out preliminary cleaning with argon-ion bombardment [Ru-4, Ru-12, Ru-14].

(110) *Surface* [Ru-1, Ru-17, Ru-18]

Repeated cycles of heating in oxygen ($\sim$1500 K, $\sim 10^{-5}$–$\sim 10^{-3}$ Pa) and annealing ($\sim$1400–1550 K) have been found to be effective for cleaning this plane.

(10$\overline{1}$0) *Surface* [Ru-2, Ru-19]

Two methods for cleaning this surface have been published. In the first method, the sample was heated to 1470 K, sputtered with argon, and annealed below 1270 K. This process yielded an AES spectrum with $R \approx 1.85$ at 470 K and sharp LEED patterns [Ru-2]. The second method consisted of a preliminary outgassing at 770 K and cycles of argon-ion bombardment and annealing at 1270–1470 K [Ru-19].

Samarium (Sm): rhdr, T_m = 1350 K

Polycrystalline samples of samarium have been cleaned by repeated argon-ion bombardment and annealing (conditions not specified) [Sm-1]. AES analysis of the sample disclosed "very small" amounts of carbon and oxygen after this treatment.

Scandium (Sc): fcc, T_m = 1814 K

Polycrystalline and (100) *Surfaces* [Sc-1 to Sc-3]

Preparation of an annealed surface that remains clean at room temperature has not been reported. Repeated cycles of argon-ion bombardment and annealing at temperatures above 1100 K produced a surface that remained clean only as long as the temperature was held above 1100 K. Oxygen, carbon, sulfur, and chlorine reappeared on the surface when the sample was cooled to ambient temperature. This phenomenon was found to occur for polycrystalline [Sc-1], (100) [Sc-2, Sc-3], and (001) [Sc-3] surfaces. In one AES study, the sulfur AES peak increased to a maximum at 850 K, but carbon and scandium peaks decreased slightly to a minimum near 850 K with only small changes in the oxygen and chlorine peaks [Sc-2]. Chlorine disappeared above 900 K, sulfur was not observed above 950 K, and oxygen and carbon vanished at 1100 K.

(0001) *Surface* [Sc-4]

Krypton-ion bombardment at room temperature and at elevated temperatures was used to reduce contamination by oxygen, carbon, sulfur, and chlorine. After 15 h of ion bombardment (1 keV, 20 μA/cm^2, 873–1173 K), carbon, sulfur, and chlorine were removed from the surface. Ion bombardment at 1173 K for a few hours was necessary to remove accumulations of sulfur and chlorine. However, even after 50 h of bombardment at 573–873 K, 5% of a monolayer of oxygen was still present. Subsequent cleaning of carbon and oxygen was achieved easily by sputtering at room temperature and at 773 K and by annealing for 40 min at 773 K.

Selenium (Se): hex, T_m = 490 K

For the $(10\bar{1}0)$ and $(1\bar{2}10)$ surfaces, the most important contaminant was carbon. High-purity (99.999% pure) crystals were cleaned by several cycles of argon-ion bombardment (500 eV, 1 μA/cm^2, 0.5 h) and annealing at 423 K for 0.5 h [Se-1].

Silicon (Si): cub-dia, T_m = 1683 K

Carbon and oxygen, both adsorbed from the atmosphere, are the major impurities on high-purity, single-crystal silicon surfaces. Cleaning these surfaces has been accomplished in the electronics industry almost exclusively by wet chemical processes [Si-1], while UHV studies have generally used wet chemistry as a preliminary step that was followed by procedures that can be performed only under vacuum conditions, such as heating and sputtering [Si-2]. However, these latter procedures frequently induce undesirable structural and chemical changes on the surface [Si-3].

An excellent summary of the knowledge about semiconductor surfaces and their preparation with the "conventional UHV" techniques has been given by Hanemann [Si-4]. Carbon is the most difficult to remove in vacuum because of its low sputter yield and its high thermal stability. Wet chemical techniques have as their objectives: (1) to etch away the natural oxide layer and all structural damage caused by prior mechanical treatments in such a way that a predictably smooth crystalline surface remains and (2) to terminate the surface with a hydrophilic (oxide terminated) [Si-5 to Si-13] or a hydrophobic (hydrogen terminated) [Si-14 to Si-21] layer. Both of these layers are so unreactive that they can be transferred to processing equipment without picking up atmospheric contaminants and so lightly bonded to the substrate that they can be removed by low-temperature heating. In both hydrophilic and hydrophobic layer formation, dry (e.g., ozone) oxidation processes may be used as intermediate steps [Si-22 to Si-24]. Silicon surfaces prepared by such "wet chemical" means can be made cleaner than those prepared conventionally in UHV. Moreover, hydrogen-terminated silicon surfaces exhibit a unique surface morphology that can be controlled over a wider range than is usually accessible by UHV preparation methods. Coupling this with the fact that the surfaces are unreconstructed suggests that H-terminated silicon surfaces prepared by wet chemistry are ideal for conventional *in situ* experiments in UHV [Si-25]. Indeed, UHV systems have been built with a chamber attached for wet-chemical pretreatment [Si-26]. Given the clear advantages of hydrogen-terminated surfaces, many dry processes have been developed to produce hydrogen-terminated surfaces *in situ* using electron-cyclotron-resonance-generated hydrogen plasma ions [Si-27, Si-28], RF-generated hydrogen atoms [Si-29], or photoexcited fluorine gas diluted with hydrogen [Si-30]. In summary, the four most common methods for preparing a clean silicon surface are [Si-31] (1) hydrophobic or hydrophilic surface pretreatment and heating between 1000 K to above 1473 K, (2) surface pretreatment and *in situ* UV/ozone beam oxidation or XeF or H-ion or H-atom reaction and heating, (3) surface pretreatment followed by argon-ion bombardment and annealing (~1000 K), and (4) *in situ* cleaving of a (111) surface.

Polycrystalline Surface [Si-32]

Polycrystalline surfaces have been cleaned by brief argon-ion sputtering (5 keV, 2.5 mA/cm^2, 10 s) without creating any surface texture. However, changes in the surface texture were observed after 45 s of sputtering.

(100) *and Vicinal Surfaces* [Si-4, Si-24, Si-26, Si-27, Si-29, Si-30, Si-33 to Si-43]

The (100) crystal face was cleaned by repeated cycles of argon-ion bombardment (500 eV) and annealing (1173 K, UHV) [Si-33, Si-34] or by simply heating the crystal in UHV to 1473–1523 K for 15 to 150 s [Si-35 to Si-37]. Although the atomic structures of the (100) vicinal surfaces are quite similar to that of the (100) surface, a different procedure has been used to clean the vicinal surfaces. Repeated cycles of argon-ion bombardment (350 eV, 20 min) and heating (1373 K, 2 min; then 1223 K, 30 min) produced a clean AES spectrum and an optimal RHEED pattern for a (100) vicinal surface [Si-38]. A HF-dipped (hydrophobic) surface rinsed with deionized water and cleaned of remaining impurities by an approximately 10-min exposure to XeF gas at 10^{-2} Pa was found to be clean but amorphous after heating to 723 K. Only after heating at 1023 K for 2 min was a perfect (2 × 1) LEED structure observed [Si-39]. An oxidized surface was dipped in HF and rinsed in isopropanol, but gave a clean LEED pattern only after repeated cycles of ion bombardment and annealing at 1200 K [Si-40]. A hydrogen-plasma-based technique for carbon removal has been combined with a modest anneal for oxide desorption at 1000 K to produce an atomically clean Si(100) (2 × 1) surface [Si-27]. Exposure to RF-generated hydrogen atoms at 623 K removed all carbon contamination from surfaces inserted into UHV after hydrophilic or hydrophobic pretreatment [Si-29]. Exposure to photoexcited fluorine diluted with hydrogen and annealing produced a clean surface [Si-30]. A carefully prepared hydrophobic surface quickly introduced into UHV shows a simple (1 × 1) pattern [Si-41 to Si-43].

(110) *Surface* [Si-4, Si-44 to Si-50]

Clean surfaces have been prepared by ion bombardment or by heating in vacuum; details were not given [Si-44]. In another study, heating produced a surface with a ratio of AES signals C(KLL)/Si(LVV) of 1/100. Bombardment with 500-eV argon ions and annealing did not reduce the carbon concentration. Subsequent bombardment with oxygen ions (500 eV, 1.3×10^{-3} Pa) led to formation of a thin oxide layer. Flashing to 1300 K evaporated the oxide and yielded a silicon surface with C(KLL)/Si(LVV) = 1/2000 [Si-45]. After hydrophobic pretreatment and quick introduction into UHV, a clean LEED pattern appeared [Si-46]. After hydrophilic pretreatment, cycles of simultaneous argon-ion bombardment and annealing (SIBA [Si-46]) (20 min, 10 min at 873 K, 800 eV, 2 μA/cm^2, 45°, and 60 min, 1173 K annealing), then slow cooling to room temperature produced good LEED patterns [Si-47]. After long and meticulous hydrophilic pretreatment [Si-49], heating for 20 s at 1173 K and 10 nPa produced a clean surface [Si-50].

(111) *Surface* [Si-4, Si-26, Si-29, Si-31, Si-35, Si-36, Si-41, Si-51 to Si-57]

A clean, reconstructed surface has been prepared by cleaving in UHV; the resultant (2 × 1) LEED structure changed to a (1 × 1) structure on heating to 700 K, and to a (7 × 7) structure on heating to 1000 K [Si-31]. The clean (111)–(7 × 7) reconstructed surface was also achieved by argon-ion bombardment (2 keV, 4 μA/cm^2) and annealing (1073 K, 30 min) [Si-51]. In most of the surface studies, rapid or flash heating to 1473–1523 K was used to prepare a clean surface. Conflicting reports exist on carbon removal: heating in vacuum to 1200 K and slowly cooling to room temperature resulted in a Si(111)–(7 × 7) surface that had less than 5% of a monolayer of carbon [Si-52], whereas heating to only 1183 K in UHV did not remove carbon but produced a carbide structure [Si-53]. A novel cleaning procedure has been used on surfaces preoxidized in an oxidizing etch to form a 2–3-nm-thick layer of SiO$_2$ on silicon [Si-54]. Once under vacuum, the carbon contamination was desorbed almost completely upon heating to 1073 K. Holding the crystal at this temperature while exposing it to an atomic gallium flux of ~0.1 monolayer/s for 15 min results in a clean (111) surface without detectable (by AES) carbon, oxygen, or gallium. Presumably, volatile SiO and GaO$_2$ are the products of the Ga + SiO reaction with the excess gallium (vapor pressure ≈1 mPa) being evaporated. Although heating at 1073 K can also result in a clean surface, the time required can be much longer than that for the gallium treatment [Si-55]. A hydrophilic oxide layer was desorbed at 1173 K and the remaining SiC particles were removed by repeated flash heating to 1523 K; the result was a clean surface with the (7 × 7) structure [Si-56]. Clean (111) and (111) vicinal surfaces have been produced by forming a 1-μm-thick layer of oxides, dipping in HF, rinsing with deionized water, and then annealing in UHV (10–20 nPa, 1473 K, 1–2 min) [Si-57]. Carefully prepared hydrophobic surfaces show barely detectable oxygen and carbon contamination and clear LEED patterns. Hydrogen desorbs at ~800 K [Si-41].

(210), (211), (311), (320), (331), (510), (511) *(High-Miller-Index) Surfaces*

Clean surfaces resulted from oxidation to form a 1-μm-thick layer of oxide, dipping in HF, rinsing with deionized water, and annealing in UHV (10–20 nPa, 1473 K, 1–2 min) [Si-58].

Silver (Ag): fcc, T_m = 1235 K

Polycrystalline and single-crystalline samples of silver (99.99%–99.9999% pure) have been found to contain mainly sulfur, carbon, oxygen, and chlorine impurities. The most commonly used method for establishing surface cleanliness has been AES. However, the C KVV AES transition at ~272 eV overlaps the Ag MNN transitions at ~260 and 266 eV. In order to establish the absence of carbon, most investigators relied on minimizing the AES peak intensity ratio R, where R = (C_{272} + $Ag_{260 + 266}$)/Ag_{303}, because the Ag MNN AES transition at ~303 eV is well separated from the carbon AES features. Values of R for clean silver have been recorded in the range of 0.42–0.55, depending on the modulation voltage used to obtain the spectrum. There is also an overlap of the N KLL peaks at ~348 and 360 eV with the Ag MNN peaks at ~351 and 356 eV. The cleaning of silver samples in UHV was usually accomplished by a combination of argon-ion sputtering and annealing.

Polycrystalline Surface [Ag-1 to Ag-3]

Silver foils have been cleaned by argon-ion bombardment [Ag-1] at 2 keV [Ag-2] or by prolonged heating to 1000 K in UHV [Ag-3]. Free evaporation of silver took place in the latter case, and the resulting surface consisted mainly of crystallites oriented in such a way as to expose the (111) surfaces.

(100) Surface [Ag-4]

The initial cleaning of (100) surfaces has been accomplished by repeated cycles of argon-ion bombardment (200–500 eV) and annealing in UHV (670 K) until sharp LEED patterns and R values of 0.40–0.45 were obtained. Subsequent cleaning was carried out by heating to 620 K in UHV. Occasionally, another cycle of sputtering and annealing was required. Embedded argon was observed to desorb between 300–450 K.

(110) Surface [Ag-3 to Ag-13]

The (110) surface has been cleaned by cycles of argon-ion sputtering and annealing in UHV. Most of the authors reviewed preferred to use a comparatively gentle ion bombardment (300–600 eV, 1–3 μA/cm^2, 300 K) [Ag-3 to Ag-7]; however, others executed this operation at higher ion energies (1 keV [Ag-10], 5 keV [Ag-11]). The annealing temperatures ranged from 670 to 720 K [Ag-4 to Ag-6, Ag-8], although some groups did anneal their samples at a higher temperature (800 K [Ag-9, Ag- 10]; 1000 K [Ag-7]). Oxygen dissolved into the bulk upon heating a (110) sample in this gas [Ag-13]. Tellurium segregation from the bulk induced a $c(2 \times 2)$ surface structure after heating one sample above 570 K for the first time [Ag-4].

(111) Surface [Ag-4, Ag-14 to Ag-20]

Cycles of inert-gas-ion sputtering followed by UHV annealing were commonly used to clean this crystal face. Most authors used argon ions [Ag-4, Ag-14 to Ag-17]; however, one did use xenon ions [Ag-18]. Sputtering was carried out with ion energies ranging from 200–500 eV [Ag-4, Ag-14, Ag-18] to 2 keV [Ag-15] and with an ion current density of 1 μA/cm^2 [Ag-14] or total ion currents of 3–40 μA [Ag-18, Ag-15]. The sample temperature during bombardment was 300 K [Ag-4, Ag-14], 420 K [Ag-17], or 750 K [Ag-15]. Annealing was performed in UHV at 570 to 900 K. Heating samples in oxygen and then annealing in UHV also produced clean surfaces [Ag-19, Ag-20]. Tellurium segregation to the surface created a $(\sqrt{3} \times \sqrt{3}) - 30°$ structure after the crystal was heated above 570 K for the first time; however, the tellurium was easily removed by argon-ion bombardment [Ag-4].

(331) Surface [Ag-21]

Clean (331) surfaces have been prepared by many repetitions of argon-ion sputtering (300 eV, 1 μA/cm^2, 300 K) and annealing (750 K, UHV). The (311) surface was unreconstructed and was stable to <900 K, the temperature at which evaporation became significant.

Sodium (Na): bcc, T_m = 371 K

We found no surface studies on bulk sodium. However, pure sodium films have been routinely prepared by evaporation techniques.

Strontium (Sr): fcc; hcp > 488 K; bcc > 878 K, T_m = 1042 K

Since a thick oxide layer formed on polycrystalline strontium after a few seconds of air exposure, the target was mechanically cleaned in a vessel filled with tetrachlorethylene, then inserted into the vacuum system under a nitrogen atmosphere. The surface was cleaned *in situ* by argon-ion bombardment (total ion dosage: ~6 × 10^{18} Ar/cm^2 at 3 keV) [Sr-1]. Alternatively, a clean surface was generated and maintained by continuous *in situ* scraping [Sr-2].

Tantalum (Ta): bcc, T_m = 3269 K

The most frequently found impurities in outgassed tantalum samples were sulfur, carbon, and oxygen. Heating tantalum single crystals in vacuum to within 10% of their melting point is reported to produce clean surfaces. However, many groups also used a high-temperature treatment in oxygen.

Polycrystalline Surface [Ta-1, Ta-2]

Carbon and sulfur were removed from polycrystalline tantalum by heating in oxygen (>2000 K, 10^{-5} Pa) [Ta-1]. Sulfur, titanium, and scandium contaminants were removed by argon-ion bombardment, and carbon was eliminated by heating to >1400 K in vacuum [Ta-2].

(100) *Surface* [Ta-3 to Ta-6]

The (100) crystal plane was cleaned by heating alternately to 2800–3000 K and 1800–2000 K for 1 min each [Ta-3 to Ta-5]. Another method consisted of heating the crystal to 2300 K, first in vacuum and then in 1.3×10^{-4} Pa oxygen, and finally flashing in UHV (>2300 K) to remove the last of the oxygen [Ta-6], presumably by desorption of oxides.

(110) *Surface* [Ta-7, Ta-8]

This crystal face has also been cleaned successfully by heating in UHV to ~2670 K. It was held at this temperature until the vacuum system reached its base pressure.

Technetium (Tc): unknown, T_m = 2445 K

Polycrystalline surfaces have been cleaned by repeatedly heating to 2273 K in a vacuum of <1.3×10^{-7} Pa [Tc-1].

Tellurium (Te): hex, T_m = 723 K

Reaction of tellurium with atmospheric oxygen and water vapor is an activated process requiring temperatures in excess of 333 K [Te-1]. Thus, it is possible to prepare tellurium surfaces in air using polishing or cleaving techniques, and then to transfer the crystal quickly to a UHV vacuum system with minimal contamination of the surface by oxygen [Te-1, Te-2]. The main surface impurity was carbon which has been removed by either argon-ion sputtering and annealing or sublimation of the surface layers of the tellurium (the vapor pressure of tellurium is about 1.3×10^{-4} Pa at 473 K). Care must be taken when subliming tellurium because some thermal etching has been observed after heating at 443 K for 15 min, and marked thermal etching occurred after heating at 563 K for 30 min [Te-3]. The only polycrystalline surfaces studied were thin films evaporated on substrates.

(0001) *Surface* [Te-2 to Te-4]

By limiting air exposure to one quarter of an hour after electropolishing and rinsing, these surfaces were found to be clean to the 0.1 monolayer level for oxygen and to the 0.05 monolayer level for carbon using electron-excited X-ray analysis [Te-2]. A similarly prepared surface was cleaned by heating at 423 K in UHV for short periods without thermally etching the surface [Te-3]. Another electrolytically polished surface did not yield an LEED pattern after heating to 473 K for 1 h, but successive argon-ion sputtering (150 eV, 1 μA/cm^2, 15 min) and annealing (473 K, 30 min) gave surfaces with well-defined LEED patterns; however, no element-specific analysis was performed to verify the cleanliness [Te-4].

(10$\overline{1}$0) *Surface* [Te-2 to Te-6]

This surface has been prepared by cleaving in air or in UHV. Cleaving in UHV produced surfaces that gave good LEED patterns [Te-4]. Air-cleaved surfaces were found to have ~0.25 monolayer of carbon and ≤0.05 monolayer of oxygen after 15 min of air exposure [Te-2]. The carbon was removed by either (a) heating in UHV (at 423 K [Te-3] or 548 K [Te-5]) for a short time or (b) repeated argon-ion sputtering (150–250 eV, 1 μA/cm^2, 1–15 h) and annealing at 473 K from a few minutes to 1 h [Te-4, Te-6]. In one study, the electron energy loss spectrum continued to change with time after AES indicated a carbon- and oxygen-free surface. The cleaning process was repeated until the energy loss spectrum remained constant and AES continued to indicate a clean surface [Te-6].

(1$\overline{2}$10) *Surface* [Te-7]

A chemically polished surface was annealed at temperatures for which subli-mation was small yet sufficient to remove some impurities. Argon-ion bombard-ment (550 eV, 2 μA/cm^2, 1 h) cleaned the surface, and annealing (523 K, 1 h) recrystallized the surface.

Terbium (Tb): hex, T_m = 1629 K

Polycrystalline samples of terbium were found by AES to contain "small amounts" of carbon and oxygen after repeated cycles of argon-ion bombardment and annealing (conditions not specified) [Tb-1].

Thallium (Tl): hcp; bcc > 503 K, $T_m = 577$ K

Samples were cut from 99.999% pure polycrystalline ingots, and clean surfaces were produced by mechanical scraping at a pressure of less than 10^{-6} Pa [Tl-1]. Immediately after scraping, the carbon and oxygen $1s$ photoemission lines could not be detected; nevertheless, the thallium surface yielded detectable $1s$ photoemission signals from both carbon and oxygen during a 24-h period in the analyzing chamber maintained at $\sim 10^{-7}$ Pa by an ion pump.

Thorium (Th): fcc; bcc > 1673 K, $T_m = 2023$ K

The most common impurities found in high-purity thorium are sulfur, chlorine, phosphorus, oxygen, and carbon. The first three elements are relatively easy to remove, while oxygen and carbon require extensive treatment. Clean polycrystalline and single-crystal surfaces have been produced by argon-ion sputtering followed by annealing at temperatures above 675 K. Annealing temperatures should be kept below 1673 K, where a phase transformation from α-Th (fcc) to β-Th (bcc) occurs. Once a thorium surface is clean, it is difficult to maintain in this condition due to the rapid chemisorption of CO from the residual gases in vacuum systems.

Polycrystalline Surface [Th-1, Th-2]

The segregation of impurities (sulfur, carbon, and phosphorus) to the surface of bulk polycrystalline thorium has been studied in detail [Th-1]; impurities, present in the bulk in the part-per-million range, segregate rapidly to the surface at elevated temperatures. The surface was found to saturate with sulfur at 1173–1373 K; this sulfur diffused into the bulk at 1373–1443 K. Carbon and phosphorus diffused into the bulk at ~753 K and ~1060 K, respectively. Extended argon-ion bombardment (0.5–5 keV, 5 µA, 5.2×10^{-3} Pa, 6 h) removed gross contamination, and a combination of high-temperature annealing (400–775 K) and ion bombardment at temperatures above 673 K was used to remove carbon and oxygen from thorium [Th-2].

(100) and (111) Surfaces [Th-3 to Th-6]

Cleansing these crystal surfaces of carbon, oxygen, and other interstitial contaminants has been accomplished by the following procedure: (a) removal of gross surface contamination by argon-ion bombardment (0.5–4 keV, 10–50 µA/cm^2, 300 K); (b) argon-ion bombardment of the sample, beginning at a temperature near 1000 K and continuing during gradual cooling to room temperature (repeat until clean surface is produced); and (c) anneal at a temperature near 1000 K for several minutes in a vacuum of less than 10^{-8} Pa. This procedure is also applicable to polycrystalline samples.

Thulium (Tm): hcp, T_m = 1818 K

We found no surface studies on bulk thulium. Thulium has been evaporated to form films.

Tin (Sn): cub; tetr > 286 K; rhdr > 434 K, T_m = 505 K

Tin is a soft metal requiring special precautions in polishing and mounting. The transformation between the cubic and tetragonal phases is accompanied by a change in volume and can cause creation of local "warts" on the surface or disintegration of the solid to a coarse powder.

Polycrystalline Surface [Sn-1 to Sn-4]

Clean tin surfaces have been produced *in situ* by micromilling with a rotating diamond edge [Sn-1] or by scraping with a tungsten carbide blade in a preparation chamber at ~10^{-7} Pa prior to transfer into the UHV chamber [Sn-2]. Clean surfaces were also prepared from zone-refined (99.999+% pure) tin by *in situ* argon-ion sputtering (1–2 keV, 20 μA/cm^2). After sputtering, no structure was observed in the tin AES spectra indicative of the presence of impurities (carbon, oxygen, etc.) up to electron energies of 2000 eV [Sn-3, Sn-4].

(100) Surface [Sn-5]

A (100) single crystal (99.999% purity, β phase) was cleaned by ion sputtering and *in situ* annealing. Exposure of the sample to a vacuum of ~3×10^{-8} Pa for 10 h resulted in a surface having no detectable AES signal from impurities and no change from the initial LEED pattern.

Titanium (Ti): hcp; bcc > 1155 K, T_m = 1933 K

Titanium is one of the most reactive transition metals. Its ability to decompose simple gases and to form stable compounds with the products has led to the widespread use of freshly deposited titanium films as getters for vacuum pumping. This high reactivity means that titanium is very difficult to clean and to maintain in a clean state; all the studies cited below showed traces of residual carbon and oxygen on titanium surfaces. The cleaning of titanium is further complicated by the hcp-to-bcc phase transition that occurs at ~1155 K; thus, single-crystal samples must be cleaned at lower temperatures. Sulfur was found to be the most persistent surface contaminant as it segregated to hot surfaces. In many instances traces of sulfur remained on the surface despite extensive efforts to remove it [Ti-1 to Ti-4]. The most effective purification procedure involved sulfur depletion by argon-ion sputtering with the sample at elevated temperature.

Polycrystalline Surface [Ti-4 to Ti-8]

Simultaneous argon-ion bombardment and annealing at 1070 K were found to result in a surface free from sulfur in one case [Ti-6] but not in another [Ti-4]; slight amounts of carbon and oxygen contamination were detected in each case. Argon-ion sputtering at 300 K coupled with brief anneals was reported to produce a sulfur-free surface, but extended annealing caused the sulfur to reappear [Ti-7]. Argon-ion sputtering (600 eV, 10 $\mu A/cm^2$) coupled with annealing at 1020 K reduced, but did not eliminate, sulfur and chlorine contamination [Ti-5]. However, inert-ion bombardment and annealing at 1733 K were effective in producing a clean surface [Ti-8].

(0001) *Surface* [Ti-1 to Ti-3, Ti-9 to Ti-11]

The most successful procedure for cleaning the basal plane of titanium consisted of a cycle of sputtering with argon ions (600 eV, 4 $\mu A/cm^2$, 300 K) and annealing in UHV at 1020 K until 50 h of sputtering time were accumulated; this cycle was followed by sputtering with argon ions (600 eV, 4 $\mu A/cm^2$, 1020 K, ~14 h) and annealing in UHV (1020 K, 4 h) [Ti-9, Ti-10]. Subsequent recleaning was achieved by sputtering with argon ions at 300 K (600 eV, 30 min) or at 1020 K (500 eV, 1.2 h) followed by annealing (1020 K, 1 h). One author was able to obtain a sulfur-free surface by cycles of ion sputtering and annealing [Ti-11], while others were not [Ti-1 to Ti-3].

($10\overline{1}0$) *and* ($10\overline{1}1$) *Surfaces* [Ti-12, Ti-13]

One author generated a clean ($10\overline{1}0$) surface by argon-ion etching at 970–1070 K [Ti-12], while another achieved the same result on the ($10\overline{1}1$) plane by (a) argon-ion bombardment (b) annealing at 820 K (several hours), and (c) repeating (a) and (b) until no impurities could be detected by XPS [Ti-13].

Tungsten (W): bcc, T_m = 3683 K

Carbon, which originates in the bulk and segregates to the surface when tungsten is heated, is the most difficult contaminant to remove. The two most widely used techniques for the initial cleaning of a tungsten surface are (1) prolonged heating at a high temperature in UHV and (2) reaction with oxygen to remove the carbon in the form of CO [W-1] followed by flashing at high temperature in UHV to desorb the oxygen as tungsten oxides. The effectiveness of the oxygen-anneal treatment varied considerably among crystals from different suppliers [W-2]. Also, there is evidence that the anneal step removed pits that were formed at the surface during the oxygen treatment [W-2].

Polycrystalline Surface [W-3 to W-14]

Repeated flashings or prolonged heating to >2500 K in UHV have yielded clean surfaces in some instances [W-1, W-10 to W-12]. However, in most studies, such simple heating procedures have not yielded atomically clean surfaces. In fact, heating for up to 10 min at 3000 K in UHV did not remove carbon once it had segregated to the surface [W-7]. The equilibrium concentration of carbon segregated to the surface was a function of temperature, but carbon dissolved into the bulk above 2200 K [W-4]. The carbon concentration was quite large at temperatures of 1500–1800 K; efficient removal of the carbon from the sample occurred by reaction at these temperatures with oxygen at $\sim 10^{-5}$ to $\sim 10^{-4}$ Pa. The tungsten oxides remaining on the surface were removed by one of three procedures: (1) flashing to above 2400 K in UHV [W-6 to W-9, W-13], (2) heating in hydrogen at 7×10^{-5} Pa for 2 min at 2200 K [W-4], or (3) sputtering with argon ions followed by anneals at 1800 K [W-14].

(100) *Surface* [W-2, W-9, W-15 to W-39]

Some evidence exists that prolonged heating at 2200–2500 K followed by flashing to 2800–3000 K yields clean surfaces [W-15, W-17, W-21]; however, in most cases the oxygen-anneal treatment described for polycrystalline surfaces was required to initially prepare a clean surface [W-9, W-16, W-18, W-19, W-22 to W-39]. A maximum in carbon segregation onto the surface has been reported to occur at 1500 ± 100 K [W-27], which is consistent with results on polycrystalline tungsten. In one study [W-20], the oxygen-anneal treatment was not effective. In this study, 20 cycles of krypton-ion sputtering with the sample at 1300 K, with flashes to 2500 K between 1-h bombardments, were required to remove the carbon.

(110) *Surface* [W-1, W-28, W-34, W-39 to W-50]

Heating this surface to high temperatures in UHV is not sufficient for the generation of clean surfaces. The (110) and vicinal surfaces on the same samples have been cleaned by the oxygen-anneal treatment detailed for the polycrystalline surfaces [W-42, W-43, W-47]. A sputter–anneal treatment has also been reported to be effective in cleaning these surfaces [W-41].

(111) *Surface* [W-39, W-51, W-52]

Carbon was removed successfully from these surfaces by heating in UHV at 2500 K [W-51]. However, the oxygen-anneal treatment was preferred [W-39, W-52].

(112) *Surface* [W-48, W-53, W-54]

This face has been cleaned by either sputter–anneal or oxygen-anneal treatments as outlined above.

Uranium (U): orth; tetr > 941 K; bcc > 1048 K, T_m = 1405 K

All the clean-surface studies reviewed were performed using polycrystalline material. Interstitial impurities (i.e., sulfur, chlorine, phosphorus, oxygen, and carbon) have been the most common contaminants, but small amounts of calcium and iron were often present. The most difficult contaminants to remove were carbon and oxygen. Heating to 1073 K in UHV was not sufficient to eliminate surface impurities [U-1 to U-4]. Many investigators found that uranium could be cleaned in two stages: (a) argon-ion sputtering (0.5–5 keV, 10–40 μA/cm^2, 300 K, 1–3 days) and (b) cycles of argon-ion sputtering at high temperatures (300–850 K) and annealing in UHV (800–1170 K) [U-5 to U-9]; however, this procedure was usually repeated several times. One group created a clean surface by argon-ion bombardment alone (5 keV, 10 μA/cm^2, 300 K, several h) [U-10]. In another study, a small amount of iron (~0.1 monolayer) was observed to segregate to the surface upon cooling from 850 to 300 K; this iron was removed by a gentle argon-ion bombardment (500 eV, 1 μA/cm^2, 300 K, 15–30 s) [U-8]. A different procedure consisted of a chemical treatment in oxygen (10^{-5} Pa, 1070 K, 30 min) and subsequent reduction in hydrogen (10^{-5} Pa, 1070 K, 10 min); this procedure effectively removed phosphorus and sulfur, but some oxygen remained [U-11]. Clean surfaces have also been produced by *in situ* abrasion with a diamond file [U-12]. For single-crystal alpha-uranium, we recommend the two-stage sputter–annealing cycle outlined above with the sample temperature never exceeding 850 K during annealing (this procedure avoids the orthohombic to tetragonal phase transition at ~941 K). Maintaining a clean surface on uranium has been difficult because of rapid chemisorption of carbon monoxide, a common residual gas in UHV systems.

Vanadium (V): bcc, T_m = 2163 K

Inert-gas-ion bombardment at high temperatures has been necessary to produce clean vanadium surfaces on single crystals of nominal 99.99% purity [V-1, V-2]. A majority of the investigators reporting on this metal were unable to produce surfaces free of oxygen, sulfur, or carbon [V-3 to V-10]; these elements were suspected of stabilizing the (100)–(1 × 1) structure. Interpretation of AES spectra of vanadium surfaces is complicated by the overlap of the vanadium LVV transition at ~509 eV with the oxygen KVV transition at ~512 eV.

(100) *Surface* [V-1, V-5 to V-9]

A clean surface was obtained after approximately 200 h of neon-ion bombardment with the sample at 800 K; a (5 × 1) reconstructed surface was produced. The (5 × 1) structure undergoes a reversible phase transition to the normal (1 × 1) structure at 630 K [V-1]. Previous LEED investigations [V-5 to V-9] reported normal (1 × 1) structures but also significant amounts of sulfur on the surface, which may have stabilized this structure.

(110) *Surface* [V-2, V-10]

Clean surfaces have been prepared by sputtering with argon ions (2 keV, 20 $\mu A/cm^2$, 670 K, ~80 h) and annealing in UHV (970–1070 K). Shortened sputter–anneal cycles were repeated until evidence of oxygen segregation (by AES or the appearance of the oxygen-induced (6 × 2) LEED pattern) was not observed after prolonged anneals [V-2].

Ytterbium (Yb): fcc, T_m = 1092 K

The only procedure documented to yield clean surfaces on bulk polycrystalline ytterbium was mechanical removal of a macroscopic surface layer in UHV with a tungsten carbide blade [Yb-1]. Argon-ion sputtering and annealing have been tried but some oxygen was left on the surface [Yb-2]. However, repeated cycles of this last procedure should eventually lead to a clean surface.

Yttrium (Y): hcp, T_m = 1795 K

The major contaminants found in yttrium foil samples were sulfur, carbon, chlorine, and oxygen [Y-1]. These could be removed by argon-ion bombardment, but chlorine always reappeared on the surface after annealing [Y-2]. One author found yttrium to be very reactive toward oxygen and reported that it was not possible to clean this metal by argon-ion etching (900 eV, 11 $\mu A/cm^2$) alone [Y-3]. In another study, cycles of inert-ion bombardment and annealing at 1595 K reduced the concentration of all nonmetallic impurities, except oxygen, below the AES detection limits; the residual oxygen level was about 5 at% [Y-4].

Zinc (Zn): hcp, T_m = 693 K

The low melting point of zinc, coupled with its high vapor pressure (~10^{-5} Pa at 420 K), causes it to be incompatible with accepted UHV bakeout procedures. This problem was circumvented by electroplating a zinc single crystal with a heavy nickel coating which prevented evaporation even at a bakeout temperature of 520

K. The crystal was cleaved after the bakeout was terminated and after the vacuum system reached its base pressure [Zn-1].

Polycrystalline Surface [Zn-2 to Zn-7]

Treatments using argon-ion etching (900–1500 eV, ~200 μA/cm^2, 300 K) were sufficient to produce clean surfaces.

(0001) *Surface* [Zn-1, Zn-8 to Zn-11]

The basal plane is commonly prepared by cleaving in UHV [Zn-1, Zn-8, Zn-9] or air [Zn-9]. In the latter case, argon-ion bombardment was necessary to remove the surface oxide layer [Zn-9 to Zn-11]. The cleaving itself has been done at both ambient temperature [Zn-1, Zn-8] and at 77 K [Zn-9]. Annealing at 390–425 K was reported to produce a well-ordered surface as determined by LEED [Zn-9].

Zirconium (Zr): hex, T_m = 2125 K

Zirconium is an efficient gettering agent for common gases and, in this respect, is chemically similar to titanium. Consequently, even zone-refined material contains appreciable levels of oxygen, carbon, and sulfur. Segregation of these bulk impurities to the surface during heating constitutes the most troublesome aspect of cleaning zirconium surfaces. This problem is further complicated by the high solubility of oxygen (29 at% at 700 K), which increases with temperature.

Polycrystalline Surface [Zr-1 to Zr-4]

After insertion into the vacuum system and brief heating to 900 K, the AES spectrum showed that the polycrystalline surface was contaminated with sulfur, chlorine, nitrogen, and oxygen. Sputter cleaning (1 μA/cm^2, 500 eV, 300 K, 1 h) produced a clean surface, but subsequent annealing led to the reappearance of sulfur, presumably due to diffusion from the bulk. It is difficult to ascertain the completeness of sulfur removal using AES because there is a significant overlap of the sulfur transition at 150 eV and the zirconium transition at 145 eV. Argon-ion sputtering at 900 K for 10 h reduced the sulfur level to the extent that the ratio of the 145-eV AES peak (Zr + S) to the 95-eV peak (Zr only) reached a minimum value of 1.3 [Zr-1]. Other authors who did not anneal *in situ* but who cleaned only by argon-ion bombardment [Zr-2] or by *in situ* micromilling with a rotating diamond edge [Zr-3] easily eliminated the sulfur. The latter authors mentioned that it was not possible to keep the zirconium surface free of oxygen at a pressure of 10^{-7} Pa in the analyzing chamber.

(0001) *Surface* [Zr-5]

An AES spectrum taken after an initial anneal at 873 K revealed large quantities of carbon and oxygen together with smaller amounts of nitrogen, boron, and/or chlorine. The carbon contamination proved most difficult to remove; it could not be reduced below the AES detection limits. The cleanest surface obtained corresponded to an AES peak ratio C_{274}/Zr_{170} of about 0.05 to 0.1. This ratio was achieved after approximately 50 h of argon-ion bombardment at room temperature followed by a number of cycles of bombardment at 823–873 K for several hours and 30-min anneals at the same temperature. The annealing temperature of 823–873 K appeared to be optimum because more carbon segregated to the surface at both higher and lower temperatures.

3. Discussion and Recommendations

Assessment of the information presented in the review section leads to several observations. Carbon, oxygen, and sulfur were most often the difficult impurities to remove from elemental surfaces. For the elements considered here, the relative occurrence of these impurities was approximately C:O:S = 6:4:3, with carbon being a key impurity for more than 30 of the reviewed elements. Often, more than one of these impurities proved to be the main barriers to achieving a clean surface. Generally, carbon and sulfur contamination resulted from segregation of bulk impurities to the surface during heating; in contrast, the most troublesome source of oxygen contamination was adsorption of oxygen-containing gases during, and immediately after, the cleaning process. Elimination of carbon and sulfur was usually accomplished by heating in a reactive gas or heating during ion bombardment. The heating ensured a continuous flow of the impurities from the bulk to the surface; thus, the bulk impurities were steadily depleted. Ion bombardment was generally used to remove oxygen, but repeated flashings in UHV to temperatures sufficient to desorb volatile oxides were sometimes effective.

The reviewed information, tempered by our own knowledge and experience in surface cleaning, was condensed into a set of recommended procedures (Table 1). These recommendations have been arranged alphabetically by element with differences for crystallographic planes detailed wherever appropriate. The procedures recommended are those we would use if faced with the need to prepare a clean surface of a particular element. Generally, the selected procedures are a consensus of the reviewed literature; however, possibilities for widespread implementation of a procedure were also considered. Careful adherence to the details of a recommended procedure does not per se ensure the creation of an atomically clean surface. Surface analysis with an element-specific technique is still required to determine the level of surface cleanliness.

Several aspects of the entries in Table 1 deserve comment. Since most of these elements can be evaporated in UHV to form films with clean surfaces, evaporation is listed as a recommended procedure only if a procedure was not found for bulk specimens. Specific ion energies and current densities for the ion bombardment steps should be considered nominal values, because sputtering phenomena are not strong functions of these parameters in the ranges of interest. Unless otherwise specified, the ion bombardment should be performed using argon ions at an energy of approximately 1 keV, a current density of a few $\mu A/cm^2$, and ambient temperature. All unspecified annealing temperatures can be assumed to be approximately two-thirds of the melting-point temperature (K) for the elements. All annealings and heatings should be performed under UHV conditions, except where information to the contrary is given.

A variety of conclusions can be drawn from the information presented in Table 1. As expected, repeated cycles of ion bombardment and annealing were recommended more often than any other type of procedure; in fact, such cycles were recommended for about 70% of the elements. Heating in reactive gas(es) and annealing in UHV was the recommendation for five elements (Ir, Mo, Os, Re, W). For five other elements (Ni, Pd, Pt, Rh, Ru), a combination of bombardment–annealing cycles and reaction-annealing cycles was recommended. Only five elemental surfaces [As, C (graphite and diamond), Nb, Ta, Tc] have been unambiguously cleaned by simply heating in UHV. *In situ* scraping was the recommended procedure for four elements (Ca, Li, Tl, Yb); however, cycles of ion bombardment and annealing would probably yield clean surfaces for these elements. Unfortunately, we did not find any documented surface cleaning studies on bulk specimens for 15 of the elements reviewed. Nevertheless, application of the procedure recommended for chemically similar elements should produce atomically clean surfaces for these 15 elements. Only one clearly systematic classification has been noted from Table 1; all elements having gas reaction–annealing cycles as part of the recommended procedure are grouped near the center of the periodic table of elements.

Although a variety of procedures may eventually produce a clean surface on those elements for which we did not list a recommended procedure, we suggest the following strategy for cleaning bulk specimens in the absence of a recommended procedure. Begin by outgassing the sample at temperatures slightly below the melting point. Using the nominal ion bombardment conditions given above (1 keV, few $\mu A/cm^2$, 300 K), sputter with argon ions until *in situ* element-specific analysis reveals that the surface is atomically clean. Anneal the sample at two-thirds of its melting point while monitoring the surface composition. If any contaminant (e.g., carbon or sulfur) segregates to the surface, determine the temperature corresponding to the maximum concentration of contamination. Then incorporate this temperature into the ion bombardment conditions and repeat the bombardment–annealing cycle until the annealed surface is atomically clean.

Since most of the elements can be evaporated in UHV to form films with clean surfaces, evaporation is listed as a recommended procedure in Table 1 only if a procedure was not found for bulk specimens. Unless otherwise specified, the ion bombardment should be performed using argon ions at an energy of approximately 1 keV, a current density of a few $\mu A/cm^2$, and ambient temperature. All unspecified annealing temperatures can be assumed to be approximately two-thirds of the melting-point temperature (K) for the elements. All annealing, heating, and scraping steps should be performed under UHV conditions, except where information to the contrary is given. The abbreviation "poly" refers to a polycrystalline surface.

Table 1. Recommended Surface Cleaning Procedures

Element	Surface planes	Recommended procedures
Actinium (Ac)	—	No information found.
Aluminum (Al)	Poly	Outgas in UHV (673 K, 48 h). Repeat cycles of Ar^+ bombardment (5 keV, 1.6 $\mu A/cm^2$, 300–500 K, 10 h first time then 30 min) and annealing (673 K, 30 min).
	(100), (110), (111)	Outgas in UHV (673 K, 48 h). Repeat cycles of Ar^+ bombardment (<2 keV, 5 $\mu A/cm^2$, <673 K, 15 min) and annealing ($\leq$700 K, 1 h).
Americium (Am)	—	No information found.
Antimony (Sb)	(0001), (111)	Cleave in UHV.
	(0001), (01$\bar{1}$2), (11$\bar{2}$0)	Repeat cycles of Ar^+ bombardment (200 eV, few $\mu A/cm^2$, 300 K, few min) and annealing (520 K, few h).
Arsenic (As)	(0001)	Heat air-cleaved crystal in UHV (493 K).
	(111)	Cleave in UHV.
Barium (Ba)	Poly	Repeat cycles of Ar^+ bombardment and annealing
Beryllium (Be)	Poly, (0001)	Repeat cycles of Ar^+ bombardment (500 eV, few $\mu A/cm^2$, 300 K, 1–2 h) and annealing (1000 K, 1 h).
Bismuth (Bi)	(0001), (01$\bar{1}$2)[a], (11$\bar{2}$0)[a]	Bombard with Ar^+ (150–300 eV, 1–5 $\mu A/cm^2$, 6 h) and anneal (510–520 K).
Boron (B)	Poly	Repeat cycles of Ar^+ bombardment and annealing (<1600 K).
	(100), (111)	Heat (1723 K, 1 min).
Cadmium (Cd)	Poly	Repeat cycles of Ar^+ bombardment and annealing (3 keV).
	(0001)	Bombard with Ar^+ (700 eV, 3 $\mu A/cm^2$, 295 K, 24 h).
Calcium (Ca)	Poly	*In situ* scraping.
Carbon (C)	Amorphous	Evaporate in UHV to form films.
	Glassy	Fracture in UHV.

(*continued*)

Table 1. Continued

Element	Surface planes	Recommended procedures
	Graphite (0001)	Cleave in air. Place immediately into ion- and electron-free vacuum system. Anneal (723 K, 5 h). Anneal ($T \geq 1273$ K) if ion conc. is high.
	Graphite ($10\bar{1}0$), ($11\bar{2}0$)	Fracture in UHV.
	Diamond (100), (110), (111)	Heat in UHV (1173–1573 K, 10 min).
Cerium (Ce)	Poly	Repeat cycles of inert-gas–ion bombardment (1 keV, 10 μA/cm^2) and annealing.
Chromium (Cr)	(100)	Repeat cycles of Ar$^+$ bombardment (500 eV) and annealing (900 K).
	(110)	Repeat cycles of Ar$^+$ bombardment (500 eV, 300 K) and annealing (670–870 K, 15 min).
	(111)	Repeat cycles of Ar$^+$ bombardment (2 keV, 5 μA/cm^2, 300 K) and annealing (1170 K).
Cobalt (Co)	Poly	Heat in oxygen (900 K, 10^{-4} Pa). Repeat cycles of Ne$^+$ bombardment (530 eV, 1 μA/cm^2, 600 K) and annealing (1000 K, 15 h).
	(100)	Repeat cycles of Ar$^+$ bombardment (150 eV, 1 μA/cm^2, 420 K) and annealing (420 K).
	(0001), ($10\bar{1}0$), ($10\bar{1}2$), ($11\bar{2}0$)	Repeat cycles of Ne$^+$ bombardment (500 eV, 1 μA/cm^2, 600 K) and annealing (723 K).
Copper (Cu)	Poly,[a] (100), (110), (111), (311)	Repeat cycles of Ar$^+$ bombardment (600 eV, few μA/cm^2, 300 K) and annealing (723 K).
Dysprosium (Dy)	Poly	Repeat cycles of Ar$^+$ bombardment (5 keV) and annealing (500 K).
Erbium (Er)	Poly	Evaporate in UHV to form films.
Europium (Eu)	Poly	Evaporate in UHV to form films.
Gadolinium (Gd)	Poly	Repeat cycles of inert-gas ion bombardment (1 keV, 10 μA/cm^2) and annealing.
Gallium (Ga)	Liquid	Repeat cycles of Ar$^+$ bombardment (3 eV, 20 μA/cm^2) and heating (600 K).
Germanium (Ge)	Polycrystalline	No information found.
	Amorphous	Evaporation or ion bombardment of clean single crystal surface.
	(100), (110)	Chemical pretreatment to produce thin, clean passive oxide layer. Heat to 573–773 K to remove oxide. or Repeat cycles of inert-gas-ion bombardment (500 eV) and annealing (1073 K).
	(111)	Cleave in UHV.
Gold (Au)	Poly (Ca-free)	Heat in UHV (1000 K, 6 h) and in oxygen (1000 K, 6×10^{-3} Pa, 8–24 h) and anneal (1000 K).
	Poly, (100), (111), Vicinals	Repeat cycles of inert-gas-ion bombardment (340 eV, 300 K) and annealing ($\leq$973 K).

(*continued*)

Table 1. Continued

Element	Surface planes	Recommended procedures
Hafnium (Hf)	Poly	Repeat cycles of Ar$^+$ bombardment (500 eV, 1 μA/cm^2, 1070–1370 K) and annealing (1370 K).
Holmium (Ho)	Poly	Repeat cycles of inert-gas-ion bombardment (1 keV, 10 μA/cm^2) and annealing.
Indium (In)	Poly	Repeat cycles of Ar$^+$ bombardment of liquid ($\geq$430 K) and then cool to solidify.
Iridium (Ir)	Poly	Repeat cycles of heating in oxygen (1500 K, ~10^{-4} Pa, several h) and annealing in UHV (1700 K, 20 min).
	(100)–5 × 1	Repeat cycles of heating in oxygen (1400–1500 K, ~10^{-5} Pa, 5–10 min) and annealing (1500 K, 1 min).
	(100)–1 × 1	Generate clean (5 × 1) surface structure. Heat in oxygen (475 K, ~10^{-5} Pa, 200–400 s) and anneal (660 K). Heat in hydrogen (400–700 K, ~10^{-5} Pa, 300 s) or in carbon monoxide (400 K, ~10^{-5} Pa, 300 s). Anneal (800 K).
	(110)–2 × 1, (111)	Bombard with Ar$^+$ to remove initial contamination. Repeat cycles of heating in oxygen (800 K, ~10^{-5} Pa) and annealing (1200–1600 K).
	(755)	Repeat cycles of heating in oxygen (770–870 K, ~10^{-5} Pa, several min) and flash annealing (1470 K).
Iron (Fe)	Poly, (100), (110), (111)	Bombard with Ar$^+$ (500 eV, 1–10 μA/cm^2, 300 K, 1–150 h) to remove oxide layer. Repeat cycles of Ar$^+$ bombardment (500 eV, 20 μA/cm^2, 300 or 650 K, 15–30 min) and annealing (450–700 K, ~1 h). Flash anneal (970 K).
Lanthanum (La)	Poly	Repeat cycles of Ar$^+$ bombardment (900 eV) and annealing.
Lead (Pb)	Poly, (100), (110), (111)	Repeat cycles of Ar$^+$ bombardment (400–700 eV, 570 K) and annealing (580 K).
Lithium (Li)	Poly	Scrape bulk sample in UHV. Bombard with Ar$^+$ (1 keV, 5 μA/cm^2, 300 K).
Lutetium (Lu)	Poly	Evaporate in UHV to form films.
Magnesium (Mg)	Poly, (100), (0001)	Repeat cycles of inert-ion bombardment and annealing ($\leq$435 K).
Manganese (Mn)	Poly	Repeat cycles of Ar$^+$ bombardment (3 keV, few h) and annealing (673 K).
Molybdenum (Mo)	Poly, (100), (110), (111), (112).	Repeat cycles of heating in oxygen (~1500 K, ~10^{-4} Pa, 3–6 h) and flash annealing (2200 K).
Neodymium (Nd)	Poly	No documented cleaning procedure.
Neptunium (Np)	—	No information found.
Nickel (Ni)	Poly, (100), (110), (111)	Repeat cycles of Ar$^+$ bombardment (500 eV, 5 μA/cm^2, 300 K, 15 min) and annealing (900 K, 15 min).
		Heat in oxygen (900 K, 10^{-4} Pa) if carbon remains.

(*continued*)

Table 1. Continued

Element	Surface planes	Recommended procedures
		Heat in hydrogen (1000 K, 10^{-3} Pa) if oxygen remains.
		Anneal (900 K, 15 min).
Niobium (Nb)	Poly, (100), (110), (111)	Heat in UHV (2300 K, several hours).
Osmium (Os)	Poly, (0001)	Heat in oxygen (1000 K, 10^{-5} Pa), anneal (2000 K).
Palladium (Pd)	Poly, (100), (110), (111), (210), (311)	Repeat cycles of inert-ion bombardment (~500 eV, 5 μA/cm^2, 900 K, ~300 h per mm thickness) and annealing (1100–1300 K).
		Heat in oxygen (<1000 K, ~10^{-4} Pa) and flash anneal (1300 K).
Phosphorus (P)	—	No information found.
Platinum (Pt)	Poly, (100), (110), (111), (112), (113), (133), (122), (012)	Repeat cycles of Ar$^+$ bombardment (~400 eV), heating in oxygen (1000–1300 K, ~10^{-4} Pa), and annealing (1373 K).
Plutonium (Pu)	Poly	Bombard with inert-gas ions (0.5–5 keV, 10–50 μA/cm^2) to remove gross contamination.
		Repeat cycles of inert-gas ion bombardment (676 K) and annealing (676 K, <10^{-8} Pa, several min).
Polonium (Po)	—	No information found.
Praseodymium (Pr)	Poly	Evaporate in UHV to form films.
Promethium (Pm)	—	No information found.
Protactinium (Pa)	—	No information found.
Radium (Ra)	—	No information found.
Rhenium (Re)	Poly, (0001)	Repeat cycles of heating in oxygen (2000–2500 K, 10^{-4} Pa) and annealing (2200 K, 2 min).
Rhodium (Rh)	(100), (111), (210), (331), (332), (775)	Bombard with inert-gas ions (0.5–1 keV, 1–10 μA/cm^2, 300 K) to remove gross contamination.
		Anneal (1250–1300 K).
		Repeat cycles of Ar$^+$ bombardment (500 eV, 1–10 μA/cm^2, 1250–1300 K), heating in oxygen (800–1300 K, 10^{-5}–10^{-4} Pa), and annealing (1400 K, few min).
Ruthenium (Ru)	(0001), (110)	Repeat cycles of heating in oxygen (1300–1500 K, ~10^{-5} Pa, 10–15 s) and annealing (1500–1600 K, 10–60 s).
	($10\bar{1}0$)	Repeat cycles of Ar$^+$ bombardment (500 eV, 1270 K) and annealing (1270 K).
Samarium (Sm)	Poly	Repeat cycles of inert-gas-ion bombardment (1 keV, 10 μA/cm^2) and annealing.
Scandium (Sc)	Poly, (100), (001)	Repeat cycles of Ar$^+$ bombardment (500 eV, few μA/cm^2, 850 K) and annealing (1100 K).[a]
	(0001)	Kr$^+$ bombardment (1 keV, 20 μA/cm^2, 1170 K, 50 h) and annealing (770 K).
Selenium (Se)	($10\bar{1}0$), ($1\bar{2}10$)	Repeat cycles of Ar$^+$ bombardment (500 eV, 1 μA/cm^2, 1/2 h) and annealing (420 K, 1/2 h).
Silicon (Si)	Poly	Gentle sputtering.

(continued)

Table 1. Continued

Element	Surface planes	Recommended procedures
(Careful hydropho-bic or hydrophilic pre-treatment.)	(100), (100) vicinals	Repeat anneal only (1173 K). Repeat cycles of Ar$^+$ bombardment (350–500 eV) and annealing (1200–1400 K, 2–3 min). Laser clean.
	(110)	Gentle anneal only (1173 K). Bombard with oxygen ions (500 eV, 1.3×10^{-3} Pa) and flash to 1300 K.
	(111)	Flash anneal (1473 K).
	(210), (211), (311), (320), (331), (510), (511)	Dip in HF. Rinse with dionized water. Anneal (10–20 nPa, 1473 K, 2 min)
Silver (Ag)	Poly, (100), (110), (111), (331)	Repeat cycles of Ar$^+$ bombardment (300–600 eV, 1–3 μA/cm^2, 300 K) and annealing (670–750 K).
Sodium (Na)	Poly	Evaporate in UHV to form films.
Strontium (Sr)	Poly	Repeat cycles of Ar$^+$ bombardment (3 keV, to a total dose of 1 A s/cm^2) and annealing.
Tantalum (Ta)	Poly, (100), (110)	Heat in UHV (2700–3000 K).
Technetium (Tc)	Poly	Heat in UHV (2270 K).
Tellurium (Te)	(0001), (1$\bar{2}$10), air-cleaved 10$\bar{1}$0)	Repeat cycles of Ar$^+$ bombardment (~300 eV, ~2 μA/cm^2, ~30 min) and annealing (470 K, 1 h).
	(10$\bar{1}$0)	Cleave in UHV.
Terbium (Tb)	Poly	Repeat cycles of inert-gas-ion bombardment (1 keV, 10 μA/cm^2) and annealing.
Thallium (Tl)	Poly	*In situ* scraping.
Thorium (Th)	Poly, (100), (111)	Bombard with inert-gas ions (0.5–5 keV, 10–50 μA/cm^2) to remove gross contamination.
		Repeat cycles of inert-gas-ion bombardment (1000 K) and annealing (1000 K, <10^{-8} Pa, several min).
Thulium (Tm)	—	Evaporate in UHV to form films.
Tin (Sn)	Poly, (100)	Repeat cycles of Ar$^+$ bombardment (1–2 keV, 20 μA/cm^2) and annealing.
Titanium (Ti)	Poly, (0001), (10$\bar{1}$0), (10$\bar{1}$1)	Repeat cycles of Ar$^+$ bombardment (600 eV, 4 μA/cm^2, 300 K) and flash annealing (1020 K) until 50 h of bombardment are accumulated.
		Bombard with Ar$^+$ (600 eV, 4 μA/cm^2, 1020 K, ~14 h). Anneal (1020 K, 1 h).
		Repeat cycles of Ar$^+$ bombardment (600 eV, 4 μA/cm^2, 300 K for 1/2 h and 1020 K for 1/2 h) and annealing (1020 K, 1 h).
Tungsten (W)	Poly, (100), (110), (110) vicinals, (111), (112), (210)	Heat in oxygen (1550 K, $~7 \times 10^{-5}$ Pa, 100–300 h per mm thickness).
		Repeat flash anneals (>2400 K).
Uranium (U)	Poly	Bombard with inert-gas ions (0.5–5 keV, 10–50 μA/cm^2) to remove gross contamination.

(*continued*)

Table 1. Continued

Element	Surface planes	Recommended procedures
		Repeat cycles of inert-gas-ion bombardment (1000 K) and annealing (1000 K, <10^{-8} Pa, several min).
Vanadium (V)	(100)–5 × 1	Ne$^+$ bombardment (500 eV, 5 μA/cm^2, 800 K, 200 h).
	(110)	Ar$^+$ bombardment (2 keV, 20 μA/cm^2, 670 K, 80 h). Anneal (970–1070 K).
		Repeat cycles of Ar$^+$ bombardment (2 keV, 20 μA/cm^2, 670 K, 5 min) and annealing (970–1070 K).
Ytterbium (Yb)	Poly	*In situ* scraping.
Yttrium (Y)	Poly	Repeat cycles of Ar$^+$ bombardment (900 eV, 5 μA/cm^2) and annealing (1600 K).
Zinc (Zn)	Poly	Repeat cycles of Ar$^+$ bombardment (1 keV, 20 μA/cm^2, 300 K) and annealing.
	(0001)	Cleave in UHV.
	Air-cleaved (0001)	Repeat cycles of Ar$^+$ bombardment (1 keV, 20 μA/cm^2, 300 K) and annealing (390–425 K).
Zirconium (Zr)	Poly	Bombard with Ar$^+$ (500 eV, ~1 μA/cm^2, 300 K, 1 h) to remove gross contamination.
		Repeat cycles of Ar$^+$ bombardment (900 K, 10 h) and annealing.
	(0001)	Bombard with Ar$^+$ (300 K, 50 h).
		Repeat cycles of Ar$^+$ bombardment (~850 K, several h) and annealing (~850 K, 30 min).

*"Procedure has not been documented by element-specific techniques to produce atomically clean surfaces.

4. Concluding Remarks

Two remarks are appropriate with regard to this chapter. First, a need exists for documented cleaning procedures for more than 15 elements. The recommended procedures for these elements have been listed in Table 1 as "no information found" or "evaporate in UHV to form films," or "no documented cleaning procedure." We would be most pleased to receive copies of published papers that may help eliminate the information gap in an updated version of this chapter. Second, we strongly suggest that future authors of surface studies provide either (1) details of their cleaning methods, including some measure of the degree of surface cleanliness from an element-specific analysis technique, or (2) reference to another paper that did provide details and documentation. The level of detail should be sufficient to permit reproduction of the surface conditions employed. A useful qualitative measure of surface cleanliness is specification of the peak amplitudes of the contaminants relative to those for the element under given analysis conditions. Both these remarks are consistent with achievement of orderly scientific progress in the development of improved surface cleaning procedures for the solid elements.

Acknowledgments

We gratefully acknowledge the extremely helpful assistance of Karen Sitzberger for the initial typing required to convert our previous manuscript to an electronic file and for the subsequent typing of additions, changes, and corrections. This work was performed under the auspices of the U.S. Department of Energy by the Lawrence Livermore National Laboratory under Contract Number W-7405-ENG-48.

References

1. R. W. Roberts, *Br. J. Appl. Phys.* **14**, 537 (1963).
2. R. G. Musket, W. McLean, C. A. Colmenares, D. M. Makowiecki, and W. J. Siekhaus, *Appl. Surf. Sci.* **10**, 143 (1982).
3. M. Grunze, H. Ruppender, and O. Elshazly, *J. Vac. Sci. Technol. A* **6**, 1266 (1988).

Aluminum

Al-1. D. T. Quinto and W. D. Robertson, *Surf. Sci.* **27**, 645 (1971).
Al-2. L. H. Jenkins and M. F. Chung, *Surf. Sci.* **28**, 409 (1971).
Al-3. T. Fort, Jr. and R. L. Wells, *Surf. Sci.* **32**, 543 (1972).
Al-4. G. Dufour, H. Guennou, and C. Bonnelle, *Surf. Sci.* **32**, 731 (1972).
Al-5. W. S. Lassiter, *Surf. Sci.* **47**, 559 (1975).
Al-6. P. H. Dawson, *Surf. Sci.* **57**, 229 (1976).
Al-7. C. Benazeth, M. Benazeth, and L. Viel, *Surf. Sci.* **65**, 165 (1977).
Al-8. R. J. Baird, C. S. Fadley, S. M. Goldsberg, P. J. Feibelmann, and M. Sunjic, *Surf. Sci.* **72**, 495 (1978).
Al-9. T. W. Rogers, C. T. Campbell, R. L. Lance, and J. M. White, *Surf. Sci.* **97**, 425 (1980).
Al-10. T. W. Rogers, R. L. Lance, and J. M. White, *Surf. Sci.* **100**, 388 (1980).
Al-11. G. C. Allen, P. M. Tucker, B. E. Hayden, and D. F. Klemperer, *Surf. Sci.* **102**, 207 (1981).
Al-12. C. Jordan, R. L. Michel, J. Gastaldi, and J. Derrien, *Phil. Mag. A* **41**, 443 (1980).
Al-13. A. F. Carley and M. W. Roberts, *Proc. R. Soc. London Ser. A* **363**, 403 (1978).
Al-14. J. W. Rogers, Jr., and T. M. White, *J. Vac. Sci. Technol.* **16**, 485 (1979).
Al-15. S. B. M. Hagström, R. Z. Bachrach, R. S. Bauer, and S. A. Flodström, *Phys. Scr.* **16**, 414 (1977).
Al-16. D. T. Quinto, B. W. Holland, and W. D. Robertson, *Surf. Sci.* **32**, 139 (1972).
Al-17. J. O. Porteus, *Surf. Sci.* **41**, 515 (1974).
Al-18. G. Allie, E. Blanc, and D. Dufayard, *Surf. Sci.* **47**, 635 (1975).
Al-19. B. A. Hutchins, T. N. Rhodin, and J. E. Demuth, *Surf. Sci.* **54**, 419 (1976).
Al-20. J. K. Grepstad, P. O. Gartland, and B. J. Slagsvold, *Surf. Sci.* **57**, 348 (1976).
Al-21. G. Allie, E. Blanc, and D. Dufayard, *Surf. Sci.* **62**, 215 (1977).
Al-22. W. Eberhardt and C. Kunz, *Surf. Sci.* **75**, 709 (1978).
Al-23. C. Argile and G. E. Rhead, *Surf. Sci.* **78**, 125 (1978).
Al-24. C. W. B. Martinson and S. A. Flodström, *Surf. Sci.* **80**, 306 (1979).
Al-25. P. Hoffmann, K. Horn, A. M. Bradshaw, and K. Jacobi, *Surf. Sci.* **82**, L610 (1979).

Al-26. R. Michel, J. Jourdan, J. Castaldi, and J. Derrien, *Surf. Sci.* **84**, L509 (1979).
Al-27. C. W. B. Martinson, S. A. Flodström, J. Lundgren, and P. Westrim, *Surf. Sci.* **89**, 102 (1979).
Al-28. R. Michel, J. Castaldi, C. Allasia, C. Jourdan, and J. Derrien, *Surf. Sci.* **95**, 309 (1980).
Al-29. K. G. Lynn and H. Lutz, *Phys. Rev. B* **22**, 4143 (1980).
Al-30. F. Jona, J. A. Strozier, Jr., and C. Wong, *Surf. Sci.* **30**, 225 (1972).
Al-31. M. Baines, A. Howie, and S. K. Anderson, *Surf. Sci.* **53**, 546 (1975).
Al-32. A. M. Bradshaw, P. Hoffman, and W. Wyrobisch, *Surf. Sci.* **68**, 269 (1977).
Al-33. P. Hoffmann, C. V. Muschuvitz, K. Horn, K. Jacobi, A. M. Bradshaw, K. Kambe, and M. Scheffler, *Surf. Sci.* **89**, 327 (1979).
Al-34. K. Jacobi, C. V. Muschivitz, and K. Kambe, *Surf. Sci.* **93**, 310 (1980).
Al-35. J. Pillon, D. Roptin, and C. Cailler, *Surf. Sci.* **59**, 741 (1976).

Antimony

Sb-1. L. Ley, R. A. Pollak, S. P. Kowalczyk, R. McFeely, and D. A. Shirley, *Phys. Rev. B* **8**, 641 (1973).
Sb-2. F. Jona, *Surf. Sci.* **8**, 57 (1967).
Sb-3. B. C. Martin and G. M. Roenblatt, *J. Cryst. Growth* **33**, 281 (1976).

Arsenic

As-1. W. P. Ellis, *Surf. Sci.* **41**, 125 (1974).
As-2. W. P. Ellis, *Surf. Sci.* **50**, 178 (1975).
As-3. B. C. Martin and G. M. Roenblatt, *J. Cryst. Growth* **33**, 281 (1976).

Barium

Ba-1. J. A. T. Verhoeven and H. van Dovern, *Appl. Surf. Sci.* **5**, 361 (1980).
Ba-2. W. V. Lampert, K. D. Rachocki, B. C. Lamartine, and T. W. Haas, *J. Electron Spectrosc. Rel. Phenom.* **26**, 133 (1982).
Ba-3. K. Jacobi, C. Astaldi, B. Frick, and P. Geng, *Surf. Sci.* **189/190**, 578 (1987).

Beryllium

Be-1. D. M. Zehner, N. Barbulesco, and L. H. Jenkins, *Surf. Sci.* **34**, 385 (1973).
Be-2. R. G. Musket, *Surf. Sci.* **44**, 629 (1974).
Be-3. F. Jona, J. A. Strozier, Jr., J. Kumar, and R. O. Jones, *Phys. Rev. B* **6**, 407 (1972).
Be-4. A. K. Green and E. Bauer, *Surf. Sci.* **74**, 676 (1978).

Bismuth

Bi-1. T. N. Taylor, J. W. Rogers, Jr., and W. P. Ellis, *J. Vac. Sci. Technol.* **15**, 559 (1978).
Bi-2. T. N. Taylor, C. T. Campbell, J. W. Rogers, Jr., W. P. Ellis, and J. M. White, *Surf. Sci.* **134**, 529 (1983).
Bi-3. F. Jona, *Surf. Sci.* **8**, 57 (1967).

Boron

B-1. G. Dagoury and D. Vigner, *Le Vide* **32**, 51 (1977).

B-2. W. Klein, *J. Less-Common Met.* **47**, 101 (1976).
B-3. G. Rovida and M. Maglietta, *J. Appl. Phys.* **44**, 3801 (1973).

Cadmium

Cd-1. P. Weightman, *J. Phys. C* **9**, 1117 (1976).
Cd-2. L. Braicovich, G. Rossi, R. A. Powell, and W. E. Spicer, *Phys. Rev. B* **21**, 3539 (1980).
Cd-3. R. Nyholm and N. Martensson, *J. Phys.* **13**, L279 (1980).
Cd-4. A. D. McLachlan, J. G. Jenkin, J. Liesegang, and R. C. G. Leckey, *J. Electron Spectrosc. Rel. Phenenom.* **3**, 207 (1974).
Cd-5. W. Joyner, M. W. Roberts, and G. N. Salaita, *Surf. Sci.* **84**, L505 (1979).

Calcium

Ca-1. M. Erbudak, P. Kalt, L. Schlapbach, and K. Bennemann, *Surf. Sci.* **126**, 101 (1983).

Carbon

C-1. F. R. McFeely, S. P. Kowalczyk, L. Ley, R. G. Cavell, R. A. Pollak, and D. A. Shirley, *Phys. Rev. B* **9**, 5268 (1974).
C-2. B. Lang, *Surf. Sci.* **80**, 38 (1979).
C-3. G. F. Amelio and E. J. Scheibner, *Surf. Sci.* **11**, 242 (1968).
C-4. J. Suzanne, J. P. Colomb, and M. Bienfart, *Surf. Sci.* **44**, 131 (1974).
C-5. L. Mattera, F. Rosatelli, C. Salvo, F. Tommasini, U. Valbursa, and G. Vidali, *Surf. Sci.* **93**, 515 (1980).
C-6. G. Derry, D. Wesner, G. Vidali, T. Thwaites, and D. R. Frankl, *Surf. Sci.* **94**, 332 (1980).
C-7. C. Wesner, G. Derry, G. Vidali, T. Thwaites, and D. R. Frankl, *Surf. Sci.* **95**, 367 (1980).
C-8. S. Calisti and J. Suzanne, *Surf. Sci.* **105**, L255 (1981).
C-9. J. J. Metois, J. C. Heyrand, and Y. Takeda, *Thin Solid Films* **51**, 105 (1978).
C-10. J. J. Lander and J. Morrison, *J. Appl. Phys.* **35**, 3593 (1964).
C-11. J. M. Thomas, E. L. Evans, M. Barber, and P. Swift, *Faraday Soc. London Trans.* **67**, 1875 (1971).
C-12. F. J. Himpsel, J. A. Knapp, J. A. Vechten, and D. E. Eastman, *Phys. Rev. B* **20**, 624 (1979).
C-13. P. G. Lurie and J. M. Wilson, *Surf. Sci.* **65**, 453 (1977).
C-14. H. G. Maquire and C. G. Cillie, *J. Phys. C* **9**, L135 (1976).
C-15. H. G. Maguire, *Phys. Stat. Solidi (b)* **76**, 715 (1976).

Cerium

Ce-1. M. B. Chamberlain and W. L. Baun, *J. Vac. Sci. Technol.* **12**, 1047 (1975).
Ce-2. T. L. Barr, in: *Quantitative Surface Analysis of Materials* (N. S. McIntyre, ed.), Proc. Symp., Cleveland, OH, 1977, American Society for Testing and Materials, Philadelphia, PA, ASTM-STP-643 (1978), p. 83.

Chromium

Cr-1. S. Ekelund and C. Leygraf, *Surf. Sci.* **40**, 179 (1973).
Cr-2. G. Gewinner, J. C. Peruchetti, A. Jeagle, and R. Reidinger, *Phys. Rev. Lett.* **43**, 935 (1979).

Cr-3.   J. C. Peruchetti, G. Gewinner, and A. Jaegle, *Surf. Sci.* **88**, 479 (1979).
Cr-4.   D. Tabor, J. M. Wilson, and T. J. Bastow, *Surf. Sci.* **26**, 471 (1971).
Cr-5.   H. Kato, Y. Sakisaka, M. Nishitima, and M. Onchi, *Surf. Sci.* **107**, 20 (1981).
Cr-6.   L. I. Johansson, L. G. Petersson, K. F. Berggren, and J. W. Allen, *Phys. Rev. B* **22**, 3294 (1980).
Cr-7.   G. Gewinner, J. C. Peruchetti, A. Jaegle, and A. Kalt, *Surf. Sci.* **78**, 439 (1978).

Cobalt

Co-1.   M. E. Bridge, C. M. Comrie, and R. M. Lambert, *Surf. Sci.* **67**, 393 (1977).
Co-2.   M. E. Bridge and E. M. Lambert, *Surf. Sci.* **93**, 413 (1979).
Co-3.   A. Ignatiev and T. Matsuyama, *J. Catal.* **58**, 328 (1979).
Co-4.   K. A. Prior, K. Schwata, and R. M. Lambert, *Surf. Sci.* **77**, 193 (1978).
Co-5.   T. Matsuyama and A. Ignatiev, *Surf. Sci.* **102**, 18 (1981).
Co-6.   M. P. Hooker and J. T. Grant, *Surf. Sci.* **62**, 21 (1977).
Co-7.   B. Alsenz, B. W. Lee, A. Ignatiev, and M. A. Van Hove, *Solid State Commun.* **25**, 641 (1978).
Co-8.   A. Ignatiev, B. W. Lee, and M. A. Van Hove, in: *Proc. 7th Int. Vac. Congr. and 3rd Int. Conf. on Solid Surfaces*, Vienna, 1977 (R. Dobrozemski, F. Rudenauer, and F. P. Viehbock, eds.), p. 1773.
Co-9.   G. L. Berning, *Surf. Sci.* **51**, 673 (1976).
Co-10. M. Welz, W. Moritz, and D. Wolf, *Surf. Sci.* **125**, 473 (1983).
Co-11. M. Maglietta and G. Rovida, *Surf. Sci.* **71**, 495 (1978).
Co-12. G. Rovida and M. Maglietta in: *Proc. 7th Int. Vac. Congr. and 3rd Int. Conf. on Solid Surfaces*, Vienna, 1977 (R. Dobrozemski, F. Rudenauer, and F. P. Viehbock, eds.), p. 963.

Copper

Cu-1.   R. Bowman, J. B. Mechelem, and A. A. Holscher, *J. Vac. Sci. Technol.* **15**, 91 (1978).
Cu-2.   R. W. Joyner, C. S. McKee, and M. W. Roberts, *Surf. Sci.* **26**, 303 (1971).
Cu-3.   J. H. Jenkins and M. F. Chung, *Surf. Sci.* **24**, 125 (1971).
Cu-4.   L. McDonnell and D. P. Woodruff, *Surf. Sci.* **46**, 505 (1974).
Cu-5.   C. Benndorf, B. Egert, G. K. Keller, and F. Thieme, *Surf. Sci.* **74**, 216 (1978).
Cu-6.   J. H. Onuferko and D. P. Woodruff, *Surf. Sci.* **95**, 555 (1980).
Cu-7.   A. Spitzer and H. Lüth, *Surf. Sci.* **102**, 29 (1981).
Cu-8.   S. M. Goldberg, R. J. Band, S. Kono, N. F. T. Hall, and C. S. Fadley, *J. Electron Spectrosc. Rel. Phenom.* **21**, 1 (1980).
Cu-9.   D. Westphal, D. Spanjard, and A. Goldmann, *J. Phys. C* **13**, 1361 (1980).
Cu-10. F. H. P. M. Habraken and G. A. Bootsma, *Surf. Sci.* **87**, 333 (1979).
Cu-11. H. P. Bonzel, *Surf. Sci.* **27**, 387 (1971).
Cu-12. J. R. Noonan, H. L. Davis, and L. H. Jenkins, *J. Vac. Sci. Technol.* 15, 619 (1978).
Cu-13. H. L. Davis, J. R. Noonan, H. L. Davis, and L. H. Jenkins, *Surf. Sci.* **83**, 559 (1979).
Cu-14. H. Papp and J. Pritchard, *Surf. Sci.* **53**, 371 (1975).

Dysprosium

Dy-1.   J. A. Schrefels, J. E. Deffeyes, L. D. Neff, and J. M. White, *J. Electron Spectrosc. Rel. Phenom.* **25**, 191 (1982).

Gadolinium

Gd-1. W. Farber and P. Braun, *Surf. Sci.* **41**, 195 (1974).

Gallium

Ga-1. J. Fine, S. Hardy, and T. D. Andreadis, in: *The Physics of Ionized Gases, Contributed Papers of SPIG-80* (B. Cobic, ed.), Boris Kidric Institute of Nuclear Sciences, Belgrad, Yugloslavia (1980) p. 2940.

Germanium

Ge-1. Y. Q Cai, P. Storer, A. S. Kheifets, and I. E. McCarthy, *Surf. Sci.* **334**, 276 (1995).
Ge-2. D. Steinmetz, F. Ringeisen, D. Bolmont, and J. J. Koulman, *Surf. Sci.* **237**, 135 (1990).
Ge-3. F. Meyer and J. J. Vrakking, *Surf. Sci.* **33**, 271 (1972).
Ge-4. F. Jona, *IBM J. Res. Dev.* **9**, 375 (1965).
Ge-5. K Prabhakarana, T. Ogino, R. Hull, and J. C. Bean, *Surf. Sci.* **316**, L1031 (1994).
Ge-6. X. J. Zhang, G. Xue, A. Agarwal, R. Tsu, M. A. Hasan, J. E. Greene, and A. Rockett, *J. Vac. Sci. Technol. A* **11**, 2553 (1993).
Ge-7. R. J. Culbertson, Y. Kuk, and L. C. Feldman, *Surf. Sci.* **167**, 127 (1986).
Ge-8. H. J. W. Zundvliet and A. van Silfhouts, *Surf. Sci.* **211/213**, 544 (1989).
Ge-9. B. Z. Olshanetshy, S. M. Repinsky, and A. A. Shklyaev, *Surf. Sci.* **64**, 224 (1977).
Ge-10. B. Z. Olshanetsky, S. M. Repinsky, and A. A. Shklyaev, *Surf. Sci.* **69**, 205 (1977).
Ge-11. D. Hanemann, *Adv. Phys.* **31**, 165 (1982), and references therein.
Ge-12. M. Taubenblatt, E. So, P. Sih, A. Kahn, and P. Mark, *J. Vac. Sci. Technol. A* **16**, 231 (1978).
Ge-13. M. Henzler and J. Töpler, *Surf. Sci.* **40**, 388 (1973).
Ge-14. S. Sinharoy and M. Henzler, *Surf. Sci.* **51**, 75 (1975).

Gold

Au-1. S. H. Overbury and G. A. Somorjai, *Surf. Sci.* **55**, 209 (1976).
Au-2. D. D. Eley and P. B. Moore, *Surf. Sci.* **76**, L599 (1978).
Au-3. M. E. Schrader, *Surf. Sci.* **78**, L227 (1978).
Au-4. P. Legare, L. Hilare, M. Sotto, and G. Maire, *Surf. Sci.* **91**, 175 (1980).
Au-5. R. W. Joyner and M. W. Roberts, *Chem. Soc. London, Faraday Trans. I* **69**, 1242 (1973).
Au-6. A. M. Mattera, R. M. Goodman, and G. A. Somorjai, *Surf. Sci.* **7**, 26 (1967).
Au-7. G. McElhiney and J. Pritchard, *Surf. Sci.* **60**, 397 (1975).
Au-8. J. F. Wendelken and D. M. Zehner, *Surf. Sci.* **71**, 178 (1978).
Au-9. E. Bertel and F. P. Netzer, *Surf. Sci.* **97**, 409 (1980).
Au-10. J. A. D. Mathew, F. P. Netzer, and E. Bertel, *J. Electron. Spectrosc. Rel. Phenom.* **20**, 1 (1980).
Au-11. D. Wolf, H. Jagodzinski, and W. Moritz, *Surf. Sci.* **77**, 265 (1978).
Au-12. W. Moritz and D. Wolf, *Surf. Sci.* **88**, L29 (1979).
Au-13. J. R. Noonan and H. L. Davis, *J. Vac. Sci. Technol.* **16**, 587 (1979).
Au-14. R. Feder, N. Muller, and D. Wolf, *Z. Phys. B* **28**, 265 (1977).
Au-15. P. Heimann and N. Neddermeyer, *J. Phys. F* **7**, L37 (1977).
Au-16. M. A. Chesters and G. A. Somorjai, *Surf. Sci.* **52**, 21 (1975).
Au-17. K. Besocke, B. Krahl-Urban, and J. Wagner, *Surf. Sci.* **68**, 39 (1977).
Au-18. S. A. Cochran and H. H. Farrell, *Surf. Sci.* **95**, 359 (1980).

Hafnium

Hf-1.   T. W. Haas, J. T. Grant, and G. J. Dooley III, *J. Vac. Sci. Technol.* **7**, 43 (1970).
Hf-2.   W. C. Weast (ed.), *CRC Handbook of Chemistry and Physics*, CRC Press, Boca Raton, FL, (1979) p. B11.
Hf-3.   M. P. Cox, J. S. Foord, R. M. Lambert, and R. H. Prince, *Surf. Sci.* **129**, 375 (1983).
Hf-4.   G. J. Dooley III and T. W. Haas, *J. Vac. Sci. Technol.* **7**, 590 (1970).
Hf-5.   G. J. Dooley III and T. W. Haas, *J. Metals* **22**, 17 (1970).
Hf-6.   C. Morant, L. Galan, and J. M. Sanz, *Surf. Interface Anal.* **16**, 304 (1990).

Holmium

Ho-1.   D. Chopra, H. Babb, and R. Bhalla, *Phys. Rev. B* **14**, 5231 (1976).

Indium

In-1.   M. Gettings and J. C. Riviere, *Surf. Sci.* **68**, 64 (1977).
In-2.   S. M. Rossnagel, H. F. Dylla, and S. A. Cohen, *J. Vac. Sci. Technol.* **16**, 558 (1979).
In-3.   P. Légaré, L. Hilaire, and G. Maire, *J. Microsc. Spectrosc. Electron.* **5**, 771 (1980).
In-4.   A. W. C. Lin, N. R. Armstrong, and T. Kuwana, *Anal. Chem.* **49**, 1228 (1977).
In-5.   H. F. Helbig and P. Adelmann, *J. Vac. Sci. Technol.* **14**, 488 (1977).
In-6.   A. C. Parry-Jones, P. Weightman, and P. T. Andrews, *J. Phys. C* **12**, 1587 (1979).
In-7.   R. Nyholm and N. Martensson, *J. Phys. C* **13**, L279 (1980).

Iridium

Ir-1.   T. Matsushima, *Surf. Sci.* **87**, 665 (1979).
Ir-2.   E. Beckman, B. Horn, and D. Fick, *Surf. Sci.* **147**, 263 (1984).
Ir-3.   C. M. Chan, *Surf. Sci.* **237**, L421 (1990).
Ir-4.   N. V. Gelfond, I. K. Igumenon, A. I. Boronin, V. I. Bukhtiyaron, M. Yu. Smirnov, I. P. Prosvirin, and R. I. Kwon, *Surf. Sci.* **275**, 323 (1992).
Ir-5.   D. A. Hutt and D. W. Bassett, *Surf. Sci.* **287/288**, 1000 (1993).
Ir-6.   K. D. Shiang, C. M. Wei, and T. T. Tsong, *Surf. Sci.* **301**, L573 (1994).
Ir-7.   J. Küppers and H. Michel, *Appl. Surf. Sci.* **3**, 179 (1979).
Ir-8.   J. Küppers, H. Michel, F. Mitschke, K. Wandelt, and G. Ertl, *Surf. Sci.* **89**, 361 (1979).
Ir-9.   T. N. Rhodin and G. Broden, *Surf. Sci.* **60**, 466 (1976).
Ir-10.   G. Broden, T. Rhodin, and W. Capehart, *Surf. Sci.* **61**, 143 (1976).
Ir-11.   G. Broden and T. Rhodin, *Chem. Phys. Lett.* **40**, 247 (1976).
Ir-12.   G. Broden, T. Rhodin, C. Brucker, R. Benbow, and Z. Hurych, *Surf. Sci.* **59**, 593 (1976).
Ir-13.   G. Broden and T. Rhodin, *Solid State Commun.* **18**, 105 (1976).
Ir-14.   J. Kanski and T. N. Rhodin, *Surf. Sci.* **65**, 63 (1977).
Ir-15.   A. Ignatiev, A. V. Jones, and T. N. Rhodin, *Surf. Sci.* **30**, 573 (1972).
Ir-16.   J. T. Grant, *Surf. Sci.* **18**, 228 (1969).
Ir-17.   E. Lang, K. Müller, K. Heinz, M. A. Van Hove, R. J. Koesner, and G. A. Somorjai, *Surf. Sci.* **127**, 347 (1985).
Ir-18.   G. Kisters, J. G. Chen, S. Lehwold, and H. Ibach, *Surf. Sci.* **245**, 226 (1986).
Ir-19.   C. M. Chan, K. L. Luke, M. A. Van Hove, W. H. Weinberg, and E. D. Williams, *J. Vac. Sci. Technol.* **16**, 642 (1979).
Ir-20.   C. M. Chan, M. A. Van Hove, W. H. Weinberg, and E. D. Williams, *Surf. Sci.* **91**, 440 (1980).

Ir-21. J. L. Taylor, D. E. Ibbotson, and W. H. Weinberg, *Surf. Sci.* **79**, 349 (1979).
Ir-22. J. L. Taylor, D. E. Ibbotson, and W. H. Weinberg, *J. Chem. Phys.* **69**, 4298 (1978).
Ir-23. C. M. Chan, E. D. Williams, and W. H. Weinberg, *Surf. Sci.* **82**, L577 (1979).
Ir-24. J. L. Taylor, D. E. Ibbotson, and W. H. Weinberg, *Surf. Sci.* **90**, 37 (1979).
Ir-25. T. S. Wittrig, D. E. Ibbotson, and W. H. Weinberg, *Surf. Sci.* **102**, 506 (1981).
Ir-26. B. E. Nieuwenhuys and G. A. Somorjai, *Surf. Sci.* **72**, 8 (1978).
Ir-27. K. Christmann and G. Ertl, *Z. Naturforsch.* **28A**, 1144 (1973).
Ir-28. C. M. Chan, M. A. Van Hove, W. H. Weinberg, and E. D. Williams, *Solid State Commun.* **30**, 47 (1979).
Ir-29. C. M. Chan and M. A. Van Hove, *Surf. Sci.* **171**, 226 (1986).
Ir-30. M. Copel, P. Fenter, and T. Gustafson, *J. Vac. Sci. Technol.* **A54**, 742 (1987).
Ir-31. W. Hettering and W. Heiland, *Surf. Sci.* **210**, 129 (1989).
Ir-32. T. S. Marinova and D. V. Chakarov, *Surf. Sci.* **217**, 65 (1989).
Ir-33. H. Bu, M. Shi and J. Rabalais, *Surf. Sci.* **236**, 135 (1990).
Ir-34. T. Matsushima, Y. Ohno, and K. Nagai, *J. Phys. Chem.* **94**, 704 (1991).
Ir-35. H. Bu, S. V. Teplov, V. Zastaynjuk, M. Shi, and J. Rabalais, *Surf. Sci.* **275**, 332 (1992).
Ir-36. C. M. Comrie and W. H. Weinberg, *J. Chem. Phys.* **64**, 250 (1976).
Ir-37. C. M. Chan, S. L. Cunningham, M. A. Van Hove, W. H. Weinberg, and S. P. Withrow, *Surf. Sci.* **66**, 394 (1977).
Ir-38. H. Conrad, J. Kuppers, F. Nitscke, and A. Plagge, *Surf. Sci.* **69**, 668 (1977).
Ir-39. J. Küppers and A. Plagge, *J. Vac. Sci. Technol.* **13**, 259 (1976).
Ir-40. D. I. Hagen, B. E. Nieuwenhuys, G. Rovida, and G. A. Somorjai, *Surf. Sci.* **57**, 632 (1976).
Ir-41. B. E. Nieuwenhuys, D. I. Hagen, G. Rovida, and G. A. Somorjai, *Surf. Sci.* **59**, 155 (1976).
Ir-42. V. P. Ivanov, G. K. Boreskov, V. I. Savchenko, W. F. Egglehoff, Jr., and W. H. Weinberg, *Surf. Sci.* **61**, 207 (1976).
Ir-43. P. A. Zhdan, G. K. Boreskov, A. I. Boronin, W. F. Egglehoff, Jr., and W. H. Weinberg, *Surf. Sci.* **61**, 25 (1976).
Ir-44. P. A. Zhdan, E. K. Boreskov, W. F. Egglehoff, Jr., and W. H. Weinberg, *Surf. Sci.* **61**, 377 (1976).
Ir-45. J. T. Grant, *Surf. Sci.* **25**, 451 (1971).
Ir-46. T. S. Marinova and K. L. Kostov, *Surf. Sci.* **185**, 203 (1987).
Ir-47. P. D. Szuromi, J. R. Engtrom, and W. H. Weinberg, *J. Chem. Phys.* **80**, 508 (1984).
Ir-48. K. L. Kostov and T. S. Marinova, *Surf. Sci.* **184**, 359 (1987).
Ir-49. J. A. D. Matthew, E. Bertel, and F. P. Netzer, *Surf. Sci.* **184**, L389 (1987).
Ir-50. T. S. Marinova and K. L. Kostov, *Surf. Sci.* **181**, 573 (1987).
Ir-51. O. A. Baschenko, V. I. Bukhtiyaron, and A. I. Boronin, *Surf. Sci.* **271**, 493 (1992).

Iron

Fe-1. H. J. Krebs, H. P. Bonzel, and G. Gafner, *Surf. Sci.* **88**, 269 (1979).
Fe-2. G. Ertl and K. Wandelt, *Surf. Sci.* **50**, 479 (1975).
Fe-3. G. Ertl, M. Huber, and N. Thiele, *Z. Naturforsch.* **34A**, 30 (1979).
Fe-4. C. Leygraf and S. Ekelund, *Surf. Sci.* **40**, 609 (1973).
Fe-5. K. O. Legg, F. Jona, D. W. Jepsen, and P. M. Marcus, *J. Phys. C* **10**, 937 (1977).
Fe-6. P. A. Dowben and R. G. Jones, *Surf. Sci.* **84**, 449 (1979).
Fe-7. G. Ertl, M. Grunze, and M. Weiss, *J. Vac. Technol.* **13**, 314 (1976).
Fe-8. T. Matsudiara, M. Watanabe, and M. Onchi, in: *Proc. 2nd Int. Conf. on Solid Surfaces*, Tokyo, (H. Kumagai and T. Toya, eds.), *Jpn. J. Appl. Phys.* Suppl. 2, Pt. 2 (1974), p. 181.

Fe-9.   T. Horiguchi and S. Nakanishi, in: *Proc. 2nd Int. Conf. on Solid Surfaces*, Tokyo (H. Kumagai and T. Toya, eds.), *Jpn. J. Appl. Phys.* Suppl. 2, Pt. 2 (1974), p. 89.

Fe-10. M. Watanabe, M. Miyamura, T. Matsuadaira, and M. Onchi, in: *Proc. 2nd Int. Conf. on Solid Surfaces*, Tokyo (H. Kumagai and T. Toya, eds.), *Jpn. J. Appl. Phys.* Suppl. 2, Pt. 2 (1974), p. 501.

Fe-11. A. Schulz, R. Courths, H. Schulz, and S. Hüfner, *J. Phys. F* **9**, L41 (1979).

Fe-12. J. Benzinger and R. J. Madix, *Surf. Sci.* **94**, 119 (1980).

Fe-13. C. F. Brucker, and T. N. Rhodin, *Surf. Sci.* **57**, 523 (1976).

Fe-14. C. Brucker and T. Rhodin, *J. Catal.* **47**, 214 (1977).

Fe-15. G. W. Simmons and D. J. Dwyer, *Surf. Sci.* **48**, 373 (1975).

Fe-16. C. R. Brundle, *IBM J. Res. Dev.* **22**, 235 (1978).

Fe-17. G. Brodén, G. Gafner, and H. P. Bonzel, *Appl. Phys.* **13**, 333 (1977).

Fe-18. H. D. Shihi, F. Jona, U. Bardi, and P. M. Marcus, *J. Phys. C* **13**, 3801 (1980).

Fe-19. G. Gafner and R. Feder, *Surf. Sci.* **57**, 37 (1976).

Fe-20. K. Yoshida and G. A. Somorjai, *Surf. Sci.* **75**, 46 (1978).

Fe-21. F. J. Szalkowski and C. A. Mergerle, *Phys. Lett.* **48A**, 117 (1974).

Fe-22. G. Brodén and H. P. Bonzel, *Surf. Sci.* **84**, 106 (1979).

Fe-23. M. A. Chesters and J. C. Riviere, in: *Proc. 7th Int. Vac. Congr. and 3rd Int. Conf. on Solid Surfaces*, Vienna , 1977 (R. Dobrozemski, F. Rudenauer, and F. P. Viehbock, eds.), p. 873.

Fe-24. H. P. Bonzel and H. J. Krebs, *Surf. Sci.* **91**, 499 (1980).

Fe-25. E. S. Jensen, C. W. Seabury, and T. N. Rhodin, *Solid State Commun.* **35**, 581 (1980).

Fe-26. W. G. Dorfeld, J. B. Hudson, and R. Zuhr, *Surf. Sci.* **57**, 460 (1976).

Fe-27. W. Arabczyk, H.-J. Müssig, and F. Storbeck, *Phys. Stat. Solidi (a)* **55**, 437 (1979).

Fe-28. H. D. Shih, F. Jona, D. W. Jepsen, and P. M. Marcus, *Surf. Sci.* **104**, 39 (1981).

Fe-29. W. Arabczk and H.-J. Mussig, *Thin Solid Films* **34**, 103 (1976).

Fe-30. I. D. Gay, M. Textor, R. Mason, and Y. Iwasawa, *Proc. R. Soc. London* **356**, 25 (1977).

Lanthanum

La-1.   T. L. Barr, in: *Quantitative Surface Analysis of Materials*, Proc. Symp., Cleveland, OH (N. S. McIntyre, ed.), American Society for Testing and Materials, Philadelphia, PA, ASTM-STP-643 (1978), p. 83.

Lead

Pb-1.   S. Evans and J. M. Thomas, *J. Chem. Soc. Faraday Trans. II* **71**, 313 (1975).

Pb-2.   J. F. McGilp, P. Weightman, and E. J. McGuire, *J. Phys. C* **10**, 3445 (1977).

Pb-3.   R. Nyholm and N. Martensson, *J. Phys. C* **13**, L279 (1980).

Pb-4.   R. A. Pollak, S. Kowalczyk, L. Ley, and D. A. Shirley, *Phys. Rev. Lett.* **29**, 274 (1972).

Pb-5.   R. W. Joyner, K. Kiski, and M. W. Roberts, *Proc. R. Soc. (London) A* **358**, 223 (1977).

Pb-5.   H. M. van Pinxteren and J. W. Frenken, *Surf. Sci.* **275**, 383 (1992).

Pb-6.   B. Pluis, J. M. Gay, J. W. M. Frenken, S. Gierlotka, J. F. van der Veen, J. E. MacDonald, A. A. Williams, N. Piggins, and J. Als-Nielsen, *Surf. Sci.* **222**, L485 (1989).

Pb-7.   J. W. M. Frenken, J. F. van der Veen, R. N. Barnett, U. Landman, and C. L. Cleveland, *Surf. Sci.* **172**, 319 (1986).

Pb-8.   A. Pavlovska, H. Steffen, and E. Bauer, *Surf. Sci.* **234**, 143 (1990).

Pb-9.   U. Breuer, H. P. Bonzel, K. C. Prince, and R. Lipowsky, *Surf. Sci.* **223**, 258 (1989).

Lithium

Li-1. R. E. Clausing, D. S. Easton, and G. L. Powell, *Surf. Sci.* **36**, 377 (1973).
Li-2. G. L. Powell, R. E. Clausing, and G. E. McGuire, *Surf. Sci.* **49**, 310 (1975).
Li-3. D. J. David, M. H. Fronig, T. N. Wittenberg, and W. E. Moddeman, *Appl. Surf. Sci.* **7**, 185 (1981).

Lutetium

Lu-1. J. Onsgaard, S. Tourgaard, P. Morgan, and F. Ryborg, *J. Electron Spectrosc. Rel. Phenom.* **17**, 29 (1980).

Magnesium

Mg-1. A. Benninghoven and L. Wiedmann, *Surf. Sci.* **41**, 483 (1974).
Mg-2. G. C. Allen, P. M. Tucker, B. E. Hayden, and D. F. Klemperer, *Surf. Sci.* **102**, 207 (1981).
Mg-3. L. H. Jenkins and M. F. Chung, *Surf. Sci.* **33**, 159 (1972).
Mg-4. B. E. Hayden, E. Schweizer, R. Kotz, and A. M. Bradshaw, *Surf. Sci.* **111**, 26 (1981).
Mg-5. H. Namba, J. Darville, and J. M. Gilles, *Solid State Commun.* **34**, 287 (1980).
Mg-6. H. Namba, J. Darville, and J. M. Gilles, *Surf. Sci.* **108**, 446 (1981).
Mg-7. J. Ghijsen, H. Namba, P. A. Thiry, J. J. Pireaux, and P. Caudano, *Appl. Surf. Sci.* **8**, 397 (1981).

Manganese

Mn-1. H.-K. Hu and J. W. Rabalais, *Surf. Sci.* **107**, 376 (1981).

Molybdenum

Mo-1. T. W. Haas, J. T. Grant, and G. J. Dooley III, *J. Vac. Sci. Technol.* **7**, 43 (1970).
Mo-2. T. Muira, *Jap. J. Appl. Phys.* **15**, 403 (1976).
Mo-3. J. Lecante, R. Riwan, and C. Guillot, *Surf. Sci.* **35**, 271 (1973).
Mo-4. H. M. Kenneth and A. E. Lee, *Surf. Sci.* **48**, 591 and 606 (1975).
Mo-5. A. G. Jackson and M. P. Hooker, *Surf. Sci.* **28**, 373 (1971).
Mo-6. G. J. Dooley III and T. W. Haas, *J. Vac. Sci. Technol.* **7**, S90 (1970).
Mo-7. K. Kunimori, T. Kawai, T. Knodow, T. Onishi, and K. Tamaru, *Surf. Sci.* **46**, 567 (1974).
Mo-8. A. Barrie and C. R. Brundle, *J. Electron Spectrosc. Rel. Phenom.* **5**, 321 (1974).
Mo-9. D. Tabor and J. M. Wilson, *J. Crystal Growth* **9**, 60 (1971).
Mo-10. A. Ignatiev, F. Jona, H. D. Shih, D. W. Jepsen, and P. M. Marcus, *Phys. Rev. B* **11**, 4787 (1975).
Mo-11. G. Guillot, R. Riwan, and J. Lecante, *Surf. Sci.* **59**, 581 (1976).
Mo-12. K. Hartig, A. P. Janssen, and J. A. Venables, *Surf. Sci.* **74**, 69 (1978).
Mo-13. T. E. Felter and P. J. Estrup, *Surf. Sci.* **76**, 464 (1978).
Mo-14. E. Bauer and H. Poppa, *Surf. Sci.* **88**, 31 (1979).
Mo-15. T. Muira and Y. Tuzi, in: *Proc. 2nd Int. Conf. on Solid Surfaces* (H. Kumagai and T. Toya, eds.), *Jpn. J. Appl. Phys.* Suppl. 2, Pt. 2 (1974), p. 85.
Mo-16. S. Thomas and T. W. Haas, *Surf. Sci.* **28**, 632 (1971).
Mo-17. R. M. Lambert, J. W. Linnett, and J. A. Schwarz, *Surf. Sci.* **26**, 572 (1971).
Mo-18. G. J. Dooley, III and T. W. Haas, *J. Vac. Sci. Technol.* **7**, 49 (1970).

Neodymium

Nd-1.   A. A. Atensio, O. I. Kapusta, and N. M. Omel'Yanovskaya, *Sov. Phys. Solid State* **20**, 1735 (1978).

Nickel

Ni-1.   C. J. Davisson and C. H. Germer, *Phys. Rev.* **30**, 705 (1927).

Ni-2.   T. W. Haas, G. J. Dooley III, A. G. Jackson, and M. P. Hooker, *Prog. Surf. Sci.* **1**, 155 (1971).

Ni-3.   M. P. Hooker, J. T. Grant, and T. W. Haas, *J. Vac. Sci. Technol.* **13**, 296 (1976).

Ni-4.   H. G. Tompkins, *Surf. Sci.* **62**, 293 (1977).

Ni-5.   A. M. Horgan and I. Dalins, *J. Vac. Sci. Technol.* **10**, 523 (1973).

Ni-6.   A. Benninghoven, P. Beckman, D. Greifendorf, J.-H. Muller, and M. Schemmer, *Surf. Sci.* **107**, 148 (1981).

Ni-7.   P. H. Dawson and W.-C. Tam, *Surf. Sci.* **81**, 164 (1979).

Ni-8.   H. Windawi and J. R. Katzer, *Surf. Sci.* **75**, L761 (1978).

Ni-9.   S. Andersson and J. B. Pendry, *Surf. Sci.* **71**, 75 (1978).

Ni-10.   T. T. Anh Nguyen and R. C. Cinti, *Surf. Sci.* **68**, 566 (1977).

Ni-11.   J. M. Blakely, J. S. Kim, and H. C. Potter, *J. Appl. Phys.* **41**, 2693 (1970).

Ni-12.   F. L. Baudais, A. J. Borschke, J. D. Fedyk, and M. J. Dignam, *Surf. Sci.* **100**, 210 (1980).

Ni-13.   U. Jostell, *Surf. Sci.* **82**, 333 (1979).

Ni-14.   D. J. Godfrey and D. P. Woodruff, *Surf. Sci.* **105**, 438 (1981).

Ni-15.   G. Allie, E. Blanc, and D. Dufayard, *Surf. Sci.* **57**, 293 (1976).

Ni-16.   T. Matsudaira, M. Nishijima, and M. Onchi, *Surf. Sci.* **61**, 651 (1976).

Ni-17.   E. G. McRae, D. Aberdam, R. Baudoing, and Y. Gauthier, *Surf. Sci.* **78**, 518 (1978).

Ni-18.   M. Barber, R. S. Bordoli, J. C. Vickerman, and J. Wolstenholme, in: *Proc. 7th Int. Vac. Congr. and 3rd Int. Conf. on Solid Surfaces*, Vienna (R. Dobrozemski, S. Rudenauer, and S. P. Viehbock, eds.), (1977), p. 983.

Ni-19.   C. Gaubert and Y. Gauthier, in: *Proc. 7th Int. Vac. Congr. and 3rd Int. Conf. on Solid Surfaces*, Vienna (R. Dobrozemski, S. Rudenauer, and S. P. Viehbock, eds.) (1977), p. 2427.

Ni-20.   P. D. Johnson and T. A. Delchar, *Surf. Sci.* **77**, 400 (1978).

Ni-21.   D. J. Godfrey and D. P. Woodruff, *Surf. Sci.* **89**, 76 (1979).

Ni-22.   J. Kolaczkiewicz, C. Koziol, and S. Mrox, *Acta Phys. Polon. A* **41**, 783 (1972).

Ni-23.   S. Andersson, *Solid State Commun.* **21**, 75 (1977).

Ni-24.   T. M. Buck, G. H. Wheatley, and L. K. Verheij, *Surf. Sci.* **90**, 635 (1979).

Ni-25.   J. E. Demuth and T. N. Rhodin, *Surf. Sci.* **42**, 261 (1974).

Ni-26.   R. C. Schouten, O. L. J. Gijzeman, and G. A. Bootsma, *Surf. Sci.* **87**, 1 (1979).

Ni-27.   K. Akimoto, Y. Sakisaka, M. Nishijima, and M. Onchi, *Surf. Sci.* **88**, 109 (1979).

Ni-28.   K. Akimoto, Y. Sakisaka, M. Nishijima, and M. Onchi, *Surf. Sci.* **82**, 349 (1979).

Ni-29.   Y. Gauthier, D. Aberdam, and R. Baudoing, *Surf. Sci.* **78**, 339 (1978).

Ni-30.   T. Fleisch, N. Winograd, and W. N. Delgass, *Surf. Sci.* **78**, 141 (1978).

Ni-31.   J. C. Bertolini, G. Dalmai-Imelik, and J. Rousseau, *Surf. Sci.* **67**, 478 (1977).

Ni-32.   G. Dalmai-Imelik, J. C. Bertolini, and J. Rousseau, *Surf. Sci.* **63**, 67 (1977).

Ni-33.   P. H. Holloway and J. B. Hudson, *Surf. Sci.* **43**, 123 (1974).

Ni-34.   K. Jacobi, M. Scheffler, K. Kambe, and F. Forstmann, *Solid State Commun.* **22**, 17 (1977).

Ni-35.   J. B. Benziger and R. J. Madix, *Surf. Sci.* **79**, 394 (1979).

Ni-36.   K. Horn, A. M. Bradshaw, and K. Jacobi, *J. Vac. Sci. Technol.* **15**, 575 (1978).

Ni-37.   C. L. Allyn, T. Gustafsson, and E. W. Plummer, *Chem. Phys. Lett.* **47**, 127 (1977).

Ni-38.   R. S. Bordoli, J. C. Vickerman, and J. Welstenholme, *Surf. Sci.* **85**, 244 (1979).

Ni-39.   P. H. Dawson and W.-C. Tam, *Surf. Sci.* **91**, 153 (1980).

Ni-40. J. C. Tracy, *J. Chem. Phys.* **56**, 2736 (1972).
Ni-41. G. Maire, J. R. Anderson, and B. B. Johnson, *Proc. R. Soc. (London) A* **320**, 227 (1970).
Ni-42. G. Casalone, M. G. Cattania, M. Simonetta, and M. Tescari, *Surf. Sci.* **62**, 321 (1977).
Ni-43. T. Fleisch, G. L. Ott, W. N. Delgass, and N. Winograd, *Surf. Sci.* **81**, 1 (1979).
Ni-44. P. R. Norton, R. L. Tapping, and J. W. Goodale, *Surf. Sci.* **65**, 13 (1977).
Ni-45. J. C. Bertolini and B. Tardy, *Surf. Sci.* **102**, 131 (1981).
Ni-46. J. C. Bertolini and J. Rousseau, *Surf. Sci.* **89**, 467 (1979).
Ni-47. D. W. Goodman, J. T. Yates, Jr., and T. E. Madey, *Surf. Sci.* **93**, L135 (1980).
Ni-48. J. McCarty, J. Falconer, and R. J. Madix, *J. Catal.* **30**, 235 (1973).
Ni-49. M. Nishijima, S. Masuda, Y. Sakisaka, and M. Onchi, *Surf. Sci.* **107**, 31 (1981).
Ni-50. H. P. Bonzel and E. E. Latta, *Surf. Sci.* **76**, 275 (1978).
Ni-51. R. Riwan, *Surf. Sci.* **27**, 267 (1971).
Ni-52. H. H. Madden and G. Ertl, *Surf. Sci.* **35**, 211 (1973).
Ni-53. J. L. Falconer and R. J. Madix, *Surf. Sci.* **46**, 473 (1974).
Ni-54. R. J. Madix and J. L. Falconer, *Surf. Sci.* **51**, 546 (1975).
Ni-55. R. J. Madix, J. L. Falconer, and A. M. Suszko, *Surf. Sci.* **54**, 6 (1976).
Ni-56. W. C. Turkenburg, R. G. Smeenk, and F. W. Saris, *Surf. Sci.* **74**, 181 (1978).
Ni-57. J. F. Van Der Veen, R. M. Tromp, R. G. Smeenk, and R. W. Saris, *Surf. Sci.* **82**, 468 (1979).
Ni-58. R. G. Smeenk, R. M. Tromp, J. F. Van Der Veen, and F. W. Saris, *Surf. Sci.* **95**, 156 (1980).
Ni-59. G. L. Price, B. A. Sexton, and B. G. Baker, *Surf. Sci.* **60**, 506 (1976).
Ni-60. R. Sau and J. B. Hudson, *Surf. Sci.* **102**, 239 (1981).
Ni-61. R. Sau and J. B. Hudson, *Surf. Sci.* **95**, 465 (1980).
Ni-62. J. A. Van Den Berg, L. J. Verheij, and D. G. Armour, *Surf. Sci.* **91**, 218 (1980).
Ni-63. J. L. Falconer and R. J. Madix, *Surf. Sci.* **48**, 393 (1975).
Ni-64. N. M. Abbas, and R. J. Madix, *Surf. Sci.* **62**, 739 (1977).
Ni-65. S. W. Johnson and R. J. Madix, *Surf. Sci.* **66**, 189 (1977).
Ni-66. J. Falconer, J. McCarty, R. J. Madix, in: *Proc. 2nd Int. Conf. on Solid Surfaces*, Tokyo (H. Kumajai and T. Toya, eds.), *Jpn. J. Appl. Phys.* Suppl. 2, Pt. 2 (1974), p. 525.
Ni-67. T. N. Taylor and P. J. Estrup, *J. Vac. Sci. Technol.* **11**, 244 (1974).
Ni-68. T. N. Taylor and P. J. Estrup, *J. Vac. Sci. Technol.* **10**, 26 (1973).
Ni-69. H. H. Madden, J. Kuppers, and G. Ertl, *J. Chem. Phys.* **58**, 3401 (1973).
Ni-70. J. Verhoeven and J. Los, *Surf. Sci.* **82**, 109 (1979).
Ni-71. R. A. Zuhr and J. B. Judson, *Surf. Sci.* **66**, 405 (1977).
Ni-72. F. C. Schouten, E. W. Kaleveld, and G. A. Bootsma, *Surf. Sci.* **63**, 460 (1977).
Ni-73. F. C. Schouten, E. Te Brake, O. L. J. Gijzeman, and G. A. Bootsma, *Surf. Sci.* **74**, 1 (1978).
Ni-74. C. Benndorf, B. Egert, C. Nöbl, H. Seidel, and F. Thieme, *Surf. Sci.* **92**, 636 (1980).
Ni-75. P. R. Mahaffy and M. J. Dignam, *Surf. Sci.* **97**, 377 (1980).
Ni-76. M. Grunze, R. K. Driscoll, G. N. Burland, J. C. L. Cornish, and J. Pritchard, *Surf. Sci.* **89**, 381 (1979).
Ni-77. K. H. Rieder, *Appl. Surf. Sci.* **2**, 74 (1978).
Ni-78. C. Varelas, K. Goltz, and R. Sizmann, *Surf. Sci.* **80**, 524 (1979).
Ni-79. P. S. Frederick and S. J. Hruska, *Surf. Sci.* **62**, 707 (1977).
Ni-80. J. C. Shelton, H. R. Patil, and J. M. Blakely, *Surf. Sci.* **43**, 493 (1974).
Ni-81. P. H. Holloway and J. B. Hudson, *Surf. Sci.* **43**, 141 (1974).
Ni-82. T. Edmonds, J. J. McCarroll, and R. C. Pitkethly, *J. Vac. Sci. Technol.* **8**, 68 (1971).
Ni-83. G. A. Sargent, G. B. Freeman, and J. L.-R. Chao, *Surf. Sci.* **100**, 342 (1981).
Ni-84. G. Casalone, M. G. Cattania, M. Simonetta, and M. Tescari, *Surf. Sci.* **72**, 739 (1978).
Ni-85. J. E. Demuth and H. Ibach, *Surf. Sci.* **85**, 365 (1979).
Ni-86. J. E. Demuth and H. Ibach, *Surf. Sci.* **78**, L238 (1978).
Ni-87. J. E. Demuth and H. Ibach, *Chem. Phys. Lett.* **60**, 395 (1979).

Ni-88. S. Lehwald, H. Ibach, and J. E. Demuth, *Surf. Sci.* **78**, 577 (1978).
Ni-89. W. Erley, K. Besocke, and H. Wagner, *J. Chem. Phys.* **66**, 5269 (1977).
Ni-90. J. C. Campuzano, R. Dus, and R. G. Greenler, *Surf. Sci.* **102**, 172 (1981).
Ni-91. T. W. Capehart and T. N. Rhodin, *Surf. Sci.* **82**, 367 (1979).
Ni-92. A. M. Horgan and I. Dalins, *Surf. Sci.* **36**, 526 (1973).
Ni-93. J. C. Bertolini and J. Rousseau, *Surf. Sci.* **83**, 531 (1979).
Ni-94. J. C. Bertolini and B. Imelik, *Surf. Sci.* **80**, 586 (1979).
Ni-95. P. Klimesch and M. Henzler, *Surf. Sci.* **90**, 57 (1979).

Niobium

Nb-1. J. M. Dickey, *Surf. Sci.* **50**, 515 (1975).
Nb-2. P. H. Dawson and W. C. Tam, *Surf. Sci.* **81**, 464 (1979).
Nb-3. S. Usami, N. Tominaga, and T. Nakajima, *Vacuum* **27**, 11 (1977).
Nb-4. H. H. Farrell and M. Strongin, *Surf. Sci.* **38**, 18, 31 (1973).
Nb-5. J. Jupille and A. Cassuto, *Surf. Sci.* **60**, 177 (1976).
Nb-6. J. Jupille, B. Bigeard, J. Fusy, and A. Cassuto, *Surf. Sci.* **84**, 190 (1979).
Nb-7. R. Pantel, M. Bujor, and J. Bardolle, *Surf. Sci.* **62**, 589 (1977).
Nb-8. K. H. Rieder, *Appl. Surf. Sci.* **4**, 183 (1980).

Osmium

Os-1. Y. Fakuda and J. W. Rabalais, *Chem. Phys. Lett.* **76**, 47 (1980).
Os-2. H. Saalfeld, S. Tougaard, K. Bolwin, and M. Neumann, *Surf. Sci.* **178**, 452 (1986).
Os-3. H. H. Graen, M. Neuber, M. Neumann, G. Illing, H.-J. Freund, and F.-P. Netzer, *Surf. Sci.* **223**, 33 (1989).

Palladium

Pd-1.   J. C. Tracy and P. W. Palmberg, *J. Chem. Phys.* **51**, 4852 (1969).
Pd-2.   D. L. Weissman, M. L. Shek, and W. E. Spicer, *Surf. Sci.* **92**, L59 (1980).
Pd-3.   J. Kuppers, F. Nitschke, K. Wandelt, and G. Ertl, *Surf. Sci.* **87**, 295 (1979); **88**, 1 (1979).
Pd-4.   K. Kunimori, T. Kawai, T. Kondow, T. Onishi, and K. Tamaru, *Surf. Sci.* **59**, 302 (1976).
Pd-5.   T. Matsushima and J. M. White, *Surf. Sci.* **67**, 122 (1977).
Pd-6.   J. A. Strozier, in: *Proc. 7th Int. Vac. Cong. and 3rd Int. Conf. on Solid Surfaces*, Vienna (R. Dobrozemski, S. Rudenauer, and S. P. Viehbock, eds.), p. 857.
Pd-7.   Y. Matsumoto, M. Soma, T. Onishi, and K. Tamaru, *J. Chem. Soc. Faraday Trans. I* **76**, 1122 (1980).
Pd-8.   G. J. Slusser and N. Winograd, *Surf. Sci.* **95**, 53 (1980).
Pd-9.   A. M. Bradshaw and F. M. Hoffman, *Surf. Sci.* **72**, 513 (1978).
Pd-10. S. D. Bader, J. M. Blakely, M. B. Brodsky, R. J. Friddle, and R. L. Panosh, *Surf. Sci.* **74**, 405 (1978).
Pd-11. H. Conrad, G. Ertl, and E. E. Lotta, *Surf. Sci.* **41**, 435 (1974).
Pd-12. K. Christman, G. Ertl, and O. Schober, *Surf. Sci.* **40**, 61 (1973).
Pd-13. P. Legare, Y. Hall, and G. Maise, *Solid State Commun.* **31**, 307 (1979).
Pd-14. T. E. Madey, J. T. Yates, Jr., A. M. Bradshaw, and F. M. Hoffman, *Surf. Sci.* **89**, 370 (1979).

Platinum

Pt-1. J. C. Hamilton and J. M. Blakely, *J. Vac. Sci. Technol.* **15**, 559 (1978).
Pt-2. B. Lang, R. W. Joyner, and G. A. Somorjai, *Surf. Sci.* **30**, 440 (1972).
Pt-3. K. Christmann, G. Ertl, and T. Pignet, *Surf. Sci.* **54**, 365 (1976).
Pt-4. W. Erley, *Surf. Sci.* **94**, 281 (1980).
Pt-5. M. Wilf and T. P. Dawson, *Surf. Sci.* **65**, 399 (1977).
Pt-6. J. N. Miller, D. T. Ling, M. L. Shek, D. L. Weissman, P. M. Stefan, I. Lindan, and W. E. Spicer, *Surf. Sci.* **94**, 16 (1980).
Pt-7. W. Heegemann, K. H. Merster, E. Bechtold, and K. Hayek, *Surf. Sci.* **49**, 161 (1975).
Pt-8. P. R. Norton, *Surf. Sci.* **44**, 624 (1974).
Pt-9. J. L. Gland and G. A. Somorjai, *Surf. Sci.* **41**, 387 (1974).
Pt-10. H. P. Bonzel and G. Pirug, *Surf. Sci.* **62**, 45 (1977).
Pt-11. G. Brodin, G. Pirug, and H. P. Bonzel, *Surf. Sci.* **72**, 45 (1978).
Pt-12. G. Kneringer and F. P. Netzer, *Surf. Sci.* **49**, 125 (1975).
Pt-13. F. P. Netzer and J. A. D. Matthew, *Surf. Sci.* **51**, 352 (1975).
Pt-14. K. Schwaha and E. Bechtold, *Surf. Sci.* **66**, 383 (1977).
Pt-15. S. R. Kelemen, T. E. Fisher, and J. A. Schwarz, *Surf. Sci.* **81**, 440 (1979).
Pt-16. M. A. Barteau, E. J. Ko, and R. J. Madix, *Surf. Sci.* **102**, 99 (1981).
Pt-17. C. M. Comrie, W. H. Weinberg, and R. M. Lambert, *Surf. Sci.* **57**, 619 (1976).
Pt-18. M. Salmeron and G. A. Somorjai, *Surf. Sci.* **91**, 373 (1980).
Pt-19. D. W. Blakely and G. A. Somorjai, *J. Catal.* **42**, 181 (1976).

Plutonium

Pu-1. D. T. Larson and R. O. Adams, *Surf. Sci.* **47**, 413 (1975).
Pu-2. D. T. Larson, *J. Vac. Sci. Technol.* **17**, 55 (1980).

Rhenium

Re-1. R. Ducros, M. Housley, M. Alnot, and A. Cassuto, *Surf. Sci.* **71**, 433 (1978).
Re-2. M. Alnot, B. Weber, J. J. Ehrhardt, and A. Cassuto, *Appl. Surf. Sci.* **2**, 578 (1979).
Re-3. J.-L. Philippart, B. Bigeard, B. Weber, and A. Cassuto, *Surf. Sci.* **45**, 457 (1974).
Re-4. M. Alnot, J. J. Ehrhardt, J. Fusy, and A. Cassuto, *Surf. Sci.* **46**, 457 (1974).
Re-5. J. Fusy, B. Bigeard, and A. Cassuto, *Surf. Sci.* **46**, 177 (1974).
Re-6. R. R. Ford and D. Lichtman, *Surf. Sci.* **25**, 537 (1971).
Re-7. Y. Fukuda, F. Honda, and J. W. Rabalais, *Surf. Sci.* **93**, 338 (1980).
Re-8. P. D. Schulze, D. L. Utey, and R. L. Hance, *Surf. Sci.* **102**, L9 (1981).
Re-9. Y. Fukuda, F. Honda, and J. W. Rabalais, *Surf. Sci.* **99**, 289 (1980).
Re-10. R. Pantel, M. Bujor, and J. Bardolle, *Surf. Sci.* **83**, 228 (1979).
Re-11. M. Housley, R. Ducros, G. Piquard, and A. Cassuto, *Surf. Sci.* **68**, 277 (1977).
Re-12. R. S. Zimmer and W. D. Robertson, *Surf. Sci.* **43**, 61 (1974).
Re-13. W. Braun, G. Meyer-Ehmsen, M. Neumann, and E. Schwarz, *Surf. Sci.* **89**, 354 (1979).
Re-14. G. J. Dooley III and T. W. Haas, *Surf. Sci.* **19**, 1 (1970).

Rhodium

Rh-1. K. A. R. Mitchell, F. R. Shepherd, P. R. Watson, and D. C. Frost, *Surf. Sci.* **64**, 737 (1977).
Rh-2. D. G. Castner, B. A. Sexton, and G. A. Somorjai, *Surf. Sci.* **71**, 519 (1978).

Rh-3. C. von Eggeling, G. Schmidt, G. Besold, L. Hammer, K. Heinz, and K. Muller, *Surf. Sci.* **221**, 11 (1989).

Rh-4. R. J. Barid, R. C. Ku, and P. Wynblatt, *Surf. Sci.* **97**, 346 (1980).

Rh-5. R. A. Marbrow and R. M. Lambert, *Surf. Sci.* **67**, 489 (1977).

Rh-6. C.-M. Chan, P. A. Thiel, J. T. Yates, Jr., and W. H. Weinberg, *Surf. Sci.* **76**, 296 (1978).

Rh-7. F. R. Shepherd, P. R. Watson, D. C. Frost, and K. A. R. Mitchell, *J. Phys. C* **11**, 4591 (1978).

Rh-8. M. P. Cox and R. M. Lambert, *Surf. Sci.* **107**, 547 (1981).

Rh-9. L. A. DeLouise and N. Winograd, *Surf. Sci.* **138**, 417 (1984).

Rh-10. M. Rebholz, R. Prins, and N. Kruse, *Surf. Sci.* **259**, L797 (1991).

Rh-11. D. G. Castner and G. A. Somorjai, *Surf. Sci.* **83**, 60 (1979).

Rh-12. G. Hoogers and D. A. King, *Surf. Sci.* **286**, 306 (1993).

Ruthenium

Ru-1. P. D. Reed, C. M. Comrie, and R. M. Lambert, *Surf. Sci.* **59**, 33 (1976).

Ru-2. T. W. Orent and R. S. Hansen, *Surf. Sci.* **67**, 325 (1977).

Ru-3. J. T. Grant and T. W. Haas, *Surf. Sci.* **21**, 76 (1970).

Ru-4. T. E. Madey, H. A. Engelhardt, and D. Menzel, *Surf. Sci.* **48**, 304 (1975).

Ru-5. K. Wadelt, J. Hulse, and J. Küppers, *Surf. Sci.* **104**, 212 (1981).

Ru-6. T. E. Madey and D. Menzel, in: *Proc. 2nd Int. Conf. on Solid Surfaces* Tokyo (H. Kumagai and T. Toya, eds.), *Jpn. J. Appl. Phys.* Suppl. 2, Pt. 2 (1974), p. 229.

Ru-7. L. R. Danielson, M. J. Dresser, E. E. Donaldson, and J. T. Dickinson, *Surf. Sci.* **71**, 599 (1978).

Ru-8. G. E. Thomas and W. H. Weinberg, *J. Chem. Phys.* **70**, 954 (1979).

Ru-9. G. E. Thomas and W. H. Weinberg, *J. Chem. Phys.* **70**, 1437 (1979).

Ru-10. E. D. Williams and W. H. Weinberg, *Surf. Sci.* **82**, 93 (1979).

Ru-11. S. R. Kelemen and T. E. Fisher, *Surf. Sci.* **87**, 53 (1979).

Ru-12. J. A. Schwarz and S. R. Kelemen, *Surf. Sci.* **87**, 510 (1979).

Ru-13. E. Umbach, S. Kulkarni, P. Feulner, and D. Menzel, *Surf. Sci.* **88**, 65 (1979).

Ru-14. S.-K. Shi and J. M. White, *J. Chem. Phys.* **73**, 5889 (1980).

Ru-15. J. C. Fuggle, T. E. Madey, M. Steinkilberg, and D. Menzel, *Surf. Sci.* **51**, 521 (1975).

Ru-16. J. C. Fuggle, T. E. Madey, M. Steinkilberg, and D. Menzel, *Chem. Phys. Lett.* **33**, 233 (1975).

Ru-17. D. W. Goodman, T. E. Madey, M. Ono, and J. T. Yates, Jr., *J. Catal.* **50**, 279 (1977).

Ru-18. G. B. Fisher, T. E. Madey, G. J. Waclawski, and J. T. Yates, Jr., in: *Proc. 7th Int. Vac. Congr. and 3rd Int. Conf. on Solid Surfaces*, Vienna (R. Dobrozemski, F. Rudenauer, and F. P. Viehbock, eds.), p. 1071.

Ru-19. R. Ku, N. A. Gjostein, and H. P. Bonzel, *Surf. Sci.* **64**, 465 (1977).

Samarium

Sm-1. W. Farber and P. Braun, *Surf. Sci.* **41**, 195 (1974).

Scandium

Sc-1. M. F. Chung and L. H. Jenkins, *Surf. Sci.* **28**, 637 (1971).

Sc-2. J. Onsgaard, S. Tougaard, and P. Morgen, *Appl. Surf. Sci.* **3**, 113 (1979).

Sc-3. J. Onsgaard, S. Tougaard, P. Morgen, and F. Ryborg, *J. Electron Spectrosc. Rel. Phenom.* **18**, 29 (1980).

Sc-4. S. Tougaard and A. Ignatiev, *Surf. Sci.* **115**, 270 (1982).

Selenium

Se-1. A. Auboiroux, D. Marchand, A. P. Legrand, and J. C. Thuillier, *Surf. Sci.* **78**, 104 (1978).

Silicon

Si-1. *Handbook of Semiconductor Wafer Cleaning Technology: Science, Technology, and Applications* (W. Kern, ed.), Noyes, Park Ridge, NJ (1993).

Si-2. F. Jona, *IBM J. Res. Dev.* **9**, 375 (1965).

Si-3. T. Bekkay, E. Sacher, and A. Yelon, *Surf. Sci.* **217**, L377 (1989).

Si-4. D. Hanemann, *Adv. Phys.* **31**, 165 (1982).

Si-5. M. Grundner, D. Gräff, P. O. Han, and A. Schnegg, *Solid State Technol.* **34**, 69 (1991).

Si-6. G. J. Pietsch, U. Köhler, and M. Henzler, *Mat. Res. Soc. Symp. Proc.* **315**, 497 (1993).

Si-7. M. Grundner and H. Jacob, *Appl. Phys. A* **39**, 37 (1986).

Si-8. W. Kern and D. A. Puotinen, *RCA Rev.* **31**, 1871 (1970); W. Kern, *Semicond. Int'l.* (April 1984), 94; W. Kern, *J. Electrochem. Soc.* **137**, 1887 (1990).

Si-9. T. Takahagi, I. Nagai, A. Ishitani, H. Kuroda, and Y. Nagasawa, *J. Appl. Phys.* **64**, 3516 (1988).

Si-10. F. J. Grunthaner, P. J. Grundthaner, R. P. Vasquez, B. F. Lewis, and J. Maserijan, *J. Vac. Sci. Technol.* **16**, 1443 (1979).

Si-11. B. F. Phillips, *J. Vac. Sci. Technol. A* **1**, 646 (1983).

Si-12. A. Ishizaka and Y. Shiraki, *J. Electrochem. Soc.* **1333**, 666 (1986).

Si-13. R. Williams and A. M. Goodmann, *Appl. Phys. Lett.* **25**, 531 (1974).

Si-14. M. Grundner, D. Gräf, P. O. Hahn, and A. Schnegg, *Solid State Technol.* **34**, 69 (1991).

Si-15. G. S. Higashi, Y. J. Chabal, K. Raghavachari, R. S. Becker, M. P. Green, K. Hanson, T. Boone, J. H. Eisenberg, S. F. Shive, G. N. DiBello, and K. L. Fulford, Proc. 1993 Spring Meeting Electrochem. Soc., 1817.

Si-16. G. J. Pietsch, U. Köhler, and M. Henzler, *Mat. Res. Soc. Symp. Proc.* **315**, 497 (1993).

Si-17. M. Grundner and H. Jacob, *Appl. Phys. A* **39**, 37 (1986).

Si-18. E. Mendel, *Solid State Technol.* **10**, 27 (1967).

Si-19. P. A. M. van der Heide, M. J. Bean, and H. J. Ronde, *J. Vac. Sci. Technol. A* **8**, 266 (1990).

Si-20. R. C. Henderson, *J. Electrochem. Soc.* **119**, 772 (1972).

Si-21. S. I. Raider, R. Flitsch, and M. J. Palmer, *J. Electrochem. Soc.* **122**, 413 (1975).

Si-22. M. Suemitsu, T. Kaneko, and N. Miyamoto, *Jpn. J. Appl. Phys. Part 1* **28**, 2421 (1989).

Si-23. T. Kaneko, M. Suemitsu, and N. Miyamoto, *Jpn. J. Appl. Phys. Part 1* **28**, 2425 (1989).

Si-24. S. Baunack and A. Zehe, *Phys. Stat. Sol. A* **115**, 223 (1989).

Si-25. G. J. Pietsch, *Appl. Phys. A* **60**, 347 (1995).

Si-26. T. Bitzer and H. J. Lewerenz, *Surf. Sci.* **269/270**, 886 (1992).

Si-27. R. E. Thomas, M. J. Mantini, R. A. Rudder, and D. P. Malta, *J. Vac. Sci. Technol. A* **10**, 817 (1992).

Si-28. Heung-Sik Tae, S. J. Par, S. H. Hwang, and K. H. Kwang, *J. Vac. Sci. Technol. B* **13**, 908 (1995).

Si-29. A. Crossley, C. J. Sofield, S. Sugden, and R. Clampitt, *Vacuum* **46**, 667 (1995).

Si-30. T. Aoyama, T. Yamazaki, and T. Ito, *J. Electrochem. Soc.* **140**, 1704 (1993).

Si-31. B. A. Joyce, *Surf. Sci.* **35**, 1 (1973).

Si-32. L. L. Kazmerski, O. Jamjoum, P. J. Ireland, and R. L. Whitney, *J. Vac. Sci. Technol.* **18**, 960 (1981).

Si-33. S. J. White and D. P. Woodruff, *Surf. Sci.* **64**, 131 (1977).

Si-34. S. J. White, D. P. Woodruff, B. W. Holland, and R. S. Zimmer, *Surf. Sci.* **74**, 34 (1978).

Si-35. F. Meyer and J. J. Vracking, *Surf. Sci.* **33**, 271 (1972).

Si-36. C. Chang, *Surf. Sci.* **23**, 283 (1970).

Si-37. T. D. Poppendieck, T. C. Hgoc, and M. B. Webb, *Surf. Sci.* **75**, 287 (1978).

Si-38. R. Kaplan, *Surf. Sci.* **93**, 145 (1980).

Si-39. V. S. Aliev, M. R. Baklanov, and V. I. Bukhtinyarov, *Appl. Surf. Sci.* **90**, 191 (1995).

Si-40. J. Westermann, H. Nienhaus, and W. Mönch, *Surf. Sci.* **311** 101 (1994).

Si-41. J. M. C. Thornton and R. H. Williams, *Phys. Scripta* **41**, 1047 (1990).

Si-42. A. E. Dolbak, B. Z. Olshanetsky, S. I. Stenin, and S. A. Teys, *Surf. Sci.* **247**, 32 (1991).

Si-43. G. J. Pietsch, in *Fortschritt-Berichte VDI Reihe* 9: *Electronik*, 148, VDI, Dusseldorf (1992), p. 23.

Si-44. B. Z. Olshanetsky and A. A. Shklyaev, *Surf. Sci.* **67**, 581 (1977).

Si-45. S. Tougaard, P. Morgan, and J. Onsgaard, *Surf. Sci.* **111**, 545 (1981).

Si-46. A. E. Dolbak, B. Z. Olshanetsky, S. I. Stenin, and S. A. Teys, *Surf. Sci.* **247**, 32 (1991).

Si-47. E. G. Keim, L. Wolterbeek, and A. van Silfhout, *Surf. Sci.* **180**, 565 (1987).

Si-48. H. Wormeester, A. M. Molenbroek, C. M. J. Wijers, A. Vansilfhout, *Surf. Sci.* **260**, 31 (1992).

Si-49. C. A. Sébenne, J. P. Lacharme, I. Andria, and M. Khial, *Appl. Surf. Sci.* **41/42**, 352 (1989).

Si-50. N. Safta, J. P. Lacharme, and C. A. Sébenne, *Surf. Sci.* **287**, 312 (1993).

Si-51. Y. W. Chung, W. Siekhaus, and G. A. Somorjai, *Surf. Sci.* **58**, 341 (1976).

Si-52. D. Morgan and F. Ryborg, *J. Vac. Sci. Technol.* **17**, 578 (1980).

Si-53. T. Narusawa, S. Shimizu, and S. Komiza, *Surf. Sci.* **85**, 572 (1979).

Si-54. J. M. Charing and D. K. Skinner, *Surf. Sci.* **15**, 277 (1969).

Si-55. S. Wright and H. Kroemer, *Appl. Phys. Lett.* **36**, 210 (1980).

Si-56. Y. Homma, M. Tomita, and T. Hayashi, *Surf. Sci.* **258**, 147 (1991).

Si-57. B. Z. Olshanetsky and S. A. Teys, *Surf. Sci.* **230**, 184 (1990).

Si-58. B. Z. Olshanetsky, A. E. Solovyov, A. E. Dolbak, and A. A. Maslov, *Surf. Sci.* **306**, 327 (1994).

Silver

Ag-1.   E. L. Evans, J. M. Thomas, M. Barber, and R. J. M. Griffiths, *Surf. Sci.* **38**, 245 (1973).

Ag-2.   W. Gettings and J. P. Coad, *Surf. Sci.* **53**, 636 (1975).

Ag-3.   R. A. Marbrow and R. M. Lambert, *Surf. Sci.* **651**, 317 (1976).

Ag-4.   H. A. Englehardt and D. Menzel, *Surf. Sci.* **57**, 591 (1976).

Ag-5.   R. A. Marbrow and R. M. Lambert, *Surf. Sci.* **61**, 329 (1976).

Ag-6.   C. Backx, C. P. M. de Groot, and P. Biloen, *Surf. Sci.* **104**, 300 (1981).

Ag-7.   M. Bowker, M. A. Barteau, and R. J. Madix, *Surf. Sci.* **92**, 528 (1980).

Ag-8.   M. Alff and W. Moritz, *Surf. Sci.* **80**, 24 (1979).

Ag-9.   I. E. Wachs and R. J. Madix, *Surf. Sci.* **76**, 531 (1978).

Ag-10. B. A. Sexton and R. J. Madix, *Surf. Sci.* **105**, 177 (1981).

Ag-11. D. Briggs, R. A. Marbrow, and R. M. Lambert, *Surf. Sci.* **65**, 314 (1977).

Ag-12. W. Heiland, F. Iberl, E. Taglauer, and D. Menzel, *Surf. Sci.* **53**, 383 (1975).

Ag-13. G. Rovida and F. Pratesi, *Surf. Sci.* **52**, 542 (1975).

Ag-14. P. J. Goddard, J. West, and R. M. Lambert, *Surf. Sci.* **71**, 447 (1978).

Ag-15. T. E. Felter, W. H. Weinberg, P. A. Zhdan, and G. K. Boreskov, *Surf. Sci.* **97**, L313 (1980).

Ag-16. H. H. Farrell, M. M. Traum, N. V. Smith, W. A. Royer, D. P. Woodruff, and P. D. Johnson, *Surf. Sci.* **102**, 527 (1981).

Ag-17. G. McElhiney, H. Papp, and J. Pritchard, *Surf. Sci.* **54**, 617 (1976).

Ag-18. Y.-Y. Tu and J. M. Blakeley, *J. Vac. Sci. Technol.* **15**, 563 (1978).

Ag-19. G. Rovida, F. Pratesi, M. Maglietta, and E. Ferroni, *Surf. Sci.* **43**, 230 (1974).

Ag-20. W. Berndt, in: *Proc. 2nd Int. Conf. on Solid Surfaces* Tokyo (H. Kumagai and T. Toya, eds.), *Jpn. J. Appl. Phys.* Suppl. 2, Pt. 2 (1974), p. 653.

Ag-21. R. A. Marbrow and R. M. Lambert, *Surf. Sci.* **71**, 107 (1978).

Strontium

Sr-1. A. Benninghoven and L. Wiedmann, *Surf. Sci.* **41**, 483 (1974).
Sr-2. B. M. Hartley, *Phys. Lett.* **27A**, 499 (1968).

Tantalum

Ta-1. N. Pacia, J. A. Dumesic, B. Weber, and A. Cassuto, *J. Chem. Soc. Faraday Trans. I* **72**, 1919 (1976).
Ta-2. C. Boiziau, V. Dose, and H. Scheidt, *Phys. Stat. Sol. B* **93**, 197 (1979).
Ta-3. M. A. Chesters, B. J. Hopkins, and M. R. Leggett, *Surf. Sci.* **43**, 1 (1974).
Ta-4. A. G. Elliot, *Surf. Sci.* **51**, 489 (1975).
Ta-5. A. Titov and W. Moritz, *Surf. Sci.* **123**, L709 (1982).
Ta-6. S. M. Ko and L. D. Schmidt, *Surf. Sci.* **47**, 557 (1975).
Ta-7. T. W. Haas, A. G. Jackson, and M. P. Hooker, *J. Chem. Phys.* **46**, 3025 (1967).
Ta-8. T. W. Haas, J. T. Grant, and G. J. Dooley, *Phys. Rev. B* **1**, 1449 (1970).

Technitium

Tc-1. T. P. Chen, E. L. Wolf, and A. L. Giorgi, *Surf. Sci.* **122**, L613 (1982).

Tellurium

Te-1. R. G. Musket, *Surf. Sci.* **74**, 423 (1978).
Te-2. P. B. Sewell and D. F. Mitchell, *Surf. Sci.* **55**, 367 (1976).
Te-3. M. El Azab, P. B. Sewell, and C. H. Champness, *J. Electron Mater.* **5**, 381 (1976).
Te-4. S. Andersson, D. Andersson, and I. Marklund, *Surf. Sci.* **12**, 284 (1968).
Te-5. D. L. Miller, *J. Vac. Sci. Technol.* **13**, 1138 (1976).
Te-6. R. J. Meyer, W. R. Salaneck, C. B. Duke, A. Paton, C. H. Griffiths, L. Kovnat, and L. E. Meyer, *Phys. Rev. B* **21**, 4542 (1980).
Te-7. M. Auboiroux, D. Marchand, P. Tougne, and A. P. Legrand, *Surf. Sci.* **59**, 218 (1976).

Terbium

Tb-1. W. Farber and P. Braun, *Surf. Sci.* **41**, 195 (1974).

Thallium

Tl-1. J. F. McGilp, P. Weightman, and E. J. McGuire, *J. Phys.* **10**, 3445 (1977).

Thorium

Th-1. W. P. Ellis, *J. Vac. Sci. Technol.* **9**, 1027 (1972).
Th-2. R. Bastasz and C. A. Colmenares, *J. Vac. Sci. Technol.* **15**, 791 (1978).
Th-3. T. N. Taylor, C. A. Colmenares, R. L. Smith, and G. A. Somorjai, *Surf. Sci.* **54**, 317 (1976).
Th-4. R. Bastasz, C. A. Colmenares, R. L. Smith, and G. A. Somorjai, *Surf. Sci.* **67**, 45 (1977).
Th-5. R. Bastasz, C. A. Colmenares, R. L. Smith, and G. A. Somorjai, *Surf. Sci.* **71**, 397 (1978).

Th-6. W. McLean, C. A. Colmenares, R. L. Smith, and G. A. Somorjai, *Phys. Rev. B* **25**, 8 (1982).

Tin

Sn-1. R. Nyholm and N. Martensson, *J. Phys. C* **13**, L279 (1980).
Sn-2. A. C. Parry-Jones, P. Weightman, and P. T. Andrews, *J. Phys. C* **12**, 1587 (1979).
Sn-3. R. A. Powell, *Appl. Surf. Sci.* **2**, 397 (1979).
Sn-4. A. W. C. Lin, N. R. Armstrong, and T. Kuwana, *Anal. Chem.* **49**, 1228 (1977).
Sn-5. Y. C. Jean, K. G. Lynn, and M. Carroll, *Phys. Rev. B* **21**, 4935 (1980).

Titanium

Ti-1.   G. W. Simmons and E. J. Scheibner, *J. Mater.* **5**, 933 (1970).
Ti-2.   T. Smith, *Surf. Sci.* **27**, 45 (1971).
Ti-3.   I. H. Kahn, *Surf. Sci.* **40**, 723 (1973).
Ti-4.   J. T. Grant, T. W. Haas, and.J. E. Houston, *Surf. Sci.* **42**, 1 (1974).
Ti-5.   H. E. Bishop, J. C. Riviere, and J. P. Coad, *Surf. Sci.* **24**, 1 (1971).
Ti-6.   L. Porte, M. Demosthenous, G. Hollinger, Y. Jugnet, P. Pertosa, and T. M. Duc, in: *Proc. 7th Int. Vac. Congr. and 3rd Int. Conf. on Solid Surfaces*, Vienna (R. Dobrozemski, F. Rudenauer, and F. P. Viehbock, eds.) (1977), p. 923.
Ti-7.   M. L. den Boer, P. I. Cohen, and R. L. Park, *Surf. Sci.* **70**, 643 (1978).
Ti-8.   M. P. Cox, J. S. Foord, R. M. Lambert, and R. H. Prince, *Surf. Sci.* **129**, 375 (1983).
Ti-9.   H. D. Shih, F. Jona, D. W. Jepsen, and P. M. Marcus, *J. Phys. C* **9**, 1405 (1976).
Ti-10. H. D. Shih, F. Jona, D. W. Jepsen, and P. M. Marcus, *Surf. Sci.* **60**, 445 (1976).
Ti-11. Y. Fukuda, F. Honda, and J. W. Rabalais, *Surf. Sci.* **91**, 165 (1980).
Ti-12. I. H. Kahn, *Surf. Sci.* **48**, 537 (1975).
Ti-13. Y. Fukuda, G. M. Lancaster, F. Honda, and J. W. Rabalais, *J. Chem. Phys.* **69**, 3447 (1978).

Tungsten

W-1.   R. M. Stern, *Appl. Phys. Lett.* **5**, 218 (1964).
W-2.   E. G. McRae and G. H. Wheatley, *Surf. Sci.* **29**, 342 (1972).
W-3.   A. Benninghoven, E. Loebach, and N. Treitz, *J. Vac. Sci. Technol.* **9**, 600 (1972).
W-4.   R. W. Joyner, J. Rickman, and M. W. Roberts, *Surf. Sci.* **39**, 445 (1973).
W-5.   T. E. Madey, J. T. Yates, Jr., and N. E. Erickson, *Chem. Phys. Lett.* **19**, 487 (1973).
W-6.   J. T. Yates, Jr., T. E. Madey, and N. E. Erickson, *Surf. Sci.* **43**, 257 (1974).
W-7.   E. B. Bas and U. Banninger, *Surf. Sci.* **41**, 1 (1974).
W-8.   M. L. Yu, *Surf. Sci.* **64**, 334 (1977).
W-9.   M. L. Yu, *Appl. Phys. Lett.* **30**, 654 (1977).
W-10. O. K. T. Wu and R. P. Burns, *Surf. Sci.* **77**, 626 (1978).
W-11. P. D. Johnson and T. A. Delchar, *Surf. Sci.* **82**, 237 (1979).
W-12. C. Nyberg, *Surf. Sci.* **82**, 165 (1979).
W-13. D. P. Woodruff, M. M. Traum, H. H. Farrell, N. V. Smith, P. D. Johnson, D. A. King, R. L. Benlow, and Z. Hurych, *Phys. Rev. B* **21**, 5642 (1980).
W-14. F. Honda, Y. Fukuda, and J. W. Rabalais, *Chem. Phys.* **47**, 59 (1980).
W-15. B. J. Hopkins and G. D. Watts, *Surf. Sci.* **45**, 77 (1974).
W-16. G. D. Watts, A. R. Jones, and B. J. Hopkins, *Surf. Sci.* **45**, 705 (1974).
W-17. S. Usami and T. Nakagima, in: *Proc. 2nd Int. Conf. on Solid Surfaces* Tokyo (H. Kumagai and T. Toya, eds.), *Jpn. J. Appl. Phys.* Suppl. 2, Pt. 2 (1974), p. 237.

W-18. M. R. O'Neill, M. Kalisvaart, F. B. Dunning, and G. K. Walters, *Phys. Rev. Lett.* **34**, 1167 (1975).

W-19. J. T. Yates, Jr., T. E. Madey, N. E. Erickson, and S. D. Worley, *Chem. Phys. Lett.* **39**, 113 (1976).

W-20. B. W. Lee, A. Ignatiev, S. Y. Tong, and M. Van Hove, *J. Vac. Sci. Technol.* **14**, 291 (1977).

W-21. P. E. Luscher and F. M. Propst, *J. Vac. Sci. Technol.* **14**, 400 (1977).

W-22. P. E. Luscher, *Surf. Sci.* **66**, 167 (1977).

W-23. M. Housley and D. A. King, *Surf. Sci.* **62**, 81, 93 (1977).

W-24. M. K. Debe, D. A. King, and F. S. Marsh, *Surf. Sci.* **68**, 437 (1977).

W-25. S. P. Withrow, P. E. Luscher, and F. M. Propst, *J. Vac. Sci. Technol.* **15**, 511 (1978).

W-26. M. L. Yu, *Surf. Sci.* **71**, 121 (1979).

W-27. K. J. Rawlings, B. J. Hopkins, and S. D. Foulias, *Surf. Sci.* **77**, 561 (1978).

W-28. J. B. Benziger, E. I. Ko, and R. J. Madix, *J. Catal.* **54**, 414 (1978).

W-29. M. Kalisvaart, M. R. O'Neill, T. W. Riddle, F. B. Dunning, and G. K. Walters, *Phys. Rev. B* **17**, 1570 (1978).

W-30. R. A. Zuhr, L. C. Feldman, R. L. Kauffman, and P. J. Silverman, *Nucl. Instrum. Methods* **149**, 349 (1978).

W-31. T. W. Riddle, A. H. Mahan, F. B. Dunning, and G. K. Walters, *J. Vac. Sci. Technol.* **15**, 1686 (1978).

W-32. K. Griffiths and D. A. King, *J. Phys. C* **12**, L755 (1979).

W-33. M. K. Debe and D. A. King, *Surf. Sci.* **81**, 193 (1979).

W-34. S. Zuber, Z. Sidorski, and J. Polanski, *Surf. Sci.* **87**, 375 (1979).

W-35. P. Heilmann, K. Heinz, and K. Müller, *Surf. Sci.* **89**, 84 (1979).

W-36. P. M. Stefan, C. R. Helms, S. C. Perino, and W. E. Spicer, *J. Vac. Sci. Technol.* **16**, 577 (1979).

W-37. A. K. Bhattacharya, J. Q. Broughton, and D. L. Perry, *J. Chem. Soc. Faraday Trans. I* **75**, 850 (1979).

W-38. Z. Hussain, C. S. Fadley, S. Kono, and L. F. Wagner, *Phys. Rev. B* **22**, 3750 (1980).

W-39. J. Kolaczkiewica and Z. Sidorski, *Surf. Sci.* **63**, 501 (1977).

W-40. R. G. Musket and J. Ferrante, *J. Vac. Sci. Technol.* **7**, 14 (1970).

W-41. T. Smith, *Surf. Sci.* **27**, 45 (1971); *J. Appl. Phys.* **43**, 2964 (1972).

W-42. K. Besocke and H. Wagner, *Surf. Sci.* **52**, 653 (1975).

W-43. K. Besocke and S. Berger, in: *Proc. 7th Int. Vac. Congr. and 3rd Int. Conf. on Solid Surfaces*, Vienna (R. Dobrozemski, F. Rudenauer, and F. P. Viehbock, eds.) (1977), p. 893.

W-44. E. Unback, J. C. Fuggle, and D. Menzel, *J. Electron Spectrosc. Rel. Phenom.* **10**, 15 (1977).

W-45. R. Butz and H. Wagner, *Surf. Sci.* **63**, 448 (1977).

W-46. P. D. Augustus and J. P. Jones, *Surf. Sci.* **64**, 713 (1977).

W-47. C. Somerton and D. A. King, *Surf. Sci.* **89**, 391 (1979).

W-48. T. E. Madey, *Surf. Sci.* **94**, 483 (1980).

W-49. J. C. Tracy and J. M. Blakely, *Surf. Sci.* **15**, 257 (1969).

W-50. N. R. Avery, *Surf. Sci.* **33**, 107 (1972) 107; **41**, 533 (1974).

W-51. J. T. Yates, Jr., and N. E. Erickson, *Surf. Sci.* **44**, 489 (1974).

W-52. W. F. Winters, *IBM J. Res. Dev.* **22**, 260 (1978).

W-53. J. Chen and C. A. Papageorgopoulous, *Surf. Sci.* **20**, 195 (1970).

W-54. B. J. Hopkins and G. D. Watts, *Surf. Sci.* **44**, 237 (1974).

Uranium

U-1. G. C. Allen and R. K. Wild, *Chem. Phys. Lett.* **15**, 279 (1972).

U-2. G. C. Allen and P. M. Tucker, *J. Chem. Soc. Dalton Trans.* **5**, 470 (1973).

U-3. R. L. Park and J. E. Houston, *Phys. Rev. A* **7**, 1447 (1973).

U-4. J. H. Verbist, J. Riga, J. J. Pireaux, and R. Caudano, *J. Electron Spectrosc. Rel. Phenom.* **5**, 193 (1974).
U-5. W. P. Ellis, *Surf. Sci.* **61**, 31 (1976).
U-6. S. B. Nornes and R. G. Meisenheimer, *Surf. Sci.* **88**, 191 (1979).
U-7. R. Bastasz and T. E. Felter, *Phys. Rev. B* **26**, 3259 (1982).
U-8. W. McLean, C. A. Colmenares, R. L. Smith, and G. A. Somorjai, *Phys. Rev. B* **25**, 8 (1982).
U-9. W. McLean and H.-L. Chen, *J. Appl. Phys.* **58**, 4682 (1985).
U-10. P. R. Norton, R. L. Tapping, D. K. Creber, and W. J. L. Buyers, *Phys. Rev. B* **21**, 2572 (1980).
U-11. G. C. Allen and R. K. Wild, *J. Chem. Soc. Dalton Trans.* **6**, 493 (1974).
U-12. H. Grohs, H. Höchst, P. Steiner, and S. Hüfner, *Solid State Commun.* **33**, 573 (1980).

Vanadium

V-1. P. W. Davies and R. M. Lambert, *Surf. Sci.* **95**, 571 (1980).
V-2. D. L. Adams and H. B. Nielsen, *Surf. Sci.* **107**, 305 (1981).
V-3. A. Benninghoven, K. H. Müller, C. Plog, M. Schemmer, and P. Steffens, *Surf. Sci.* **63**, 403 (1977).
V-4. A. Benninghoven, K. H. Müller, and M. Schemmer, *Surf. Sci.* **78**, 565 (1978).
V-5. F. J. Szalkowski and G. A. Somorjai, *J. Chem. Phys.* **56**, 6097 (1972).
V-6. F. J. Szalkowski and G. A. Somorjai, *J. Chem. Phys.* **61**, 2064 (1974).
V-7. F. J. Szalkowski and G. A. Somorjai, *J. Chem. Phys.* **64**, 2985 (1976).
V-8. L. Fiermens and J. Vennik, *Surf. Sci.* **24**, 541 (1971).
V-9. L. Fiermens and J. Vennik, *Surf. Sci.* **35**, 42 (1973).
V-10. E. A. K. Nemeh and R. C. Cinti, *Surf. Sci.* **40**, 583 (1973).

Ytterbium

Yb-1. G. K. Wertheim and S. Hufner, *Phys. Rev. Lett.* **35**, 53 (1975).
Yb-2. G. Martin and D. Chopra, *IEEE Trans. Nucl. Sci. NS* **26**, 1169 (1979).

Yttrium

Y-1. T. W. Haas, J. T. Grant, and G. J. Dooley III, *J. Vac. Sci. Technol.* **7**, 43 (1970).
Y-2. T. W. Haas, J. T. Grant, and G. J. Dooley III, *Phys. Rev. B* **1**, 1449 (1970).
Y-3. T. L. Barr, in: *Quantitative Surface Analysis of Materials, Proc. Symp.* Cleveland, OH (N. S. McIntyre, ed.), American Society for Testing and Materials, Philadelphia, PA, ASTM-STP-643 (1978), p. 83.
Y-4. M. P. Cox, J. S. Foord, R. M. Lambert, and R. H. Prince, *Surf. Sci.* **129**, 375 (1983).

Zinc

Zn-1. M. B. Ives and W. A. J. Carson, *Mat. Res. Bull.* **6**, 1151 (1971).
Zn-2. V. Mizutani, T. Kondow, and M. Uda, *Phys. Stat. Sol.* **91B**, 693 (1979).
Zn-3. L. Yin, T. Tsang, I. Alder, and E. Yellin, *J. Appl. Phys.* **43**, 3464 (1972).
Zn-4. L. Yin, I Alder, M. H. Chen, and B. Crasemann, *Phys. Rev. A* **7**, 897 (1973).
Zn-5. S. P. Kowalczyk, R. A. Pollak, F. R. McFeely, L. Ley, and D. A. Shirley, *Phys. Rev. B* **8**, 2387 (1973).
Zn-6. J. M. Mariot and G. Dufour, *J. Phys. C* **10**, L213 (1977).
Zn-7. D. D. Sarma, M. S. Hedge, and C. N. R. Rao, *Chem. Phys. Lett.* **73**, 443 (1980).

Zn-8. I. Abbati, L. Braicovich, R. A. Powell, and W. E. Spicer, in: *Proc. 7th Int. Vac. Cong. and 3rd Int. Conf. on Solid Surfaces*, Vienna (R. Dobrozemski, F. Rudenauer, and F. P. Viehbock, eds.) (1977), p. 919.

Zn-9. W. N. Unertl and J. M. Blakely, *Surf. Sci.* **69**, 23 (1977).

Zn-10. I. Abbati, L. Braicovich, B. DeMichelis, and A. Fasana, *Solid State Commun.* **26**, 515 (1978).

Zn-11. F. J. Himpsel, D. E. Eastman, and E. E. Kock, *Phys. Rev. Lett.* **44**, 214 (1980).

Zirconium

Zr-1. J. S. Foord, P. J. Goddard, and R. M. Lambert, *Surf. Sci.* **94**, 339 (1980).

Zr-2. J. H. Weaver, R. L. Benbow, and Z. Hurich, *Solid State Commun.* **22**, 173 (1977).

Zr-3. R. Nyholm and N. Martensson, *J. Phys. C* **13**, L279 (1980).

Zr-4. G. N. Krishnan, B. J. Wood, and D. Cubicciotti, *J. Electrochem. Soc.; Solid State Sci.* **128**, 191 (1981).

Zr-5. W. T. Moore, P. R. Watson, D. C. Frost, and K. A. R. Mitchell, *J. Phys. C* **12**, L887 (1979).

5

Specimen Treatments: Surface Preparation of Metal Compound Materials (Mainly Oxides)

Ulrike Diebold

1. Introduction

Compared to elemental solids, the surface preparation of compounds containing more than one element is considerably more difficult. For any material it is desirable to achieve surface impurity concentrations below the detection limit of standard analytical techniques. Well-ordered single-crystalline surfaces need to be prepared and are characterized by low-energy electron diffraction (LEED) and, more recently, scanning tunneling microscopy (STM). For compounds, several additional issues must be considered.

First, there is the question of *surface stoichiometry.* The surface composition of a compound material may be very different from the bulk. This can be due to energetic reasons—surfaces enriched with one component are often thermodynamically more stable, and surface segregation is induced by annealing. A variety of other reasons can lead to nonstoichiometry of compound surfaces. One component might be more volatile than the others; for example, annealing causes the loss of oxygen on many metal oxide surfaces[1] and the enrichment of C in SiC.[2] Even if this is not the case, commonly used preparation techniques such as sputtering, and sometimes even polishing, may cause the enrichment of one component at the

Ulrike Diebold • Department of Physics, Tulane University, New Orleans, LA 70118.

Specimen Handling, Preparation, and Treatments in Surface Characterization, edited by Czanderna *et al.* Kluwer Academic / Plenum Publishing, New York, 1998.

surface as compared to the bulk. It is not always easy to judge with electron spectroscopic techniques whether the topmost layer is truly stoichiometric, but small deviations from stoichiometry may affect surface properties dramatically. In certain fortunate cases of compounds with a highly ionic bonding character, a deviation from surface stoichiometry is reflected in different oxidation states (see Sec. 3.1). Then the spectral shapes of X-ray photoelectron spectroscopy (XPS) core levels or Auger electron spectroscopy (AES) transitions are often a good indication of surface stoichiometry, although quantification is not always straightforward.

Second, for materials with a strongly ionic bonding character, such as metal oxides or halides, *surface stability* becomes an issue. Metallic single crystals can be cut in any direction, but cutting ionic materials along a particular crystallographic axis might produce a polar surface where the numbers of positive and negative charges are not the same.[3,4] While charge-neutral surfaces are usually stable, these polar surfaces tend to reconstruct, or they form microscopic facets. The (111) face of the rocksalt structure is an example of such a polar surface. Rocksalt is a very common structure and is found in alkali halides, many metal monoxides (NiO, MgO, FeO, TiO, etc.), and transition-metal carbides and nitrides of the fourth through the sixth group of the periodic table. In addition to surface instabilities due to the energetics of the system, one has to be aware of electron-induced desorption effects during experiments. These are particularly strong in alkali halides[5] and "maximum valency" oxides,[6] i.e., oxides where the metal cation has donated all its valence electrons to the neighboring oxygen anions in an ionic bond. Electron bombardment causes electronic excitations that can induce desorption of surface halogen and oxygen ions from halides and oxides, respectively. This process might generate unwanted artifacts if Auger electron spectroscopy is used to monitor surface composition or if LEED is used for characterization of geometric structure.

Lastly, many metal compound materials are *insulators*. Working with poorly conducting materials poses an additional challenge. Most surface characterization techniques rely on the emission or scattering of charged particles, and charging phenomena can severely hamper spectroscopic measurements. Poor electric and thermal conductivity go hand-in-hand, and one has to be careful to avoid thermal stress during heating or cooling of insulators since this might lead to fracturing of the sample. There is also the sometimes nontrivial question of how to mount a specimen that cannot be spot-welded onto a holder, and, equally important, how to accurately measure its temperature. Readings from a thermocouple can be off by as much as 100 K when it is attached to the sample holder instead of the ionic crystal itself.[7] Since sample introduction load-locks and sample transfer systems have become increasingly popular, there are no limits to experimental skill and imagination to overcome this challenge.

For several semiconductor compounds, for example the extensively studied GaAs, established growth and cleaning recipes exist. For others, such as the far less investigated metal oxide surfaces, ". . . the surface preparation of each separate crystal face of each different compound is a research program in itself."[1] In this chapter, we concentrate on these latter materials. Clearly, it is well beyond the scope of this work to discuss all compounds, i.e., metal and semiconductor alloys, carbides, halides, oxides, selenides, sulfites, etc. A selection has to be made. The chosen focus on metal oxides is somewhat subjective as it reflects the interest of the author. The recently surging interest in metal oxide surfaces, their technological relevance, and the fact that instructive case studies can be drawn from this class of materials, are more general reasons for this choice. In addition, surface preparation appears to be one of the most crucial aspects that has hampered progress of surface-related research on this exciting class of materials. For example, titanium dioxide, discussed in Sec. 3.1, has been studied extensively. One reason for the popularity of this material in surface science lies in the availability of treatment recipes for producing and reproducing a well-ordered, stoichiometric surface *in situ*.

We first give an overview of preparation techniques. While particular emphasis is put on issues important for metal oxides, this section is relevant for other materials as well. We discuss *ex situ* preparation methods, cleavage, preferential sputtering, annealing, and film deposition. We then give some examples for metal oxide surfaces. The examples have been selected in order to review a variety of problems encountered and their solutions. Care has been taken to discuss mainly materials where agreement exists among different authors about surface preparation recipes. The literature up to 1992 is mainly based on the comprehensive book on oxide surfaces by Henrich and Cox;[1] later references were obtained by a combination of searches in electronic databases[8] and the master index of the journal *Surface Science*.

2. Overview of Preparation Methods

2.1. *Ex Situ* Preparation: Sample Handling, and Mounting

Many metal oxides occur as minerals in nature, but the purity of such natural single crystals is often not sufficient for many experiments. Some metal oxides can be purchased from commercial crystal growers, especially materials which are popular substrates for epitaxial film growth. MgO, Al_2O_3, and $SrTiO_3$ (which is extensively used for thin-film growth of high-T_c superconductors) can be bought in cut and polished form from Commercial Crystal Laboratories (U.S.A.), Kelpin Kristallhandel (Germany), Goodfellow Ltd. (U.K.), Earth Jewelry (Japan), and many others. The same is true for the other materials discussed in the next section, namely NiO, TiO_2, and ZnO. Others are not sold commercially and are hard to come by. The difficulty in getting high-quality single-crystalline samples that are large

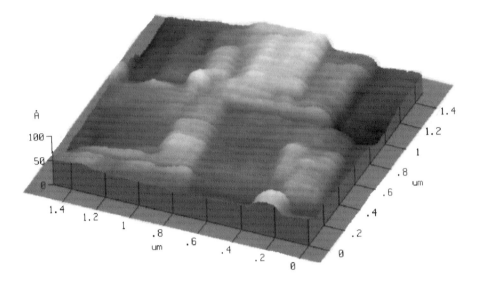

Figure 1. Cleavage of ionic crystals often results in stoichiometric, well-ordered, but not atomically smooth surfaces. This figure shows an atomic force microscopy image of multiatomic steps on a NiO(100) surface, produced by cleaving in air. The steps are approximately 2.5 nm high and oriented along <100> type directions. (Reprinted with permission from M.D. Antonik and R.J. Lad, *Journal of Vacuum Science and Technology A* **10**(4), 669–673 (1992). Copyright 1992 American Vacuum Society.)

enough in size for surface science experiments is certainly one reason why some metal oxides have not been studied extensively.

Some ionic crystals fracture or cleave easily along high-stability crystallographic directions; this is the preparation technique of choice when a "truly pristine" surface is desired which deviates little from its ideal form. Cleavage is usually quite straightforward. The form of the single crystal itself or test cleavages immediately indicate the right orientation. When samples are cleaved in air, some *in situ* cleaning technique has usually to be applied in order to get rid of contamination accumulated in the ambient atmosphere. Cleavage in ultrahigh vacuum (UHV) certainly results in the cleanest surfaces possible. Crystals are usually aligned and notched along the desired crystallographic direction before they are mounted onto a sturdy sample holder and inserted into UHV. Samples cleaved in ultrahigh vacuum render a surface stoichiometry and geometry that is closest to the equilibrium structure of the crystal. However, even in the case where crystals cleave quite easily, this does not always result in a microscopically smooth surface. Figure 1 shows an atomic force microscopy (AFM) image of a NiO(001) surface that was cleaved in air.[9] The figure shows flat terraces separated by 2.5-nm-high steps. Cleavage steps ranging from 10 to 1000 nm in height and atomically flat terraces up to 500 nm in width were also seen on the same sample. MgO, another material with rocksalt structure that cleaves easily along (001), showed similar features.[9] These high cleavage steps are one reason why cut and polished MgO substrates are preferred for epitaxial growth, compared to cleaved ones.

Polishing cut or cleaved samples can result in surfaces with a very high degree of flatness. "Superpolishing" using chemical agents with a residual root-mean-square (rms) roughness of less than 1 nm is reported for some materials.[10] This is also referred to as epitaxial polish, and several oxide samples with such a finish are commercially available.* Even in the case of a very hard sample, e.g., sapphire (Al$_2$O$_3$), superpolishing in direct contact with a rotating tin lap using silica aquasol is reported to result in very smooth surfaces with a high degree of perfection.[11] Selective etching of the SrO layer on a SrTiO$_3$(001) surface is also reported to result in very smooth surfaces.[12]

As mentioned, sample mounting, especially when the sample has to be heated in order to get rid of unwanted carbon, is often problematic. Attaching samples with Ta strips on a Ta sample holder is the preferred technique in this laboratory, where a sample transfer system is used. A more sophisticated mounting technique is the deposition of a thick metal coating on the back of the sample, onto which support wires can be spot-welded. Notched single crystals for mounting support wires are especially useful if accurate thermal desorption experiments are planned.[13] Temperature measurement poses a problem, as has been mentioned. Common ways to avoid this problem include spring-loading a thermocouple against the sample,[7] attaching it with ceramic cement (e.g., Ceramaseal 105), and improving the thermal contact through polishing the holder and both sides of the specimen[14] while measuring the temperature at the sample holder. When samples are to be moved between different experimental stages in connected UHV chambers, calibration measurements have to be made involving comparison of one thermocouple that is pressed onto the sample and a second one on the sample holder. This is a cumbersome, but certainly accurate, way to achieve reliable temperature measurements.

2.2. *In Situ* Preparation: Sputtering and Annealing

For cleaning metallic substrates, heating to high temperatures or heating in different gaseous atmospheres such as hydrogen or oxygen can be sufficient to remove impurity atoms from the surface.[15] Similarly, heating in oxygen may be used to "burn off" carbon contaminations from oxide surfaces. An example is shown in Fig. 2.

---

*One vendor (Commercial Crystal Laboratories, U.S.A.) sells oxide single crystals with an rms roughness of 0.4–0.7 nm. A roughness of 0.2 nm is quoted for MgO(001) samples. The exact preparation recipe is proprietary, but grinding the sample and then polishing each side for 30 h with a chemomechanical treatment is reported.

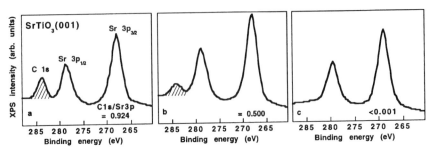

Figure 2. Carbon can often be removed efficiently from oxide surfaces by heating in an oxygen atmosphere. This figure shows X-ray photoelectron spectra of the Sr $2p$ region of a $SrTiO_3(100)$ surface. The sample was (a) rinsed with acetone in a supersonic cleaner, followed by heating at 650°C for 1 h in ultrahigh vacuum; (b) like (a), but rinsed in ethanol; and (c) rinsed with ethanol, and heated in 1 × 10^{-6} Torr oxygen at 650°C for 1 h. (Reprinted with permission from M. Yoshimoto, H. Okhubo, N. Kanda, and H. Koinuma, *Japanese Journal of Applied Physics* **31**, 3664–3666 (1992). Copyright 1992 *Japanese Journal of Applied Physics*.)

One of the most-often used *in situ* sample cleaning procedures is sputtering, i.e., bombardment with rare-gas ions (often argon) in the energy range from 500 eV to 10 keV. This procedure always causes a certain degree of undesirable radiation damage, but in many cases sputtering cannot be avoided. It is the only way to remove many contaminants other than C or overlayers that have been deposited during the course of an experiment.

For compound systems that consist of more than one element, ion bombardment often leads to preferential sputtering; i.e., the different components are not sputtered in proportion to their surface concentration. This leads to an "altered layer," a near-surface region with a composition different from the bulk stoichiometry. The thickness of the altered layer can range from less than 10 to several hundred Ångstroms, depending on the composition of the material, and on bombardment parameters such as ion beam energy, angle, ion species, and the sample temperature. Because sputtering is a dynamic process and because many effects occur at the same time, it is often hard to predict which component will be enriched at the surface and within the altered layer. The masses of the constituents of the sample and their relative surface binding energies are certainly important factors. For example, the heavier component will usually be enriched when there is a big mass difference. On the other hand, the surface binding energy determines which component will be sputtered preferentially when the masses of the sample constituents are similar. Additionally, ion beam mixing, recoil implantation, and thermal- or radiation-induced segregation or diffusion also play a role. A good overview of processes that occur during preferential sputtering, as well as an extended compendium of experimental data of surface composition changes, of

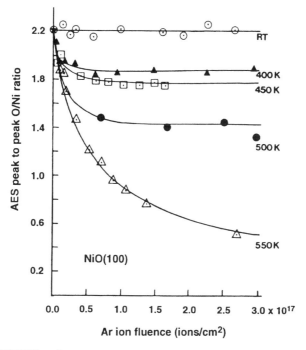

Figure 3. The NiO(100) surface oxygen concentration in terms of O/Ni AES ratio is shown as a function of 2-keV Ar-ion fluence. Congruent sputtering is found at room temperature, but preferential sputtering at slightly elevated temperatures. (Reprinted with permission from M. A. Langell, *Surface Science* **186**, 323–338 (1987). Copyright 1987 Elsevier Science B.V.)

altered layer thicknesses, and of total sputtering yield coefficients for metal alloys, binary semiconductors, and compound materials was given by Betz and Wehner.[16]

For compounds with one high-vapor pressure component, e.g., oxides or halides, the volatile component will almost always be sputtered preferentially. Oxygen is removed from many oxide surfaces and a metal-rich surface is created. (No preferential sputtering of O was observed for BeO, MgO, Al_2O_3, VO, V_2O_3, Cr_2O_3, Ga_2O_3, and SnO_2.)[16,17] The volatile component probably desorbs spontaneously once a certain, critical composition is reached. This phenomenon explains the pronounced temperature-dependent sputtering yield that has been found for NiO (Fig. 3).[18] In addition to preferential sputtering of oxygen, structural transformations take place for many oxide materials, from crystalline to amorphous, or to other crystalline phases with a different stoichiometry.[19]

Sputter damage can usually be healed through heating, and stoichiometry in oxides can be restored by annealing in oxygen. Heating of metal compounds or alloys may lead to the above-mentioned segregation of one component, however.

Also, quite high temperatures may be necessary to regain long-range order, flat surfaces, and stoichiometry in refractory materials.

2.3. Film Deposition

Metal compound films on metallic or semiconducting substrates are commonly used to circumvent many of the problems that occur when working with poorly conducting bulk materials—charging effects, difficulties in mounting and handling, reliability in temperature measurements, etc. In particular, sample charging can be avoided if films are only a few monolayers thick and electrons can tunnel through these "ultrathin" overlayers. This is also a way to synthesize materials that are not available in the form of high-quality single-crystalline specimens.

Metal oxides can be grown simply by oxidizing the relevant metal, and an abundance of literature exists on the initial interaction of oxygen with well-defined transition-metal surfaces. (The interested reader is referred to the review article by Brundle and Broughton.[20]) However, the thickness of the oxidized region is hard to control, and oxidation almost never results in a film with high crystalline quality—the lattices of the metals and their respective oxides are generally too dissimilar. Oxidation of alloys can lead to good films in certain cases; for example, it has been observed that a well-ordered Al_2O_3 overlayer is formed through oxidation of NiAl(110).[21] Two kinds of substrate materials have commonly been used for the growth of epitaxial metal–oxide films, refractory metals (molybdenum,[22] tungsten,[23] rhenium[24]), and noble metals (platinum,[25,26] silver,[27] gold[28,29]); see recent review articles.[22,30,31] Refractory materials are convenient to work with because the metal oxide overlayer can be desorbed at high temperatures, thus avoiding the necessity of time-consuming sputter–annealing cycles. Noble metals, on the other hand, are not oxidized themselves, and it is reasoned that a sharper and better-defined interface to the metal oxide overlayer is formed. In both cases, a good lattice match between substrate and metal overlayer is required for a high-quality film geometry. The deposition mode and parameters (temperature, deposition rate, oxygen background during deposition versus oxidation of a deposited metal overlayer) that result in a film with optimum quality have to be carefully optimized in trial experiments.

For ultrathin oxide films on foreign substrates, substrate-mediated phases can be formed that are not stable in the bulk. These phases can be rather complex. One example is shown in Fig. 4. Chromium oxide has been deposited by evaporation of Cr in an oxygen atmosphere of 2×10^{-6} Torr. The substrate was a Pt(111) sample, held at a temperature of 600 K.[32] The figure shows the LEED patterns of the resulting chromium oxide overlayers. The $p(2\times2)$ pattern is assigned to a Cr_3O_4 phase that is formed at the interface and the $(\sqrt{3}\times\sqrt{3})R30°$ structure to the (0001) face of Cr_2O_3. (These assignments and the supporting XPS measurements are discussed in Ref. 32.) The two structures are formed below and above a coverage

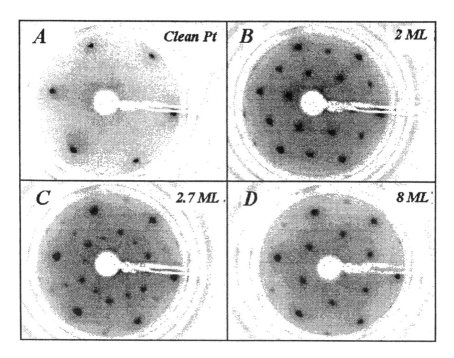

Figure 4. Ultrathin metal oxide films on single-crystalline metal substrates are often used in order to prevent charging and to synthesize specimen for materials that are not easily accessible as single-crystalline bulk materials. Ultrathin film growth may result in the formation of phases that are not stable in the bulk because of the strong interaction between the metal oxide overlayer and substrate at the interface. Here, LEED patterns of chromium oxide overlayers with different film thicknesses grown on Pt(111) are shown: (a) clean Pt(111), (b) two equivalent monolayers (ML), $p(2\times2)$ structure, (c) 2.75 equivalent monolayers, superposition of (b) and (d), and (d) five equivalent monolayers, $(\sqrt{3}\times\sqrt{3})R30°$ structure. The $p(2\times2)$ structure is assigned to the formation of $Cr_3O_4(111)$, which is a thermodynamically unstable phase. The $(\sqrt{3}\times\sqrt{3})R30°$ structure reflects the (0001) face of Cr_2O_3. (Reprinted with permission from L. P. Zhang, M. Kuhn, and U. Diebold, *Surface Science* **375** (1), 1–12 (1997). Copyright 1997 Elsevier Science B.V.)

of two monolayers, respectively. Cr_3O_4 is not known to occur in the bulk, but is stabilized through interaction with the Pt interface. While the influence of the substrate is not always as dramatic as in this example, one has nevertheless to be cautious; ultrathin films may exhibit properties quite different from the bulk because of strain effects.

Many binary oxides have been prepared by oxidation or oxide thin-film growth on a foreign substrate.[22,30] In Sec. 3.5, we discuss one example of this preparation technique, the formation of NiO thin films on nickel single crystals as well as on other materials.

3. Preparation of Metal Oxide Surfaces: Some Examples

3.1. Titanium Dioxide

Titanium dioxide (rutile*) is by far the most frequently studied sample in the surface science of metal oxides, partly because of the relative ease of surface preparation. The most stable face is the charge-neutral (110) orientation; this is also the surface that has been investigated most extensively. $TiO_2(110)$ terminates in a (1×1) structure and reconstructs to a (1×2) surface upon annealing to high temperatures. The (001) surface tends to facet,[33] while the (100) surface forms (1×3), (1×5), and (1×7) reconstructions.[34]

It is convenient to work with TiO_2 from an experimental point of view. First, bulk single crystals can be doped (n-type) in a high-temperature anneal so that charging problems are virtually absent on this surface. Second, a stoichiometric surface is easily prepared by sputtering followed by annealing in oxygen, and third, defects (mainly oxygen vacancies) can be introduced easily in a controlled fashion. It may sound surprising that easy defect creation is counted as an advantage, but defects *are* very important for many surface science studies. There is overwhelming evidence that the surface chemistry of this and many other oxide materials is heavily influenced by defects. A large number of surface science investigations have concentrated on the relationship between defects and surface properties of oxides. Often, TiO_2 has been used as a prototype material for these studies.

Three preparation procedures are discussed here: initial "conditioning" through n-type doping, reproducible preparation of a stoichiometric surface, and defect creation.

The doping through intrinsic defects of TiO_2 bulk single crystals is easy to perform and can be monitored simply by visual inspection. TiO_2 has a 3-eV band gap, and fresh TiO_2 samples are colorless and transparent. When such a crystal is heated in UHV, its color changes to yellowish because of the creation of temperature-induced color centers in the bulk. This color change is reversible, and the crystal becomes transparent again after cooling to room temperature. However, during annealing to approximately 1000 K, a slightly greenish color is observed. The original colorless state is not restored upon cooling, but a bluish hue remains. This permanent color change is due to the creation of bulk oxygen vacancies that induce defect states in the bulk band gap. Repeated heating, or heating to higher temperatures, darkens the sample further. Depending on the thermal history of a crystal, colors from light blue to opaque and metallic gray have been reported. A light blue color indicates sufficient bulk conductivity for any kind of electron

---

*Titanium dioxide exists in three modifications: brookite, anatase, and rutile. To the author's knowledge only rutile has been studied in surface science.

spectroscopic measurements and even STM. For strongly reduced, opaque rutile crystals, a carrier concentration of $10^{26}/m^3$ has been calculated from weight loss and Hall coefficient data.[35]

Almost all experimental work to date involves samples that have been conditioned in such a way.* Although the bulk is now slightly defective and semiconducting, the surface can be reoxidized by annealing in oxygen, and an insulating film with the bulk band gap is produced. The thickness of this film is unknown but is well above the information depth of most surface spectroscopies.

It has been shown by Henrich et al.[38] that the surface electronic and geometric structure of a sputtered and annealed $TiO_2(110)$ sample does not differ from a cleaved one. Various preparation recipes have been reported, but a procedure similar to the one reported by Pan et al.[39] has been repeatedly used by the author and results in perfectly stoichiometric $TiO_2(110)$ surfaces. In the first preparation step, the sample is sputtered. The ion energy and species are not critical for TiO_2 surface preparation. Higher energies and heavier ions result in more "reduced", i.e., oxygen-deficient surfaces due to preferential sputtering of oxygen, but the sputter-induced damage can be annealed in any case. After sputtering, the sample temperature is slowly increased to 1000 K in 2×10^{-6} Torr oxygen, and this temperature is held for several minutes. The sample is slowly cooled to room temperature and is exposed to the same oxygen atmosphere for approximately 30 min more before reducing the background pressure.

This procedure results in a stoichiometric surface,† as judged by XPS (Fig. 5). The sharp peaks in spectrum a in Fig. 5 are typical for fully oxidized TiO_2, where all Ti cations are in their 4+ oxidation state. The Ti $2p$/O $1s$ intensity ratio of a such a stoichiometric surface is also consistent with a 1:2 ratio of Ti:O.[41] The chemical shift of Ti^{4+} as compared to metallic Ti^0 is 5.2 eV;[42] hence the different oxidation states can clearly be distinguished in XPS spectra. Note the broad Ti $2p$ feature that results from the different oxidation states on a sputtered, oxygen-deficient surface (spectrum c in Fig. 5). Annealing at 1000 K, but in UHV instead of in oxygen, results in a low binding energy shoulder in the Ti $2p$ XPS spectrum, indicative of the presence of oxygen vacancies (Fig. 5, spectrum b). These vacancies are also visible in STM images of a surface annealed in UHV (Fig. 6).[43] This STM image

---

*There is recent evidence[36] that the presence of subsurface defects changes the behavior of TiO_2 surfaces. While Pt overlayers on conditioned $TiO_2(110)$ surfaces are encapsulated in a layer of reduced TiO_x upon heating to above 450 K in UHV,[37] this effect is not observed on a $TiO_2(100)$ sample that has no bulk oxygen vacancies.[36] The latter sample was doped with Nb in order to provide enough conductivity for surface science experiments.
†Recent STM measurements indicate that annealing in oxygen causes considerable roughening of a sample with bulk oxygen vacancies[40] and that annealing in UHV is necessary for a flat surface morphology.

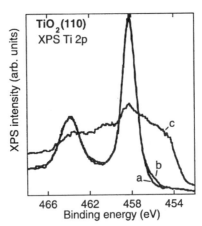

Figure 5. X-ray photoelectron spectra of $TiO_2(110)$ surfaces subjected to different sample treatments. The stoichiometric surface (a) was obtained after sputtering and annealing in 2×10^{-6} Torr oxygen at 1000 K. Point defects (oxygen vacancies), are produced when the surface is annealed at 1000 K for 3 min in ultrahigh vacuum. These defects cause the presence of a Ti^{3+} oxidation state and a shoulder in spectrum (b). Sputtering produces a highly defective, oxygen-deficient surface with several oxidation states visible in the XPS Ti $2p$ spectrum (c).

Figure 6. Scanning tunneling microscopy image (50×50 nm) of a $TiO_2(110)$ surface annealed at 920 K in ultrahigh vacuum.[40] Large terraces with monoatomic steps are visible. The inset shows a magnified area of one of the terraces. White features (one is circled in the inset) are indicative of point defects that are created through the high-temperature anneal. These defects result in the Ti^{3+} shoulder in the XPS spectrum of Fig. 5b.

also shows the flat surfaces with relatively large terraces that can be obtained through sputtering and annealing in ultrahigh vacuum.

As mentioned, many "maximum valency" oxides, i.e., oxides with cations in the highest possible oxidation state, are prone to electron-stimulated desorption (ESD) processes. Titanium dioxide is a classic example for such a material, and ESD experiments on TiO_2 have led to a model that explains this effect.[6] The onset for electron-stimulated desorption of O^+ ions is around the ionization energy of the Ti $3p$ core level.[6] There is a certain probability that this core hole decays through an interatomic Auger process with a neighboring oxygen atom. If this Auger process leads to the formation of O^+ ions, they will be repelled by the sudden reversal of the Madelung potential.

3.2. Strontium Titanate

$SrTiO_3$ has the cubic perovskite structure. The charge-neutral (100) surface and, to some extent, the polar (111) surface have been studied. $SrTiO_3$ is similar to TiO_2 in many ways. The electronic structure near the Fermi level is mainly Ti $3d$ and O $2p$ derived; the width of the band gap (3.2 eV) is comparable to that of TiO_2, and the bulk also becomes n-type semiconducting upon annealing at high temperatures (≥ 1000 K) for several hours.* This high-temperature reduction is again accompanied by a color change from transparent to blue or even dark gray, depending on the amount of color centers produced by creation of bulk oxygen vacancies. The semiconducting bulk prevents charging and makes $SrTiO_3$ a popular system for surface science investigations, where satisfactory scanning tunneling microscopy images can even be achieved.[44,45] The surface preparation is more complex, however, because of the presence of a third component. The $SrTiO_3$ crystal structure has stacked two kinds of nonpolar atomic planes, having either TiO_2 or SrO stoichiometry. The (100) surface can be terminated by a TiO_2 layer or a SrO layer, or a combination of both.

Because $SrTiO_3$ has a close lattice match to high-temperature superconducting perovskites, there has been an increased interest in finding and investigating *ex situ* and *in situ* surface preparation procedures that lead to high-quality substrates for MBE growth of these materials.[12] It was mentioned in Sec. 2.1 that $SrTiO_3$ single crystals can be bought from many different vendors. Commercial single-crystalline oxide wafers are prepared by mechanochemical polishing in a solution containing colloidal silicon; aqueous NH_3 with pH = 10.5–11.0 is reported as the polishing solution by one group.[46] This procedure results in surfaces with mostly TiO_2 termination;[46] a ~5% SrO surface content has been determined with CAICISS

---

*Doping with Nb (0.01%[44]) has been reported as a way to produce conducting samples without a high-temperature anneal.

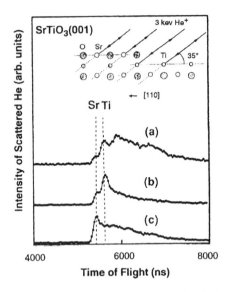

Figure 7. A SrTiO$_3$ crystal with (001) orientation may have TiO$_2$ or SrO stoichiometry or a combination of both. Using appropriate preparation procedures, the surface termination can be tailored to the desired stoichiometry. This figure shows CAICISS (*coaxial impact-collision ion scattering spectroscopy*) spectra taken at an incident angle of 35° along the [110] direction of a SrTiO$_3$(001) surface (inset) after different pretreatments. (a) as-supplied substrate; the surface is mainly TiO$_2$-terminated due to the polishing procedure (SrO content ~5%). (b) after annealing at 1000°C for 10 h in 1 atm oxygen (SrO content ~1%), and (c) after growth of a homoepitaxial, 20-nm-thick SrTiO$_3$ film with deposited under 10^{-6} Torr oxygen by pulsed laser deposition (SrO content 100%). (Reprinted with permission from M. Yoshimoto, T. Maeda, K. Shimkonzono, H. Koinuma, M. Shinohara, O. Ishiyama, and F. Ohtani, *Applied Physics Letters* **65** (2), 3197–3199 (1994). Copyright 1994 American Institute of Physics.)

(coaxial impact-collision ion scattering spectroscopy) on a polished sample without any *in situ* treatment (Fig. 7).[47] The dominant termination with the TiO$_2$ plane may be related to a report that simply rinsing a SrTiO$_3$ sample in distilled water preferentially removes Sr ions from the surface.[48] When polished SrTiO$_3$ samples are immersed in a buffered NH$_4$F-NH$_3$ solution (10 M NH$_4$F with a pH of 4.4–4.6), atomically smooth surfaces are obtained without any further cleaning, and the topmost layer is 100% TiO$_2$ (Fig. 8).[12] The acidic solution is thought to dissolve preferentially the basic oxide SrO and to leave intact the acidic TiO$_2$ layer.

Annealing with or without oxidizing gases has been investigated as further preparation step for polished SrTiO$_3$ substrates. Annealing at 700°C in UHV for 1 h removes C, S, P, K, and Cl surface impurities but not Ca, which may cause a p(2×2) reconstruction on SrTiO$_3$(100).[49] (The Ca had segregated to the surface upon annealing in air at 890°C for 30 min.) Oxygen annealing further increased the TiO$_2$ content on polished (100) surfaces; annealing at 1000°C for 10 h in 1

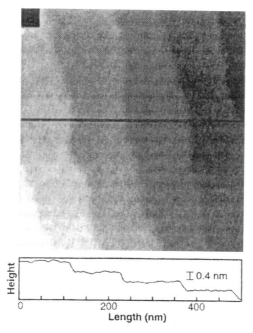

Figure 8. AFM image and line profile of a $SrTiO_3(100)$ surface etched in a buffered $NH_4F–NH_3$ solution. This surface exhibits wide, atomically flat surfaces, and 0.4-nm-high steps. The surface contains 100% TiO_2. (Reprinted with permission from M. Kawasaki, K. Takahashi, T. Maeda, R. Tsuchiya, M. Shinohara, O. Ishiyama, T. Yonezawa, M. Yoshimoto, and H. Koinuma, *Science* **266**, 1540–1542 (1994). Copyright 1994 American Association for the Advancement of Science.)

atmosphere O_2 produced a surface that was 99% TiO_2 terminated (Fig. 7).[47] In order to remove a small amount of C contamination, annealing in 1×10^{-4} Pa NO_2 at 673 K is also successful and results in a mostly TiO_2 terminated surface.[46] Relatively high temperatures (>1000°C in UHV) are needed before initially rough surfaces show wide and smooth terraces in STM.[45,50] Annealing to 1200°C for 2 min has been reported by Tanaka *et al.*[45] to result in ($\sqrt{5}\times\sqrt{5}$)R26.6° reconstruction with an ordered array of missing oxygen atoms. Annealing above 1300°C causes Sr enrichment.[50]

SrO-terminated surfaces can be obtained by growing a monolayer of SrO on a TiO_2-terminated surface. Slow deposition of Sr in an oxygen atmosphere (~1×10^{-5} Pa O_2 or NO_2) at moderate sample temperatures (600–900 K) is reported.[46,51] Homoepitaxial $SrTiO_3(001)$ thin films deposited on a TiO_2-terminated sample at 630°C under 10^{-6} Torr oxygen with laser MBE also result in 100% SrO terminated surfaces (Fig. 7).[47]

Argon bombardment severely reduces $SrTiO_3$ surfaces. This reduction is not completely removed upon annealing in UHV to 1100 K.[52] In addition to sputtering-induced loss of oxygen, preferential loss of Sr has been reported for the (001) surface,[52] but Sr enrichment occurs for the (111) surface,[53] and no preferential removal of either cation is seen for $SrTiO_3(110)$.[54] It seems unlikely that the surface orientation would account for these enormous differences in sputtering behavior. Similarly controversial results have been reported for NiO and have been resolved by investigating the temperature dependence of sputtering behavior (Fig. 3).[18] Possibly, similar effects cause these confusing results for $SrTiO_3$.

$SrTiO_3$ does not cleave well. Fractured (100) surfaces have been investigated by Henrich et al.[52] The fractured surface appears to have a relatively high defect density and exhibits a fuzzy LEED pattern that improves upon annealing in UHV to 1100 K. The defect concentration decreases upon annealing and upon exposure to oxygen; both effects are expected for oxygen vacancies. Temperature-created defects are only seen on a mostly TiO_2-terminated surface, not on a SrO-terminated surface.[51] Similar to TiO_2, $SrTiO_3$ is also very prone to loss of oxygen by electron-stimulated desorption when exposed to an electron beam.

3.3. Magnesium Oxide

For decades, magnesium oxide has been one of the most popular substrates for epitaxial thin-film growth. MgO has a high melting point (2800°C). It is readily available as a single crystal, is inexpensive, and has a good lattice match with a variety of cubic materials, including fcc metals and high-temperature superconductors. Originally, MgO was used because it is well-suited for the production of electron micrographs of thin-film structures.*

It was recognized early that surface pretreatment is an important ingredient for epitaxy on MgO, almost equally important as the lattice misfit between the substrate and the overlayer.[55,56] Consequently, quite a few papers have compared the growth on differently prepared surfaces (see Refs. 55, 57–60). Because MgO can be cleaved quite easily with a knife edge along the [001] direction, cleavage in UHV, cleavage in air (with or without subsequent heat treatment to remove the impurities that accumulate during exposure to the ambient), as well as the usual sputtering and annealing techniques have been applied as surface preparation techniques. Although the surface morphology and the atomic geometry are similar for cleavage in air and in UHV,[59] there are differences in the metal overlayer growth on these two surfaces. For example, it has been observed that the epitaxial relationships of several metal overlayers to MgO are different for these two surface pretreatments[55,58] and that the size distributions of Pd clusters grown on MgO also depend on the cleaving environment.[60] Probably, impurities like hydrocarbons and water

---

*A metal overlayer is deposited onto MgO, coated with a thick film of carbon, and the substrate is then dissolved in an acidic solution. The metal overlayer on the C film is inspected with a transmission electron microscope.

that adsorb on the surface during exposure to the ambient play a role and modify the surface diffusion and nucleation processes that affect metal-overlayer growth phenomena. While cleavage results in wide terraces and a very good geometric order, i.e., LEED patterns with low background, the mesoscopic morphology is dominated by steps. It has been mentioned that very flat surfaces can be achieved through mechanochemical superpolishing.

When a surface must be regenerated for several experiments, only repeated *in situ* cleavages or sputtering and annealing cycles are feasible. No preferential sputtering occurs on this surface, but a MgO surface was still rough after annealing above 1000°C (Fig. 9).[59] This residual roughness is presumably due to incomplete diffusion of this high-melting-point material. A higher annealing temperature would probably result in smoother surfaces. However, several authors have reported segregation of trace impurities (e.g., Ca, Sr, and Fe) when commercial, high-purity MgO is annealed to temperatures above 1000°C.[57,59,61]

Electron beam damage is not a major problem when employing surface spectroscopic techniques.[62] Although MgO has a very wide band gap ($E_{gap} = 8$

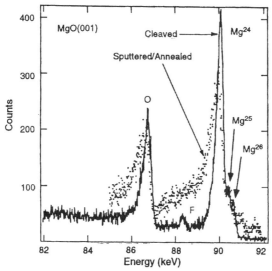

Figure 9. While cleavage of MgO samples results in (001) surfaces with a high degree of geometric ordering, sputtering leads to a surface roughening that is still visible after annealing above 1000°C. Medium-energy ion scattering (MEIS) spectra of O and Mg taken in the (010) scattering plane for a UHV-cleaved (solid line) and a sputtered–annealed (dots) MgO(001) surface. Several Mg isotopes are resolved. The small peak between the O and the Mg peaks results from F in the residual gas. An rms surface roughness of ~1.5–2.0 nm is estimated for the sputtered surface. (Reprinted with permission from J. B. Zhou, H. C. Lu, T. Gustafsson, and P. Häberle, *Surface Science* **302**, 350–362 (1994). Copyright 1994 Elsevier Science B.V.)

eV), charging effects are not as severe as for other materials. Because the secondary-electron coefficient of MgO is very high for electron energies above 100 eV, the sample floats to a *positive* potential under electron bombardment for an AES measurement. Consequently, peaks will be shifted in energy, but the sample will not charge to the potential of the primary-electron beam as is the case for other insulators when operating without a charge-compensation technique. This high secondary-electron coefficient at high energies has been put to use in a high-resolution electron energy loss spectroscopy (HREELS) experiment, where a *high*-energy flood gun (2800 eV) was used[63] to compensate for sample charging with the low-energy HREELS electron beam.

3.4. Zinc Oxide

Zinc oxide has the wurtzite structure. Its surface chemistry has been extensively investigated because of its importance in heterogeneous catalysis.[64] Cleavage perpendicular to the crystallographic c-axis results in two different polar surfaces with very different chemical properties; one surface is exclusively Zn-terminated and the other is O-terminated. In the literature, these polar faces are usually labeled (0001)–Zn and $(000\bar{1})$–O, respectively. In addition to these surfaces, the $(10\bar{1}0)$ prism face and the $(11\bar{2}0)$ faces of ZnO have been investigated. The nonpolar $(10\bar{1}0)$ face is structurally the most stable surface, followed by (0001)–Zn and $(000\bar{1})$–O.[65] Often, the $(000\bar{1})$–O face exhibits LEED patterns of poorer quality than the (0001)–Zn face, with some authors reporting streaking and beam splitting for both faces.[65–67]

Once cut, the sides of a (0001) crystal can be distinguished because they exhibit very different behavior toward etching with various acids (HCl, HF, nitric acid, etc.). The Zn face is unaffected but the O face is rapidly etched and visibly roughened when dipping the sample in acid (Fig. 10).[68,69] This procedure is used to assign the (0001)–Zn and $(000\bar{1})$–O surfaces and for orientation of the crystal before polishing. Single crystals grow in needlelike form with the O end planar and the Zn end pyramidal.[70] Klein describes how to distinguish the orientation of the crystallographic c-axis without cutting. Upon etching of needle-shaped crystals with HF, he observed a characteristic triangular-shaped etch pattern on the $(10\bar{1}0)$ prism faces with the tip of the triangle pointing into the direction of the positive c-axis.[68]

Undoped, untreated crystals exhibit strong electrical charging. No charging problems are usually reported after high-temperature "conditioning," such as annealing at 700–750 K for several hours.[71] In an early work, charging was suppressed by irradiating the sample with UV light, which leads to photoinduced conductivity in this wide-band-gap semiconductor.[65]

Most authors report sputtering and annealing as an *in situ* preparation procedure. This yields (1×1) LEED patterns for the polar (0001) and $(000\bar{1})$ faces as long

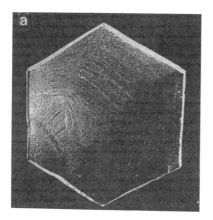 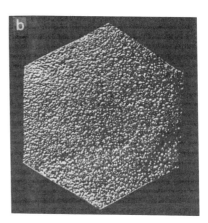

Figure 10. An etching procedure is used to distinguish between the two polar surfaces that are formed upon cleaving ZnO crystals perpendicular to the crystallographic c-axis. The (0001)–Zn face is unaffected by the etching process (20 min in HF 40%) (a), while the (000$\bar{1}$)–O face is visibly roughened (b). The crystal diameter is on the order of 1 mm. (Reprinted with permission from A. Klein, *Zeitschrift für Physik* **188**, 352–360 (1965). Copyright 1965 Springer-Verlag.)

as the annealing temperature is kept below 870 K.[72] The same LEED result has been obtained for surfaces prepared by cleavage in UHV Sputtering is usually performed with Ar^+ at relatively low energies (<1000 eV).[66,69,73] Ion bombardment is not reported to change the surface stoichiometry by preferential sputtering.[17]

Annealing has to be done at moderate temperatures (~700–800 K), mostly in UHV. Sublimation of the material occurs at relatively low temperatures (region d in Fig. 11).[74] Apparently, ZnO is very sensitive toward morphological changes through heating. Prolonged annealing of ZnO(0001)–Zn at 800 K creates hexagonal pits by thermal etching.[75] Göpel and Lampe[74] used the low sublimation temperature of the material as a way to prepare samples without sputtering. Controlled sublimation at 1050 K for 180 s and subsequent oxygen treatment (10^{-4} Pa, 700 K, 10^4 s) has been reported to yield a good LEED pattern from ZnO (10$\bar{1}$0),[74] although this procedure may also lead to Ca impurities. Annealing to 970 K without sputtering has led to a small amount of detected potassium on the (0001)–Zn face in one instance.[76]

Generally, ZnO is very easily reducible. A detailed study on the formation of point defects has been performed for ZnO (10$\bar{1}$0).[77] Defects cannot be detected with XPS or AES, but they have been monitored by observing surface conductivity changes with O_2 adsorption.[74] Point defects also lead to significant changes in UPS valence-band spectra.[77] Defects are believed to influence heavily the surface reactivity of ZnO crystals;[78] e.g., they have been reported to act as specific sites for strong CO_2 chemisorption.[77]

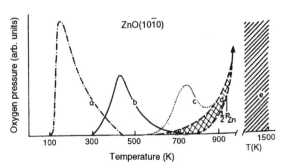

Figure 11. Residual oxygen gas pressure during heating of a ZnO (10$\bar{1}$0) sample. (a) Desorption of physisorbed oxygen that was adsorbed at 100 K; (b) desorption of chemisorbed O^{2-} that was adsorbed at 300 K; (c) desorption from nonideal (edge and kink) surface positions during the first heating cycle of an Ar^+-bombarded ZnO(10$\bar{1}$0) surface; (d) sublimation of the crystal. P_{O_2} in the hatched range is obtained on surfaces with different point-defect concentrations. (e) High-temperature range of crystal preparation determining bulk defects in ZnO samples. (Reprinted with permission from W. Göpel and U. Lampe, *Physical Review B* **22** (12), 6447–6462 (1980). Copyright 1980 American Physical Society.)

Care must be taken when annealing the sample in oxygen, since oxygen has been reported to adsorb in the form of O_2^- on ZnO (10$\bar{1}$0) at 300 K and to desorb at ~450 K (region b in Fig. 11).[74] For example, annealing in 1×10^{-6} Torr O_2 was performed at 850 K if the ion scattering spectroscopy (ISS) signal from a ZnO (000$\bar{1}$)–O sample was below a certain value in repeated sputtering/–850-K UHV annealing cycles. In this case, the oxygen valve was closed before cooling the sample to ensure that any nonlattice oxygen was desorbed, although cooling in O_2 did not significantly alter the ISS Zn:O signal ratio.[79]

3.5. NiO Thin Films on Nickel and Other Metal Single-Crystal Surfaces

The oxidation of Ni single crystals has been studied extensively over the past decade,[20] and oxidation of the three low-Miller-index faces (100), (110), and (111) has been investigated. The lattice mismatch between nickel and NiO is 16%.[80] Hence, the oxidized films are not very well ordered; they contain a large number of defects and are not very smooth. Experimentally, formation of a nickel oxide thin film is quite straightforward. After preparing a clean Ni surface (see Chapter 5, this volume), the surface is simply exposed to oxygen, and the oxygen uptake can be followed with a spectroscopic technique (e.g. AES). (NiO is not sensitive to irradiation with electrons in the low-energy (<<100 keV) regime.)[81] In an important work by Holloway and Hudson,[82] the basic processes have been described and the ground has been laid for most of the later research.

The interaction of oxygen with Ni can roughly be divided into three regimes that occur with increasing oxygen exposure. For all the Ni faces studied, the reaction

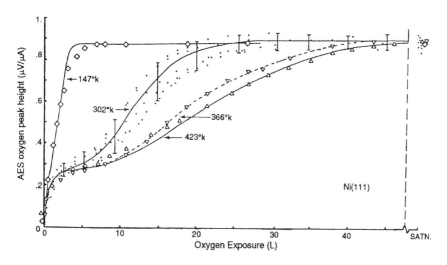

Figure 12. Oxidation of metal single crystals is a popular way of preparing metal oxide thin films, although the structural quality of the films is rarely comparable to surfaces of oxide bulk crystals. This figure shows the height of the 507-eV oxygen Auger signal during exposure of a Ni(111) single crystal to oxygen in Langmuirs (1 L = 1 × 10^{-6} Torr · s). The oxygen uptake is initially rapid, then slows for a while, and increases again until a saturation coverage is reached. The initial reaction rate is only slightly temperature sensitive, but the rate of the second reaction shows significant inverse dependence on temperature. (Reprinted with permission from P. H. Holloway and J. B. Hudson, *Surface Science* **43** (1), 141–149 (1974). Copyright 1974 Elsevier Science B.V.)

follows this route: formation of a chemisorbed layer → fast oxidation → slow oxidation (Fig. 12). In the first step, dissociative chemisorption is observed that leads to the formation of a series of ordered overlayers; a $p(2\times2)$ structure is followed by a $c(2\times2)$ layer on Ni(100),[82] a (2×2) structure is followed by a $(\sqrt{3} \times \sqrt{3})R30°$ layer on Ni(111),[83] and a $c(2\times4)$ overlayer is followed by other reconstructions at higher oxygen exposures on Ni(110).[84] Already for low oxygen exposures [10 L on Ni(100)],[80] oxide islands are formed; i.e., the oxygen atoms are incorporated in the bulk of the nickel. These islands act as nucleation sites for further oxidation. On Ni(100), the oxide islands form preferentially along step edges and are tilted to accommodate the strain that is induced by the mismatch between the nickel and the NiO lattice.[80] The oxidation rate depends on the crystal face. The (111) surface oxidizes most easily, and the (100) face has the lowest rate of oxidation at the same temperature.[83] Temperature plays a big role for the rate of oxidation as well as the oxide phase formed (the lower the temperature, the faster the oxidation process occurs;[82,83] see Fig. 12). On Ni(100), for example, three different oxide structures have been observed: NiO(111) for adsorption at temperatures <300 K, a (7×7) structure that has been assigned as a strained form of

NiO(100) around room temperature, and NiO(100) for temperatures above 500 K.[85]

NiO(100) is the most stable face of this rocksalt structure material and is always the end product of high-temperature oxidation, independent of the orientation of the Ni crystal that is oxidized. There is agreement that the oxide formed between 80 and 400 K has a limiting thickness of two to three layers, whereas the oxide is somewhat deeper when formed above 500 K.[85] Oxidation (1200 L)/annealing (670 K) cycles have been used by Bäumer *et al.* to form the best-ordered NiO(100) overlayers on Ni(100).[80] However, these films are not nearly as well ordered as cleaved bulk NiO single crystals, and high defect densities (20–25%) are reported.[80]

As mentioned in the introduction, NiO has the rocksalt structure, and the (111) face is polar and unstable. Preparation of bulk single-crystal NiO(111) surfaces is usually not successful; in the only account that has been, significant segregation of trace Si impurities from the bulk was found.[86] The NiO(111) faces that have been observed on Ni(100) and Ni(111) are possibly stabilized by OH groups that are adsorbed from the background gas during the low-temperature oxidation process.[87]

Both epitaxial NiO(100) and NiO(111) films have been prepared on foreign substrates. Well-ordered NiO(100) films have been grown on Ag(100)[27] and on Mo(100),[22] and NiO(111) films have been prepared on Au(111)[27–29] and on Mo(110).[88] The preparation recipes vary somewhat from group to group, but, usually, evaporation from a metal source (either a Knudsen-type source or simply a shielded W filament wrapped with a pure Ni wire) in an oxygen atmosphere (~10^{-6} Torr) at elevated substrate temperatures (400–700 K) is used, sometimes with annealing to somewhat higher temperatures after the deposition process.

4. Summary

The surface preparation of compound specimens is considerably more complex than that of elemental materials. Surface stoichiometry can deviate from the ideal value because of surface segregation, thermal instability, preferential sputtering, polishing, etc. Faceting and reconstructions are common, especially on polar surfaces. On maximum valency materials, one has to be aware of electron-stimulated desorption processes. Insulating samples are subject to charging during spectroscopic measurements, and sample mounting as well as reliable temperature measurements are not as straightforward as for conductive materials.

In this chapter, we have focused on surface preparation of metal oxides, but many of the discussed topics are relevant for other metal compounds as well. Only few materials are commercially available as single-crystal bulk materials. Cleaving results in surfaces that are closest to the ideal form in terms of stoichiometry and ordering. Cleaved surfaces often exhibit large steps. Superpolishing oxides with mechanical–chemical procedures results in samples with a very small rms roughness. Preferential sputtering of oxygen is very common for metal oxides; in specific cases, a strong dependence on the sample temperature is observed. Thin-film metal oxides on conductive substrates are often used for experiments when work with bulk materials is too difficult.

As examples for these and for other issues that are relevant for oxide materials, we have discussed preparation procedures for single crystals of TiO_2, $SrTiO_3$, MgO, and ZnO, as well as for NiO thin films.

References

1. V. E. Henrich and P. A. Cox, *The Surface Science of Metal Oxides*, Cambridge University Press, Cambridge (1994).
2. T. M. Parrill and Y. W. Chung, Effects of initial silicon carbide surface composition on room temperature Au/β-SiC(001) interface formation, *Surf. Sci.* **271**, 395–406 (1992).
3. P. W. Tasker, The stability of ionic crystal surfaces, *J. Phys. C: Solid State Phys.* **12**, 4977–4984 (1979).
4. J. P. LaFemina, Total energy computations of oxide surface reconstructions, *Crit. Rev. Surf. Chem.* **3** (3/4), 297–386 (1994).
5. M. Szymonski, J. Kolodziej, Z. Postawa, P. Czuba, and P. Piatkowski, Electron-stimulated desorption from ionic crystal surfaces, *Prog. Surf. Sci.* **48** (1/4), 83–96 (1995).
6. M. L. Knotek and P. J. Feibelman, Stability of ionically bonded surfaces in ionizing environments, *Surf. Sci.* **90**, 78–90 (1979).
7. M. A. Barteau, Organic reactions on well-defined oxide surfaces, *Chem. Rev.* **96** (4), 1413 (1996).
8. Uncover (http://uncweb.carl.org).
9. M. D. Antonik and R. J. Lad, Faceting, reconstruction, and defect microstructure at ceramic surfaces revealed by atomic force microscopy, *J. Vac. Sci. Technol. A* **10** (4), 669 –673 (1992).
10. J. Kühnle and O. Weis, Mechanochemical superpolishing of diamond using $NaNO_3$ or KNO_3 as oxidizing agents, *Surf. Sci.* **340**, 16–22 (1995).
11. B. Hader and O. Weis, Superpolishing sapphire: A method to produce atomically flat and damage free surfaces, *Surf. Sci.* **220**, 118–130 (1989).
12. M. Kawasaki, K. Takahashi, T. Maeda, R. Tsuchiya, M. Shinohara, O. Ishiyama, T. Yonezawa, M. Yoshimoto, and H. Koinuma, Atomic control of the $SrTiO_3$ crystal surface, *Science* **266**, 1540–1542 (1994).
13. A. L. Linsebigler, G. Lu, and J. T. Yates, Jr., Photocatalysis on TiO_2 surfaces: Principles, Mechanisms, and Selected Results, *Chem. Rev.* **95**, 735–758 (1995).
14. M. B. Hugenschmidt, L. Gamble, and C. T. Campbell, The interaction of H_2O with a $TiO_2(110)$ surface, *Surf. Sci.* **302**, 329–340 (1994).
15. M. Grunze, O. Ruppender, and O. Elshazly, Chemical cleaning of metal surfaces in vacuum systems by exposure to reactive gases, *J. Vac. Sci. Technol. A* **6** (3), 1266–1275 (1988).
16. G. Betz and G. K. Wehner, in: *Sputtering of Multicomponent Materials, Sputtering by Particle Bombardment II* (R. Behrisch, ed.), Springer, Berlin (1983), pp. 11–90.
17. J. B. Malherbe, S. Hofmann, and J. M. Sanz, Preferential sputtering of oxides: A comparison of model predictions with experimental data, *Appl. Surf. Sci.* **27**, 355–365 (1986).
18. M. A. Langell, Preferential sputtering in the 3d transition metal monoxides, *Surf. Sci.* **186**, 323–338 (1987).

19. H. M. Naguib and R. Kelly, Criteria for bombardment-induced structural changes in non-metallic solids, *Radiat. Effects* **25**, 1–12 (1975).

20. C. R. Brundle and J. Q. Broughton, in: *The Chemical Physics of Solid Surfaces and Heterogenous Catalysis*, Vol. 3, Part A, *Chemisorption Systems* (D. A. King and D. P. Woodruff, eds.), Elsevier, Amsterdam (1990), pp. 381–388.

21. R. M. Jaeger, H. Kuhlenbeck, H.-J. Freund, M. Wuttig, W. Hoffmann, R. Franchy, and H. Ibach, Formation of a well-ordered aluminium oxide overlayer by oxidation of NiAl(110), *Surf. Sci.* **259**, 235–252 (1991).

22. D. W. Goodman, Chemical and spectroscopic studies on metal oxide surfaces, *J. Vac. Sci. Technol.* **14** (3), 1526–1531 (1996).

23. G. S. Herman, M. C. Gallagher, S. A. Joyce, and C. H. F. Peden, Structure of epitaxial thin TiO_x films on W(110) as studied by low energy electron diffraction and scanning tunneling microscopy, *J. Vac. Sci. Technol. B* **14** (2), 1126–1130 (1996).

24. Y. Wu, E. Garfunkel, and T. E. Madey, Growth of ultrathin Al_2O_3 films on Ru(0001) and Re(0001), *J. Vac. Sci. Technol. A* **14** (4), 2554–2563 (1996).

25. G. H. Vurens, V. Maurice, M. Salmeron, and G. A. Somorjai, Growth, structure and chemical properies of FeO overlayers on Pt(100) and Pt(111), *Surf. Sci.* **268**, 170–178 (1992).

26. Z. Zhang and V. E. Henrich, Cation-ligand hybridization for stoichiometric and reduced TiO_2(110) surfaces determined by resonant photoemission, *Phys. Rev. B* **43** (14), 12004–12011 (1991).

27. K. Marre and H. Neddermeyer, Growth of ordered thin films of NiO on Ag(110) and Au(111), *Surf. Sci.* **287/288**, 995–999 (1993).

28. H. Hannemann, C. A. Ventrice, Jr., T. Bertrams, A. Brodde, and H. Neddermeyer, Scanning tunneling microscopy on the growth of ordered NiO layers on Au(111), *Phys. Stat. Sol. (a)* **146**, 289–297 (1994).

29. C. A. Ventrice, Jr., T. Bertrams, H. Hannemann, A. Brodde, and H. Neddermeyer, Stable reconstruction of the polar (111) surface of NiO on Au(111), *Phys. Rev. B* **49** (8), 5773–5776 (1994).

30. C. T. Campbell, Ultrathin metal films and particles on oxide surfaces: structural, electronic and chemisorptive properties, *Surf. Sci. Rep.* **27** (1–3), 1–111 (1997).

31. D. W. Goodman, Model Studies in Catalysis Using Surface Science Probes, *Chem. Rev.* **95**, 523–536 (1995).

32. L. P. Zhang, M. Kuhn, and U. Diebold, Growth, structure, and thermal properties of chromium oxide films on Pt(111), *Surf. Sci.* **375** (1), 1–12 (1997).

33. G. E. Poirier, B. K. Hance, and M. White, Identification of the facet planes of phase I TiO_2(001) rutile by scanning tunneling microscopy and low energy electron diffraction, *J. Vac. Sci. Technol. B* **10** (1), 6–15 (1992).

34. G. W. Clark and L. L. Kesmodel, A scanning tunneling microscopy study of TiO_2(001) surface reconstructions, *Ultramicroscopy* **41**, 77–82 (1992).

35. D. C. Cronemeyer, Electrical and optical properties of rutile single crystals, *Phys. Rev.* **87** (5), 876–886 (1952).

36. Y. Gao, Y. Liang, and S. A. Chambers, Thermal stability and the role of oxygen vacancy defects in strong metal support interaction—Pt on Nb-doped TiO_2(100), *Surf. Sci.* **365** (3), 638–648 (1996).

37. F. Pesty, H.-P. Steinrück, and T. E. Madey, Thermal stability of Pt films on TiO_2(110): evidence for encapsulation, *Surf. Sci.* **339** (1/2), 83–95 (1995).

38. V. E. Henrich, G. Dresselhaus, and H. J. Zeiger, Observation of two-dimensional phases associated with defect states on TiO_2, *Phys. Rev. Lett.* **36** (22), 1335–1338 (1976).

39. J.-M. Pan, B. L. Maschhoff, U. Diebold, and T. E. Madey, The interaction of water, oxygen, and hydrogen with TiO_2(110) surfaces having different defect densities, *J. Vac. Sci. Technol. A* **10**, 2470–2476 (1992).

40. M. Li, W. Hebenstreit, and U. Diebold, Oxygen–induced restructuring of rutile TiO_2 (110) surface, *Surf. Sci.* **414** (1/2), L951–L959 (1998).

41. U. Diebold and T. E. Madey, TiO_2 by XPS, *Surf. Sci. Spectra* **4** (3), 227–231 (1998).

42. J. Mayer, E. Garfunkel, T. E. Madey, and U. Diebold, Titanium and reduced titania overlayers on titanium dioxide (110), *J. Electr. Spectrosc.* **73**, 1 –11 (1995).

43. U. Diebold, J. F. Anderson, K.-O. Ng, and D. Vanderbilt, Evidence for the tunneling site on transition metal oxides: TiO_2(110), *Phys. Rev. Lett.* **77** (7), 1322–1326 (1996).

44. T. Matsumoto, T. Kawai, and S. Kawai, STM-imaging of a $SrTiO_3$(100) surface with atomic-scale resolution, *Surf. Sci. Lett.* **278**, L153–L158 (1992).

45. H. Tanaka, T. Matsumoto, T. Kawai, and S. Kawai, Surface structure and electronic property of reduced $SrTiO_3$(100) surface observed by scanning tunneling microscopy, *Jpn. J. Appl. Phys.* **32**, 1405–1409 (1993).

46. T. Hikita, T. Hanada, and M. Kudo, Structure and electronic state of the TiO_2 and SrO terminated $SrTiO_3$ surfaces, *Surf. Sci.* **287/288**, 377–381 (1993).

47. M. Yoshimoto, T. Maeda, K. Shimonzono, H. Koinuma, M. Shinohara, O. Ishiyama, and F. Ohtani, Topmost surface analysis of $SrTiO_3$(001) by coaxial impact-collision ion scattering spectroscopy, *Appl. Phys. Lett.* **65** (2), 3197–3199 (1994).

48. D. M. Tench and D. O. Raleigh, Electrochemical processes on strontium titanate electrodes, Natl. Bureau of Standards Special Publication Vol. 455, NBS, Gaithersburg, MD (1975), pp. 229–240.

49. J. E. T. Andersen and P. J. Møller, Impurity-induced 900°C (2×2) surface reconstruction of $SrTiO_3$(100), *Appl. Phys. Lett.* **56** (19), 1847 (1990).

50. Y. Liang and D. A. Bonnell, Structures and chemistry of the annealed $SrTiO_3$(001) surface, *Surf. Sci.* **310**, 128–134 (1994).

51. A. Hirata, K. Saiki, A. Koma, and A. Ando, Electronic structure of a SrO-terminated $SrTiO_3$(100) surface, *Surf. Sci.* **319**, 267–271 (1994).

52. V. E. Henrich, G. Dresselhaus, and H. J. Zeiger, Surface defects and the electronic structure of $SrTiO_3$ surfaces, *Phys. Rev. B* **17** (12), 4908–4921 (1978).

53. W. J. Lo and G. A. Somorjai, Temperature-dependent surface structure, composition, and electronic properties of the clean $SrTiO_3$(111) crystal face: Low-energy electron diffraction, Auger-electron spectroscopy, electron energy loss, and ultraviolet-photoelectron spectroscopy studies, *Phys. Rev. B* **17** (12), 4942–4950 (1978).

54. S. M. Mukhopadhay and T. C. S. Chen, Surface properties of perovskites and their response to ion bombardment, *J. Appl.Phys.* **74** (2), 872–876 (1993).

55. A. K. Green, J. Dancy, and E. Bauer, Insignificance of lattice misfit for epitaxy, *J. Vac. Sci. Technol.* **7** (1), 159–163 (1970).

56. K. Yagi and G. Hanjo, Roles of lattice fitting in epitaxy, *Thin Solid Films* **48**, 137 (1978).

57. T. Hanada, M. Asano, and Y. Mitzutani, Epitaxial growth of Ag deposited on air-cleaved MgO(100) by molecular beam deposition, *J. Crystal Growth* **116**, 243–250 (1992).

58. P. W. Palmberg, T. N. Rhodin, and C. J. Todd, Epitaxial growth of gold and silver on magnesium oxide cleaved in ultrahigh vacuum, *Appl. Phys. Lett.* **11** (7), 33–35 (1967).

59. J. B. Zhou, H. C. Lu, T. Gustafsson, and P. Häberle, Surface structure of MgO(001): a medium-energy ion scattering study, *Surf. Sci.* **302**, 350–362 (1994).

60. C. R. Henry, C. Chapron, C. Duriez, and S. Giorgio, Growth and morphology of palladium particles epitaxially deposited on a MgO(100) surface, *Surf. Sci.* **253**, 177–189 (1991).

61. M. Gajdardziska-Josifovska, P. A. Crozier, M. R. McCartney, and J. M. Cowley, Ca segregation and step modification on cleaved and annealed MgO(100) surfaces, *Surf. Sci.* **284**, 186–199 (1993).

62. D. G. Lord and M. Prutton, Electrons and the epitaxial growth of metals on alkali halides, *Thin Solid Films* **21**, 341–357 (1974).

63. P. A. Thiery, M. Liehr, J. J. Pireaux, and R. Caudano, Infrared optical constants of insulators determined by high-resolution electron energy loss spectroscopy, *Phys. Rev. B* **29** (8), 4824–4826 (1984).

64. C. N. Satterfield, *Heterogeneous Catalysis in Industrial Practice*, McGraw-Hill, New York (1991).

65. M. F. Chung and H. E. Farnsworth, Investigations of surface stability of II-VI wurtzite compounds by LEED, *Surf. Sci.* **22**, 93–110 (1970).

66. R. Davis, J. F. Walsh, C. A. Muryn, G. Thornton, V. R. Dhanak, and K. C. Prince, The orientation of formate and carbonate on ZnO, *Surf. Sci.* **298**, L196–202 (1993).

67. C. T. Campbell, D. A. Daube, and J. M. White, Cu/ZnO (000$\bar{1}$) and ZnOx/Cu(111) model catalysts for methanol synthesis, *Surf. Sci.* **182**, 458–476 (1987).

68. A. Klein, Polare Eigenschaften von Zinkoxid-Kristallen, *Z. Phys.* **188**, 352–360 (1965).

69. R. R. Gay, M. H. Nodine, V. E. Henrich, H. J. Zeiger, and E. L. Solomon, Photoelectron study of the interaction of CO with ZnO, *J. Am. Chem. Soc.* **102**, 6752–6761 (1980).

70. A. N. Mariano and R. E. Hanneman, Crystallographic polarity of ZnO crystals, *J. Appl. Phys.* **34** (2), 384–388 (1963).

71. P. J. Møller, S. A. Komolov, E. F. Lazneva, and E. H. Pedersen, CO$_2$-intermediates in the CO/ZnO(0001) interface, *Surf. Sci.* **323**, 102–108 (1995).

72. S.-C. Chang and P. Mark, The crystallography of the polar (0001)-Zn and (000$\bar{1}$)-O surfaces of zinc oxide, *Surf. Sci.* **46**, 293–300 (1974).

73. S. V. Didzilius, K. D. Butcher, S. L. Cohen, and E. I. Solomon, Chemistry of copper overlayers on zinc oxide single-crystal surfaces: Model active sites for Cu/ZnO methanol synthesis, *J. Am. Chem. Soc.* **111**, 7110–7123 (1989).

74. W. Göpel and U. Lampe, Influence of defects on the electronic structure of zinc oxide surfaces, *Phys. Rev. B* **22** (12), 6447–6462 (1980).

75. P. J. Møller, S. A. Komolov, and E. F. Lazneva, VLEED form a ZnO(0001) substructure, *Surf. Sci.* **307–309**, 1177–1181 (1994).

76. P. M. Thibado, G. S. Rohrer, and D. A. Bonnell, Experimental and simulated tunneling spectra of the polar ZnO surfaces, *Surf. Sci.* **318**, 379–394 (1994).

77. W. Göpel, R. S. Bauer, and G. Hansson, Ultraviolet photoemission studies of chemisorption and point defect formation on ZnO non-polar surfaces, *Surf. Sci.* **99**, 138–158 (1980).

78. M. A. Barteau, Site requirements of reactions on oxide surfaces, *J. Vac. Sci. Technol. A* **11** (4), 2162–2168 (1993).

79. K. H. Ernst, A. Ludviksson, R. Zhang, J. Yoshihara, and C. T. Campbell, Growth model for metal films on oxide surfaces: Cu on ZnO (000$\bar{1}$)-O, *Phys. Rev. B* **47** (20), 13782–13796 (1993).

80. M. Bäumer, D. Cappus, H. Kuhlenbeck, H.-J. Freund, G. Wilhelmi, A. Brodde, and H. Nedder-meyer, The structure of thin NiO(100) films grown on Ni(100) as determined by low-energy electron diffraction and scanning tunneling microscopy, *Surf. Sci.* **253**, 116–128 (1991).

81. M. I. Buckett and L. D. Marks, Electron irradiation damage in NiO, *Surf. Sci.* **232**, 353–366 (1990).

82. P. H. Holloway and J. B. Hudson, Kinetics of the reaction of oxygen with clean nickel single crystal surfaces. I. Ni(100) surface, *Surf. Sci.* **43**, 123–140 (1974).

83. M.-T. Liu, A. F. Armitage, and D. P. Woodruff, Anisotropy of initial oxidation kinetics of nickel single crystal surfaces, *Surf. Sci.* **114**, 431–444 (1982).

84. L. Eierdal, F. Besenbacher, E. Lægsgaard, and I. Stensgaard, Interaction of oxygen with Ni(110) studied by scanning tunneling microscopy, *Surf. Sci.* **312**, 31–53 (1994).

85. W.-D. Wang, N. J. Wu, and P. A. Thiel, Structural steps to oxidation of Ni(100), *J. Chem. Phys.* **92** (3), 2025–2035 (1990).

86. L.-C. Dufour, A. El Ansarri, P. Dufour, and M. Vareille, Effect of thermal segregation on surface properties and reactivity of chromium doped nickel oxide, *Surf. Sci.* **269/270**, 1173–1179 (1992).

87. M. A. Langell and M. H. Nassir, Stabilization of NiO(111) thin films by surface hydroxyls, *J. Phys. Chem.* **99**, 4162–4169 (1995).

88. C. Xu and D. W. Goodman, Surface chemistry of polar oxide surfaces: Formic acid on NiO(111), *J. Chem. Soc. Far. Trans.* **91** (20), 3709–3715 (1995).

89. M. Yoshimoto, H. Okhubo, N. Kanda, and H. Koinuma, Two-dimensional epitaxial growth of SrTiO$_3$ films on carbon-free surface of Nb-doped SrTiO$_3$ substrate by laser molecular beam epitaxy, *Jpn. J. Appl. Phys.* **31**, 3664–3666 (1992).

90. P. H. Holloway and J. B. Hudson, Kinetics of the reaction of oxygen with clean nickel single crystal surfaces. II. Ni(111) surface, *Surf. Sci.* **43** (1), 141–149 (1974).

6

In Situ Processing by Gas or Alkali Metal Dosing and by Cleavage

Piero A. Pianetta

1. Introduction

The purposes of this chapter are to discuss the techniques for controllably dosing surfaces in ultrahigh vacuum (UHV) with gas molecules, for the conceptually related technique of alkali metal dosing, and for preparing single-crystal surfaces by cleavage.

2. Gas Adsorption

One of the most common experimental techniques used to study the interaction of molecules with surfaces is the adsorption of gas molecules on a clean surface. In the simplest version of this technique, a clean or otherwise prepared sample is exposed to a given pressure of gas for a fixed amount of time. The amount of molecules adsorbed on the surface is then a function of the dosage and *sticking coefficient*, s, of the particular molecule with the surface of interest.

When a surface is exposed to, for example, oxygen at a pressure P, the number of gas molecules hitting the surface per unit time and per unit area is approximately

Piero A. Pianetta • Stanford University, Stanford Synchrotron Radiation Laboratory, Stanford, CA 94309.

Specimen Handling, Preparation, and Treatments in Surface Characterization, edited by Czanderna *et al.* Kluwer Academic / Plenum Publishers, New York, 1998.

$3.6 \times 10^{20} P$ (Torr) cm^{-2} s^{-1}.[1] Therefore, if a surface were exposed to a gas at a pressure of 10^{-6} Torr for 1 s, the surface would be struck by 4×10^{14} molecules/ cm^{-2}s^{-1}. The fractional monolayer coverage of the surface depends on the number of adsorption sites per unit area (approximately 10^{15} sites/cm^2). This simple consideration has led to the measurement unit called the Langmuir, which is commonly used to describe gas dosages. Exposures are typically defined in Torr-s or Langmuirs, where 1 Langmuir is equal to 10^{-6} Torr-s, so 2.5 s are required to cover 10^{15} sites/cm^2 at $s = 1$. Thus, the exposure in Langmuirs, the sticking coefficient, and the available surface sites determine the time it takes to cover a surface by a monolayer at a given base pressure of a background gas in a UHV chamber. Therefore, a base pressure of 10^{-10} Torr, for a reactive surface, will give a monolayer of contamination in 2.5×10^4 s.

The simplicity of this dosing technique accounts for its common use. Using a standard ion gauge found in all UHV systems and with the addition of a simple leak valve, it is possible to achieve dose ranges from 0.1 to over 10^9 Langmuir for common gases such as O_2, CO, H_2, etc. The base pressure of the system ultimately determines the lower limit. For a base pressure of 10^{-11} Torr, which is common for surface research systems, a dosing pressure of 10^{-9} Torr would result in a maximum of 1% contamination from the background gases for a dosing gas with $s = 1$. Again keeping to the simplest version of the possible apparatus which would use a manual leak valve, the time needed for stabilizing the exposure pressure can be about 5 s, whereas the time to close the leak valve is less than 2 s. Therefore, to keep the errors due to time within 10%, exposure times of 100 s or larger are typical. This simple consideration results in the practical lower limit given above; that is, an exposure at 10^{-9} Torr for 100 s gives 10^{-7} Torr-s or 0.1 Langmuir. Faster exposure times are possible but have not been extensively employed.[2] In addition, it is often desirable to throttle the ion pump by using a gate valve to keep the ion pump at a lower pressure than the rest of the chamber. This offers two benefits: the ion pump stays very close to its base pressure and ensures rapid pumpdown when the exposure is complete. Furthermore, the regurgitation of pumped gases by the ion pump due to the added gas load is reduced.

For surfaces such as semiconductors where the sticking coefficients are in the 10^{-3} to 10^{-6} range, the use of an ion gauge to measure pressure has been shown to significantly increase the adsorption rate of oxygen.[3,4] In such cases, cold cathode gauges have been shown to be effective at low exposure pressures; a wide range of thermocouple gauges or capacitance manometers have been used at higher exposure pressures. Typically, semiconductors require very high exposures to obtain significant fractions of a monolayer adsorbed on the surface. For example, GaAs requires exposures of over 10^6 Langmuir before significant effects can be detected in the photoemission spectra, and monolayer coverages are not achieved until an exposure of 10^{12} Langmuir is reached. The pressures required to obtain reasonable exposure times are typically greater than 10^{-3} Torr and thus require that the ion pump be completely valved off from the rest of the chamber or the sample to be dosed in an ancillary preparation chamber that can be isolated from the measurement chamber. In such cases, it is desirable to use a turbomolecular pump to evacuate the chamber rapidly upon completion of the exposure. Contamination rates from background gases can be maintained at acceptable levels often by flowing the gas through the chamber. This can be accomplished by pumping on the chamber through a conductance barrier (such as a partially closed valve or a long tube) with a turbomolecular pump and maintaining the pressure at the desired level by increasing the flow through the leak valve.

The simple valve control techniques have and are still being used in many applications. However, there are a number of disadvantages when a high chamber pressure is needed to expose the sample. The base pressure can become degraded after the whole chamber has been exposed to a relatively high gas pressure. In addition, outgassing from the chamber walls, manipulator parts, or heated surfaces such as filaments can provide additional uncontrolled gas influx. This is especially troublesome for thermally stimulated desorption experiments where desorption from the back surface of the sample provides an information signal that is not desired. As a result, a number of directional dosing techniques have been developed to enhance the exposure of the sample surface while reducing the overall pressure in the chamber.

The use of directional gas dosers to provide an enhancement of the gas flux onto the sample surface over the background gas has been explored by a number of workers. The ideal doser is a differentially pumped molecular beam;[5] however, such systems are unduly complicated, and acceptable solutions have been achieved by using simpler designs. The three designs most commonly used are a cosine emitter,[6] a single capillary,[7] and a multichannel array.[8] The primary goals of a doser are to enhance the flux and to provide a uniform dose across the surface. In this section, the characteristics of these doser designs are discussed, as are implementation and calibration methods. Finally, variations of the simple designs will be described, which allow for special experimental configurations and the use of reactive gases.

To understand the gas flow through any of the different doser designs, it is instructive to examine the basic assumptions. First, the gas is in the molecular flow regime; i.e., the mean-free path of gas molecules is much larger than the dimensions of the doser. Discussion of the high-flux collisionally opaque regime is beyond the scope of this work, and the reader is referred to the literature for that treatment.[9] In the molecular flow regime, gas molecules undergo diffuse reflection from the walls, resulting in a cosine angular distribution of flux from each element of wall area. Under these conditions, Clausing showed that while the flux distribution from a hole in the side of a thin wall has a cosine distribution, the distribution from a short tube is peaked in the forward direction.[10,11] The degree of peaking is a function of the ratio of the length, L, to radius, R, of the tube. His treatment was extended by others to include parameters of use for doser applications.[8,10,12–15] These angular distributions are shown in Fig. 1. Even a thin plate that emits with a cosine distribution provides a flux enhancement because molecules are preferentially emitted along the axis of the hole. As L/R increases, the thin plate turns into the single capillary doser and the angular distribution becomes more and more peaked, resulting in increased flux enhancement and a reduced uniformity across the surface.[7,8,16–18] Several capillaries can be arranged in an appropriate pattern to increase the uniformity.[19] The limit of using many single capillaries is a multi-capillary array, which consists of thousands of small channels and yields excellent uniformity across the sample and good flux enhancement.[8]

A typical implementation of a short-tube emitter in a dosing application is shown in Fig. 2. It is worthwhile describing this implementation in some detail because the basic design elements also apply to the other dosers. Gas is supplied from the primary source to the reservoir through valve V_1 at a pressure P_r. Closing V_1 and opening V_2 permits the gas to flow through a capillary to restrict the flow into the cylindrical effusion source, which has an orifice mounted on its end. The orifice, a short tube with $L/R = 1$, directs the gas molecules onto the sample surface. Dosing can be terminated by closing valve V_2 and opening valve V_3 for pumping. This general configuration has typically been used because the ability to select the

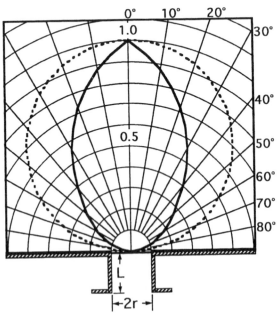

Figure 1. Cosine θ flux distribution from a small orifice (dotted line) and peaked angular distribution (solid line) from an orifice of length L and radius R.[10]

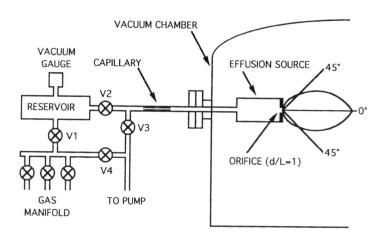

Figure 2. Arrangement for dosing with a peaked distribution from a short-tube emitter (effusion source) in an UHV chamber and attendant valves, gauges, pump, capillary flow restrictor and gas sources.[6]

pressure in the reservoir prior to dosing provides a degree of control not possible if the effusion source were fed directly through a leak valve.

To gain a quantitative understanding of the factors influencing the number of gas molecules that are emitted from the cosine emitter, it is instructive to consider the idealized example in Fig. 3. Three chambers are connected by apertures of different sizes that represent the reservoir with flow restriction, effusion source with emitter orifice, and sample chamber with a pump of speed S, respectively. Valves have been eliminated for simplicity.

The total number of molecules emitted by this source in the molecular flow regime, i.e., in which the mean-free path of the gas molecules is much greater than L or R, is given by[9]

$$N_{tot} = KA_e(N_A/2\pi Mk_bT)^{1/2}P_e \tag{1}$$

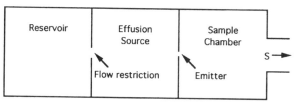

Figure 3. Schematic arrangement of a gas reservoir, flow restriction, effusion source, emitter, sample chamber, and pumping direction.

where P_e is the pressure in the effusion source (unless otherwise noted, pressures are in dyne/cm^2, where 1 Torr = 1.33×10^3 dyne/cm^2), A_e is the area of the effusion source (πR^2), M is the molecular weight of the gas molecule, N_A is Avogadro's number (6.02×10^{23}), k_b is Boltzmann's constant (1.38×10^{-16} erg/K), and T (K) is temperature. An approximate form of the Clausing factor, $K = 1/(1 + 3L/16R)$, takes into account the geometry of the emitter, so Eq. (1) can be applied to capillary sources as well. For a hole in a thin plate, $L/R << 1$ and $K \cong 1$.

The only unknown in Eq. (1) is P_e, which can be readily calculated from the system parameters:[1]

$$P_e = \frac{C_r P_r}{C_r + C_e S/(C_e + S)} \cong P_r \left(\frac{C_r}{C_e} \right) = P_r \left(\frac{A_r}{A_e} \right) = P_r \left(\frac{R_r}{R} \right)^2 \tag{2}$$

where C_r is the conductance of the flow-restricting aperture of radius R_r, C_e is the conductance of the emitter orifice of radius R, and $C_r << C_e << S$. The conductance for a simple aperture (i.e., $L = 0$), in l/s, is

$$C_x = (k_b T N_A / 2\pi M)^{1/2} A_x = 3.64 A_x (T/M)^{1/2} \quad (\text{l/s}) \tag{3}$$

where A_x is the area of the aperture. Combining the approximations of Eq. (2) with Eq. (1) gives

$$N_{\text{tot}} \cong A_r P_r (N_A / 2\pi M k_b T)^{1/2} \tag{4}$$

To complete the analysis, we calculate the enhancement factor, E, defined as the ratio of the total number of gas molecules striking the sample surface to those striking the sample which originate from the background pressure in the chamber. This is clearly an important measure of doser performance. The flux of molecules striking the sample directly from the doser is

$$F_{\text{doser}} = f N_{\text{tot}} / A \tag{5}$$

where f is the fraction of gas molecules coming from the doser that strike the sample, and A is the area of the sample. The quantity f is a measure of the efficiency of a particular doser configuration and is a function of geometrical factors such as the angular distribution of the gas molecules being emitted from the doser, the doser-to-sample distance, and the sample area. It will be seen that values of f between 0.3 and 0.5 tend to provide an acceptable balance between doser efficiency and uniformity of dose across the sample. From Eqs. (4) and (5), the corresponding pressure, P_{doser}, on the sample from this flux is

$$P_{\text{doser}} = (2\pi M k_b T/N_A)^{1/2} f N_{\text{tot}} / A \cong f(A_r / A) P_r \tag{6}$$

The portion of the flux contributing to the background pressure consists of the molecules missing the sample completely, $(1 - f)N_{\text{tot}}$, and the molecules striking the sample that are reflected, $(1 - s)fN_{tot}$. Therefore, the total number of molecules contributing to the background pressure is

$$N = (1 - sf)N_{tot} \tag{7}$$

Since the number of molecules entering the chamber must equal the number of molecules being pumped out of the chamber at a pumping speed S and pressure P, we may write[1]

$$N = PS/k_bT = (1 - sf)N_{tot} \tag{8}$$

Therefore, the background pressure, P, in the sample chamber due to dosing and the corresponding flux, F, hitting the sample are, respectively,

$$P = (1 - sf)(k_bT/S)N_{tot} \tag{9}$$

and

$$F = (N_A/2\pi Mk_bT)^{1/2}P \tag{10}$$

The effective pressure on the sample, P_{eff}, from the doser and the background pressure due to dosing is

$$P_{eff} = P_{doser} + P = (2\pi Mk_bT/N_A)^{1/2}fN_{tot}/A + (1 - sf)(k_bT/S)N_{tot} \tag{11}$$

The corresponding flux, F_{tot} is

$$F_{tot} = (N_A/2\pi Mk_bT)^{1/2}P_{eff} \tag{12}$$

We can rewrite P and P_{eff} in terms of the system parameters as

$$P \cong (1 - sf)C_rP_r/S \tag{13}$$

$$P_{eff} \cong f(A_r/A)P_r + (1 - sf)C_rP_r/S \tag{14}$$

The enhancement factor, E, which is the ratio of P_{eff} to P, can be written, from Eqs. (3), (13), and (14), as

$$E = 1 + fS(2\pi M/k_bTN_A)^{1/2}\{(1 - sf)A\}^{-1}$$

$$= 1 + 0.275fS(l/s)(M/T)^{1/2}\{A(cm^2)(1 - sf)\}^{-1} \tag{15}$$

Using typical parameters, $P_r = 1$ Torr, $R_r = 1$ μm, $R = 0.5$ cm, $A = 1$ cm^2, $S = 200$ l/s, $M = 32$, $T = 300$ K, $s = 0.2$, and $f = 0.5$, we find that $P_e \cong 4 \times 10^{-6}$ Torr, $P_{eff} \cong 1.73 \times 10^{-8}$ Torr, $P \cong 1.58 \times 10^{-9}$ Torr, $N_{tot} \cong 1 \times 10^{13}$ molecules/s, and $E = 11$. Under these conditions, a dose of 1 Langmuir is achieved in 58 s. The enhancement factor of 11 due to the higher effective pressure on the sample with respect to the background pressure in the chamber is one of the major justifications for using a doser. The enhancement factor is a very useful parameter when comparing different doser configurations used under identical conditions. However, when the system specific variables used in Eq. (15), i.e., the pumping speed, sticking coefficient, or sample area, are changed because a different sample or vacuum system is being used, the enhancement factor can vary by over a factor of 2 for identical dosers. The sticking coefficient could also change for the same sample during an experiment since it is a function of adsorbate surface coverage. Therefore, when comparing the enhancement factors cited by different researchers, it is important to ensure that the system-specific variables are identical. If they are not, meaningful comparisons can only be made after they are normalized by using Eq. (15). Note that f is a doser-specific variable that can be used to compare doser efficiencies without normalization and is, therefore, a conceptually simpler quantity than E for making comparisons.

This simple treatment was presented as a concrete example with actual numbers to give the reader a more intuitive grasp of the problem. The configuration chosen also allows one to obtain a quick assessment of the performance of a particular doser and to understand where gains can be made in the physical design of the entire dosing system. The simple treatment can also be extended to other experimental arrangements such as a capillary flow restrictor (instead of an aperture) or a single-capillary doser by using the appropriate expressions for the conductance of these components. A multicapillary array can be treated as a single capillary multiplied by the ratio of the area of the entire array to the area of a single channel.

However, to compare quantitatively the performance of the different types of dosers and determine f, which is a necessary input to the previous analysis, it is necessary to calculate the angular distribution of the flux from different dosers. The same calculation will also provide the flux uniformity across the sample surface. This analysis follows the treatment by Campbell and Valone[8] and Madey.[6] Consider Fig. 4, which illustrates a multichannel array doser of diameter d and thickness L at a distance x from a sample of diameter D. The radius and length of each individual channel are R and L, respectively. The angle θ is measured from the axis of the doser, and θ_{max} is the angle from the center of the doser to the edge of the sample. The doser to sample distance, x, is defined in terms of θ_{max}; i.e., $x = D/\{2 \tan(\theta_{max})\}$. The multichannel array is taken to consist of an infinite number of capillaries with $L/R = 40$, which has been shown to be the limiting case beyond which no further changes in the angular distributions take place.[8] Typical values of R and L are 10 and 500 μm, respectively. By varying d/D from 0 to a number greater than 1, it is possible to vary the conditions from that of a single-capillary doser to that of a multichannel array of diameter larger than the sample.

The normalized angular distribution of gas molecules emerging from an individual capillary, $F(\theta)$, can be integrated over the surface of the multicapillary array to give the total angular distribution of a particular doser configuration, $F_{tot}(\theta)$. For the details of these calculations, the reader is referred to the work of Campbell

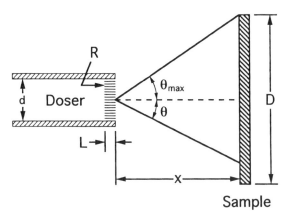

Figure 4. Arrangement for using a multichannel doser of diameter d with an "infinite" number of capillaries of length L and radius R at a distance x from a sample of diameter D.[8]

and Valone,[8] when $d \leq D$, and Winkler and Yates,[15] when $d \geq D$. Although these calculations are based on earlier work,[12,13,20-22] the more recent work developed the analysis into a form useful for actual dosing applications. The fraction of the total flux intercepted by the sample is calculated by integrating $F_{tot}(\theta)$ from zero to θ_{max} and then by dividing that result with the integral of $F_{tot}(\theta)$ from $0°$ to $90°$. These results are shown in Fig. 5, where the fraction of the total flux striking the sample, f, and the enhancement factor, E, are plotted as a function of θ_{max} and $x/D = 1/(2 \tan \theta_{max})$ for a simple cosine emitter, single capillary (i.e., $d/D = 0$), and multichannel arrays with $d/D = 1.0$ and 1.5. By making θ_{max} the independent variable, it is possible to choose an arbitrary doser-to-sample distance for any given sample size. The factors E and f are related through Eq. (15) using the following system specific variables: $S = 200$ l/s, $s = 0.2$, and $A = 1$ cm^2. Note that f is a function of only the doser and sample geometries, whereas E is highly dependent on the system-specific parameters, including the sample sticking coefficient, s. The curves in Fig. 5 show that the maximum theoretical enhancement for all dosers with $d/D \leq 1$ has a value of 21 obtained at a zero doser-to-sample distance; i.e., $\theta_{max} = 90°$. For $d/D > 1$, the maximum value of f is determined by the ratio of the sample-to-doser area, which must always be less than 1, and E is correspondingly less than the maximum value calculated for $d/D \leq 1$. For example, the value of E for $d/D = 1.5$ at $\theta_{max} = 90°$ has a value of approximately 10, which is half that obtained for $d/D = 1$. As the doser-to-sample distance is increased, θ_{max} goes to zero, f goes to zero, and E goes to 1; i.e., this is the limit of the whole chamber exposure. Although a single needle doser gives the best value of E for all doser-to-sample distances, the dose across the sample is highly nonuniform, as will be discussed. However, Glines has shown that an array of as little as seven single-needle dosers can be designed to give a high

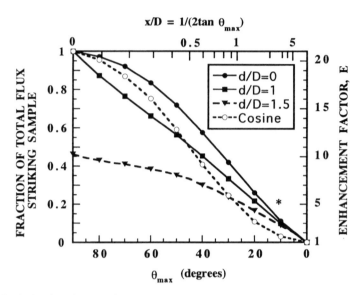

Figure 5. The fraction of the total flux striking the sample, f, and enhancement factor, E, are plotted as a function of θ_{max} and $x/D = 1/(2 \tan \theta_{max})$ for the cosine emitter, single capillary, and multichannel array with $d/D = 1.0$ and 1.5. The single f-value for the Glines emitter is plotted as an asterisk.[8,15,19]

value of E at a relatively large doser-to-sample distance with excellent uniformity.[19] This special design, which will be discussed in more detail, gives $f = 0.21$ and $E = 5$ at $\theta_{max} = 11°$ when normalized to the system-specific parameters used here ($E = 10$ for the system-specific parameters used in the original work, $S = 300$ l/s, $s = 1$, and $A = 0.8$ cm^2). This result, shown in Fig. 5, is a factor of 2 better than the single- or multiple-capillary dosers and a factor of 5 better than the cosine emitter at $\theta_{max} = 11°$. The Glines doser was specifically designed for best performance at $\theta_{max} = 11°$. At larger values of θ_{max}, f and E will approach the values of the single-needle doser becoming equal to them at $\theta_{max} = 90°$ (by definition). This example illustrates the importance of considering nonstandard doser geometries when encountering new experimental configurations.

Campbell and Valone[8] have analyzed the flux uniformity across the surface of a sample, $F_{tot}(\theta)$, for the cosine emitter, single capillary, and multichannel array with doser-to-sample distances corresponding to $\theta_{max} = 80°, 60°, 40°$, and $20°$. For a 1-cm-diameter sample, these values of θ_{max} correspond to doser-to-sample distances of 0.09, 0.29, 0.60, and 1.37 cm, respectively. These results are summarized in Fig. 6, which shows the fraction of the sample, in units of sample diameter, that is exposed to a uniform flux versus θ_{max} and x/D for the single capillary, the multichannel array with $d/D = 1$, the cosine emitter, and the seven single-capillary array of Glines. In this context, a uniform flux is defined to be a variation in flux across the sample of less than ±10%. Therefore, the vertical axis in Fig. 6 is simply the fractional sample diameter exposed to a flux greater than 80% of that at the center of the sample. In addition, the corresponding enhancement factor, E, is listed for each point in the figure. Figure 6 shows that for all plotted values of θ_{max}, the single-capillary doser gives a very nonuniform flux distribution in which most of the flux is concentrated within an area that is less than 10% that of the sample. This situation is clearly not desirable even though the enhancement factor is higher than those for other types of dosers. The uniformity can be significantly improved by forming arrays of single-capillary dosers as shown in the emitter design of Glines. As might be expected, the cosine emitter is also highly nonuniform for small doser-to-sample distances. Even at $\theta_{max} = 60°$, the fraction of the sample exposed to a uniform flux is only 0.45 the sample diameter. However, as θ_{max} becomes smaller than 45°, i.e., a sample-to-doser distance greater than one-half the sample diameter, the flux becomes uniform over more than 90% of the sample diameter while still providing an excellent enhancement factor of 11. In fact, for $\theta_{max} < 40°$, the fractional area over which the flux is uniform is greater than that of the sample. At doser to sample distances greater than the sample diameter (i.e., $\theta_{max} < 25°$), the flux uniformity becomes very high (±5%), but $E < 3$. The multicapillary array with $d/D = 1$ provides a very uniform flux across 90% of the sample diameter at a

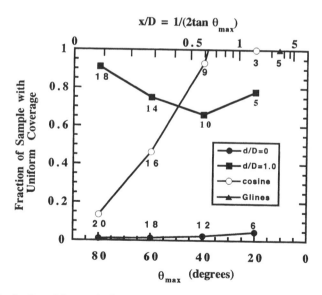

Figure 6. The fraction of the sample diameter exposed to a uniform flux is plotted as a function of θ_{max} and x/D for the single capillary, the multichannel array with $d/D = 1$, the cosine emitter, and the seven, single-capillary array of Glines. The enhancement factor, E, is listed on the graph for each point.[8,19]

doser-to-sample distance of $0.1D$ ($\theta_{max} = 80°$) with an enhancement factor of over 18. A more practical doser-to-sample distance of $0.3D$ ($\theta_{max} = 60°$) avoids multiple collisions between the doser and the sample surface but will still give a uniform flux over 75% of the sample diameter and an enhancement factor of over 14. In the range where $\theta_{max} \cong 45°$, the cosine emitter actually performs as well as the large multicapillary array, giving a larger area of uniform flux across the sample with similar enhancement factors. At larger doser-to-sample distances, the large multi-capillary array with $d/D \geq 1$ again outperforms the cosine emitter, giving a uniform flux across 80% of the sample diameter with twice the enhancement factor. The array designed by Glines outperforms both the cosine emitter and the multicapillary array at large doser-to-sample distances ($\theta_{max} \cong 11°$), giving uniform flux over the entire sample diameter with an enhancement factor twice that of the cosine emitter.

Campbell and Valone conclude that the best choice of doser design that yields a large enhancement factor and a small flux gradient is a multichannel array with a diameter slightly larger than that of the sample positioned $0.1D$ from the sample. When the experimental configuration does not allow such small distances or a large doser such as is needed for the multicapillary array, the cosine emitter at a doser-to-sample distance of $0.5D$ provides more than adequate performance.[8] Also, for large doser-to-sample distances, an array of appropriately arranged single-capillary dosers designed using the approach of Glines should be considered.[19]

Let us now examine several specific doser designs. The different designs typically share the elements discussed above in the description of the cosine emitter: a gas reservoir, a flow restrictor, and the emitter. Referring to Fig. 2, the gas reservoir is connected to the main gas supply, the gas line to the chamber, and the turbo-molecular pump by Nupro valves (Swagelok Companies, Cleveland, OH). In many cases, the reservoir is simply the volume of the gas line between the gas supply and chamber. Since the pressure in the reservoir is typically 1 Torr, a capacitance manometer is used to accurately measure this pressure. The flow restrictor serves as the conductance barrier between the reservoir and the emitter. Several different designs have been used for the flow restrictor, including stainless steel or glass capillaries and small orifices. The main characteristic of the flow restrictor is that it have a conductance between 10^{-6} to 10^{-8} l/s.

In one example of a flow restrictor, a 2-μm-diameter orifice in a 280-μm-thick Ni–Cu bimetal sheet (Buckbee Mears Co., St. Paul, MN) was mounted inside a Cajon VCR fitting (Swagelok) between two Cu or Ni gaskets.[23] This gives a conductance of 10^{-5} l/s and allows a pressure on the order of 1 Torr to be used in the reservoir. Difficulties have been experienced in the actual mounting of the aperture due to rotation of the gaskets. This has led to one scheme in which the two gaskets were riveted together.[23] In another scheme to overcome this same problem, a 5-μm aperture is used that has been laser-drilled into a solid Ag-plated stainless steel Cajon VCR gasket (SS-8-VCR-2-BL) whose center has been machined to a thickness of 250 μm. This latter method requires less skill in assembly and its only drawback is the custom fabrication of the apertures. The effective area of the aperture was determined through a calibration method that will be discussed. The second example of a flow restrictor consists of a 20-μm-internal-diameter, 40-mm-long capillary mounted in a Cajon VCR fitting. This gives a conductance of 10^{-8} l/s and thus needs higher pressures in the reservoir. Both types of orifices have been reported to operate with similar levels of success, with the one difference being that the aperture design will be somewhat simpler to fabricate. Some researchers have protected the restriction orifice with a micron filter in the gas line.[19]

The flow restrictor can be replaced with a precision leak valve. In this case, the dosing tube can be brazed into a solid copper gasket that is then mounted directly to the leak valve, which in turn is mounted on a linear translator to allow a variable doser-to-sample distance. However, this method requires that the pressure in the chamber be measured with the sample facing away from the doser to first set the conductance of the leak valve, and thus the flow rate, before the sample can be rotated into the gas flow for the actual exposure. For some applications, the operational difficulty is compensated by the fabrication simplicity. For many cases, the convenience or necessity of keeping the sample in a certain position during dosing makes the use of the restriction aperture necessary.

The actual doser is then connected to the restriction orifice. This consists of an effusion source which can be as simple as a 5-mm ID tube onto which is attached the emitter. The response time of the doser is approximately the ratio of the tubing volume downstream of the flow restriction to its conductance and scales as the square of the tubing length over its radius. Therefore the tubing should be as wide and short as possible. Acceptable response times have been obtained with 25-cm-long, 5-mm ID tubing.[19] A typical doser mounting for either the cosine emitter or the multicapillary array is shown in Fig. 7. This example illustrates the use of a restriction orifice that is mounted in a Cajon VCR fitting. In addition, the emitter is a multichannel plate of diameter greater than that of the main tubing. Also note that a baffle is installed between the supply tubing and the emitter to ensure that the supply tubing does not collimate the gas molecules. For the case of the cosine emitter, a thin foil (0.04 mm) with a small hole (0.3 mm) can be brazed onto the end of the doser tube. This configuration closely approximates a cosine emitter.[8] However, a "short tube" can also be used with $L/R \leq 1$, which is again actually a hole in a thin plate. Madey *et al.* have used such an emitter with $L = R = 0.625$ mm.[6] The choice is typically dictated by the fabrication capabilities or preferences at a particular laboratory.

Interesting variations in emitter design have been recently described that allow special configurations. The first example due to Glines has already been introduced. Glines designed a doser that can be positioned at a 45° angle to the sample normal, and that allows a mass spectrometer to be positioned directly in front of the sample during dosing.[19] This special experimental configuration requires a relatively large

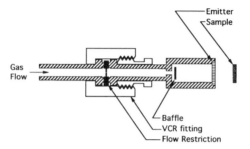

Figure 7. Typical mounting arrangement for either a cosine emitter or multicapillary array showing a restriction orifice mounted in a VCR fitting, a baffle to prevent collimation of the gas molecules by the supply tubing, and a multichannel plate emitter of diameter greater than the main tubing.[23]

doser-to-sample distance for which the standard emitter designs have a low efficiency (i.e., f value). The Glines emitter consists of six tubes, 1.5 mm in diameter, placed in a hexagonal array around a seventh tube, 0.89 mm in diameter. The distance between the tubes is approximately equal to their diameter. The precise placement and diameter of these tubes are chosen to give a calculated uniformity of ±10% over a 6-mm sample at a 15.5-mm doser-to-sample distance ($\theta_{max} = 11°$). Integrating the flux distribution over the sample area, Glines showed that $f = 0.21$, and $E = 5$ (when normalized to the system-specific parameters used here; $E = 10$ for the system-specific parameters used in the original work, $S = 300$ l/s, $s = 1$, and $A = 0.8$ cm^2). This result has also been plotted in Fig. 5. Comparing this f-value with the results for the other emitters in Fig. 5 for $\theta_{max} = 11°$, it is clear that this design is more efficient than the straight-tube emitter and multichannel array by a factor of 2 and the cosine emitter by a factor of 8. Furthermore, Fig. 6 shows that the Glines emitter also provides a high degree of dose uniformity over the whole sample area, again at $\theta_{max} = 11°$. Since this emitter design consists of single capillaries (i.e., $d/D = 0$), each with high efficiency but poor uniformity, good uniformity for an array can be achieved only at distances where there is significant overlap between the flux distributions of adjacent capillaries. Referring to the curve for $d/D = 0$ in Fig. 6 shows that the single-capillary doser maintains a narrow flux distribution until θ_{max} is smaller than 20°. However, the Glines emitter design does show that in cases where the experimental arrangement requires relatively large doser-to-sample distances, an array of single-capillary emitters can be designed to give both high efficiency and excellent uniformity. The reader is referred to the original reference for the particular details of the design.[19] Except for the novel emitter configuration, the doser is similar to the basic design in Fig. 2. In another novel emitter design, an array of 0.03-mm holes is directionally etched in a Si(100) wafer (0.3 mm thick, giving $L/R = 10$) with a density that increases radially toward the edge of the array. A numerical analysis shows that a high uniformity (<99%)

could be achieved over a large area since a 100-mm wafer could be used as the emitter.[24]

Reactive and low-vapor-pressure gases also pose special problems for doser design and operation. It is not possible to adequately treat all the possibilities that might be encountered, so a listing of representative examples will be provided. Although some materials will react with the stainless steel in the dosing system, it has been found that gases such as NO[25] and $Fe(CO)_5$[26] can still be used if the system is subjected to pretreatment. In the case of NO, the system is simply preexposed to 1×10^{-7} Torr of NO for 1 h, i.e., 360 Langmuir, and after that NO and O_2 may be used with minimal decomposition of the NO into N_2 and CO as measured by a mass spectrometer. In the case of $Fe(CO)_5$, the system is first dosed with oxygen at 10 Torr for 10 min, i.e., 6×10^9 Langmuir, then dosed with $Fe(CO)_5$ at 4 Torr for 10 min. This is found to deactivate most of the sites on the chamber and doser walls that decomposed the $Fe(CO)_5$. Even the glass multichannel arrays have been found to be highly reactive to materials such as ethers.[27] For materials such as hydrazine[28] or $TiCl_4$,[29] where passivation does not seem to work, dosing systems have been fabricated of materials consisting entirely of glass and polymers or Teflon, respectively. Finally, low-vapor-pressure gases, which may or may not be reactive, such as H_2S, cyclohexane, or benzene, also pose serious problems because they will readily deposit on the doser walls. In these cases, it is possible to continuously mix the adsorbate with an inert carrier gas such as Ar in the reservoir while pumping on it through a conductance barrier, thus ensuring that a fresh supply of the mixture is always available to the doser.[30]

Calibration of the flux directed onto the sample from the doser, F_{doser}, can be done in several ways. The method, which is conceptually the simplest, involves obtaining curves of the measured coverage as a function of exposure for both the doser method measured in terms of the reservoir pressure and the whole chamber exposure method measured in Langmuirs. From these data, a curve of reservoir pressure as a function of exposure in Langmuirs is obtained. The coverage is typically measured using X-ray photoelectron spectroscopy (XPS), Auger electron spectroscopy (AES), or temperature-programmed desorption (TPD). This method is direct and requires only a minimal knowledge of the system parameters. In addition, the uniformity in the coverage across the wafer can be determined with the AES measurements. However, the dose range over which the calibration may be performed is limited by the sticking coefficient that also varies with coverage. For example, oxygen exposures of a sample with an initial sticking coefficient of 1 such as W(100) will saturate at very low exposures, whereas samples such as GaAs require exposures of $>10^6$ Langmuir before any oxygen is detected. Therefore, calibration of the entire desired exposure range may often not be practical with this technique, and it is important to develop a calibration scheme based on the system parameters and independent of the sample.

From Eqs. (4) and (5), we see that the parameters that must be determined to get a calibration for F_{doser} are A, P_r, A_r, and f. The reservoir pressure, P_r, and sample area, A, are easily measured. In the range of 1–100 Torr, P_r can be measured to better than 2% accuracy with a capacitance manometer. The area of the flow restrictor, A_r, can be determined by plotting $\ln(P_r)$ as a function of exposure time. The slope of this curve is the inverse of the time constant, τ, which can be related to the effective area of the aperture, A_{eff}, by

$$A_{eff} = V(2\pi M/k_b N_A)^{1/2}/\tau \qquad (16)$$

where V, the volume of the reservoir, is measured by expanding a gas into it from a flask of a known volume and pressure. With a 5-μm flow restrictor, τ has been measured to be 1×10^5 s.[31]

For a given doser-to-sample distance, f can be determined either from Fig. 5 or experimentally using Eq. (13) and simple measurements.[6,8,19] In most cases, Fig. 5 is adequate since a number of comparisons in the literature for different dosing configurations have been in agreement with its results to about 10%. However, in those cases where there are no other internal calibrations to verify the exposure, and thus f needs to be determined explicitly, measurements using the oxygen on W(100) system can easily be made. Oxygen on W(100) is convenient for these measurements because an initial sticking coefficient of 1 is needed for the analysis and the oxygen can be removed after each measurement by flashing the crystal to 2500 K.[6] With the sample turned away from the doser, the valve from the reservoir is opened, thus activating the doser and quickly bringing the chamber pressure to an equilibrium value, P_{eq}. Since the sample is out of the beam, Eq. (13) will give the chamber pressure if s is set to zero; i.e., $P_{eq} = C_r P_r/S$. The sample is rotated to face the doser at a time t_0, and then the oxygen partial pressure drops immediately due to oxygen adsorption onto the sample surface. At t_0 the initial sticking coefficient of oxygen for a sample such as W(100) is 1, and Eq. (13) becomes $P(t_0) = (1 - f)C_r P_r/S$. Solving these two equations for f yields

$$f = \{P_{eq} - P(t_0)\}/P_{eq} \qquad (17)$$

The measurement is repeated for different doser-to-sample distances and can be compared to the plot of Fig. 5 for verification. Thereafter, this calibration can be used to obtain f if the doser-to-sample distance is known accurately.[8,19,32]

Now F_{doser} can be determined from these experimental values. The actual exposure, F_{tot}, is the sum of F_{doser} and F from Eqs. (5), (10), and (12). Because F is proportional to $1 - sf$ and s tends to zero as the sample surface becomes saturated, F will increase during the course of an exposure to a maximum value of $0.1F_{doser}$ for configurations where $E \cong 11$. Even for very reactive surfaces, which have the largest change in s (i.e., from 1 to 0), the errors caused by setting $s = 0$ in the expression for F are only 5%. Higher accuracy will clearly be achieved if the average value of s during the exposure is used in the expression for F. In cases where the initial sticking coefficient is less than 0.4, the effect is well within the other sources of error.

Another technique involving dynamic TPD measurements is potentially more accurate than the methods described, but it is much more complex and the reader is referred to the literature.[19]

The time-dependent sticking coefficient can be easily determined with an extension of this discussion. At a time, t, $P(t) = \{1 - s(t)f\} C_r P_r / S$ where $s(t)$ is the sticking coefficient at time t. Solving for s in the expression for $P_{eq} - P(t)$ gives[6]

$$s(t) = \{P_{eq} - P(t)\} / fP_{eq} \tag{18}$$

The average coverage, Θ at time t can then be calculated from

$$\Theta(t) = F_{tot} \int_{t_0}^{t} s(t)\, dt \tag{19}$$

This method is for $s(t) > 0.1$, which primarily applies to relatively reactive metal surfaces. For surfaces with smaller sticking coefficients such as semiconductors and metal oxide surfaces, where $s(t_0) < 10^{-3}$, the more direct route of determining actual coverage as a function of exposure is necessary.

3. Alkali Metal Adsorption

Because of their high vapor pressures, the adsorption of alkali metal atoms onto surfaces was originally treated much in the same manner as a classical gas exposure. For example, in the early studies of Cs adsorption onto GaAs,[33,34] a specific vapor pressure of Cs was established in the entire vacuum chamber and the sample exposed for enough time to achieve the desired exposure. Either Cs chromate channels or elemental sources were used to supply the Cs. The Cs chromate channels were of the type used as getter sources in vacuum tubes and were placed out of the sample's line of sight. Directional versions of these sources, called alkali metal dispensers, have become the preferred method of alkali metal dosing and will be discussed in detail. In the elemental sources, Cs was typically stored in a glass ampule, which was located within the vacuum chamber, and broken open when needed. In a very simple but effective scheme, the ampule was enclosed in copper tubing that could be isolated from the main chamber with an all-metal valve. After the tubing was evacuated and brought to a low pressure, the valve was closed and the ampule broken by compressing the copper tubing with pliers. A dose could be delivered by simply opening the valve and heating the ampule to increase the pressure in the entire chamber up to the desired value. The coverage of Cs on the sample was monitored by measuring the work function of the sample. Since the sticking coefficient of Cs to itself is very low at room temperature, the saturation coverage results in a single layer of Cs on the GaAs surface.

The technique described has obvious disadvantages because all the surfaces in the chamber were exposed and therefore coated with Cs, including the insulators. Improvements using directional dosers are possible,[35] but the necessity of breaking an ampule in the vacuum system results in a reduced level of control. Alkali metals are reactive and have high vapor pressures so that care must be taken in handling these materials. For example, the temperatures at which a given alkali metal has vapor pressures of 10^{-8}, 10^{-6}, and 10^{-4} Torr are as follows: Li (227°C, 307°C, 407°C); Na (74°C, 124°C, 192°C); K (23°C, 60°C, 125°C); Rb (−3°C, 37°C, 111°C); Cs (−16°C, 22°C, 30°C).[1,36] These data showing temperature versus pressure are plotted in Fig. 8. The high vapor pressures at room temperature of K, Rb, and Cs make it necessary that the source be sealed from the chamber when not in use. Furthermore, Li, Na, and K react violently when exposed to air, and it is recommended that gentle heating be used during outgassing to remove adsorbed gases.[36]

Fortunately, the need for controllable alkali metal sources for use in generating the photoemissive surfaces of photomultipliers, image intensifiers, camera tubes, etc., has resulted in significant developments that have significantly improved the sources available.[37] The new sources, known as alkali metal dispensers or simply dispensers, have proven invaluable to the surface science community. These developments have made the alkali metal source more like a thermal evaporator for a

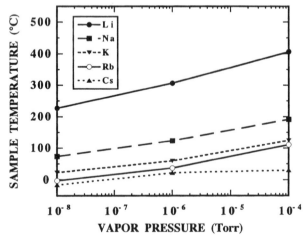

Figure 8. Plot illustrating the temperatures at which the alkali metals reach vapor pressures between 10^{-8} and 10^{-4} Torr.[1,36]

simple metal. The alkali metal dispensers consist of an alkali metal chromate (e.g., Me_2CrO_4) mixed with a reducing agent, such as Si, Zr, or Al. The alkali metal, Me, can be Li, Na, K, Rb, or Cs. A mixture of these powders will release alkali atoms when heated to temperatures between 550 and 850°C. The reducing agent now used in dispensers is a Zr84%–Al16% alloy that also acts as an effective getter to trap the gases emitted during the reduction reaction.[37–39]

The commercially available alkali metal dispensers, supplied by SAES Getters of Milan, Italy, consist of a nichrome tube with a trapezoidal cross section that holds the mixture of alkali metal chromate and reducing agent (Fig. 9). The metal tube has a slit along its length through which the alkali metal evaporates. In addition, to prevent any of the mixture from falling out of the tube and to increase the compression of the powders, a nichrome wire that partially obstructs the slit is fit inside the tube. End caps, which are welded onto each end of the tube, serve as terminals through which current can be passed to heat the tube. The tube is, in effect, an evaporation boat with precise dimensions and resistance, so the current is quite reproducible for any applied voltage. Evaporation typically takes place at 600–800°C, depending on the alkali. These temperatures are achieved when a current of 4 to 7 A passes through the dispenser.

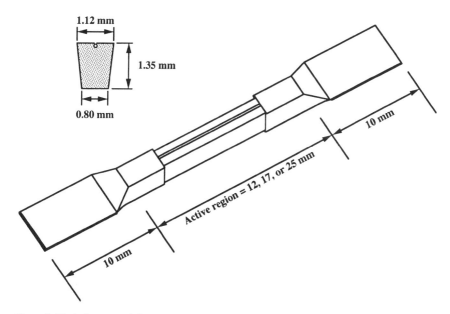

Figure 9. Typical commercially available alkali metal dispenser of "FT" type with a trapezoidal cross section and nichrome tube to hold the mixture of alkali metal and reducing agent.[37]

Although, it is possible to obtain dispenser sources of arbitrary length, the standard sources of "FT" type with integral terminals have lengths of 12, 17, or 25 mm, with the terminals adding 10 mm on each end. With these lengths, it is possible to fabricate a simple alkali metal source by spot welding one or more dispensers to a multipin current feedthrough mounted on a Conflat flange (Varian Associates, Palo Alto, CA). Given the relatively low currents needed for evaporation, the requirements for the current feedthroughs are quite modest. The dispensers can be handled in atmosphere for mounting; however, they should be stored in a dessicator because prolonged exposure to moisture will damage the dispensers. The manufacturer also recommends that latex or rubber gloves be used rather than cotton or nylon. These sources can be used in the same way as conventional evaporation sources in which shutters are used to control the flow of evaporant onto the sample surface. One important exception is that quartz crystal microbalances (QCM) are not useful for measuring the evaporation rate because of the low sticking probability of the alkali metals. As a result, calibration of the dispenser using a physical property of the sample itself such as work function, LEED patterns, temperature-programmed desorption, or XPS signal from the adsorbate or substrate should be used. These indirect methods have the advantage that the sticking coefficient of the alkali metal on the sample is automatically taken into account. It is possible to use a line-of-sight mass spectrometer to calibrate the rate of alkali metal evolution, but it leaves open the question as to how much of the metal sticks to the sample surface.

In photocathode applications, the presence of gases during or after fabrication can permanently damage the photosensitive surface being created. As a result, the evolution of gases from alkali metal dispensers has been studied extensively.[37,40] As mentioned, ZrAl alloys are used as the reducing agents primarily because of their gettering properties, which serve to eliminate most of the gas evolution during alkali metal evaporation. After proper outgassing, to be described, the actual release of gases during alkali metal evaporation is quite small. Results published by SAES Getters, Inc., show that H_2, CO, and CO_2 are released in varying degrees by the dispensers during alkali metal evaporation.[37] Other reports also indicate that very small amounts of water are evolved.[40] Although various values have been given for the evolution of gases in these studies, the most useful measure for the surface scientist is the pressure rise in the sample chamber during evaporation. For chambers with base pressures below 7×10^{-11} Torr and typical pumping speeds from 220 to 400 l/s, the maximum reported pressure rise has been 5×10^{-10} Torr, with typical values being below 1×10^{-10} Torr (R. Cao and T. Kendelwicz, personal communications; also see Refs. 40 and 41). In one case where no increase in chamber base pressure was detected at 1×10^{-10} Torr, a slight increase in the partial pressure of hydrogen was observed.[42] In another case, well-outgassed dispensers of Na, K, and Cs being used in a chamber with a base pressure of 3×10^{-11} Torr could be routinely operated for up to 7 min without raising the chamber pressure above 6×10^{-11} Torr.[43] Even the maximum reported pressure of 5×10^{-10} Torr with an exposure time of 60 s results in a dose of less than 0.03 Langmuir in additional contamination, with more typical values being less. In most cases, this level of contamination can be tolerated. In fact, the more typical pressure increases of less than 1×10^{-10} Torr result in negligible levels of contamination.

Because of the well-defined dimensions of the alkali metal dispensers, the temperature of the dispenser at a given current is reproducible and constant independent of the alkali metal. Figure 10 shows the graph of dispenser temperature versus current through the dispenser.[38]

It is necessary to thoroughly degas the dispenser to deposit clean alkali metals. The degassing procedures are usually omitted from the literature, and the product brochure simply states that the maximum degassing temperature is 500°C, which corresponds to a current of approximately 3.5 A. In what follows, the experience from our laboratory will be described, which has been shown to give clean films for all the alkali metals except Li that has not yet been used. The conditions for Li have been taken from the literature. The dispensers are outgassed only after the chamber has been baked and has achieved its base pressure. There is a concern that outgassing the dispensers when the chamber is hot, which is done with standard filaments and evaporation sources, might actually contaminate the alkali source (Cao, pers. comm.). To avoid degrading the chamber base pressure, the current through the dispenser is increased slowly up to 4 A always so the chamber pressure

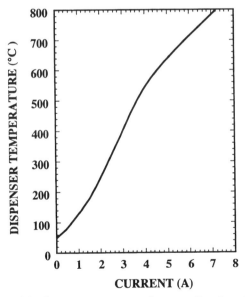

Figure 10. Dependence of the dispenser temperature on the current flow through the dispenser for an ambient temperature of 50°C (for the dispenser in Fig. 9 with an active length of 25 mm).[38]

remains below 5×10^{-10} Torr. Although the dispensers are relatively uniform in behavior, the time required to reach the target current of 4 A can take between 3 and 4 h. Note that this current gives a temperature that is somewhat above the manufacturer's "maximum degassing temperature." For the next 4 to 6 h, the current is maintained at 4 A until the pressure has stabilized below 2×10^{-10} Torr for at least 1 h. The dispensers can be successfully degassed faster by more rapidly increasing the current up to the target value of 4 A either by allowing the chamber pressure to rise to slightly higher values or by cycling the doser current to allow for the chamber pressure to recover between outgassing cycles (Cao, Kendelwicz, E. J. Nelson, pers. comm.). At this point, the dispensers are ready to use and all that remains to be done is to determine the currents and evaporation rates that are commonly used. In many cases, the application determines the exact conditions so that it will be useful to consider several specific applications.

In the production of III–V semiconductor surfaces with a negative electron affinity (NEA) for photocathode electron gun applications[44] (and possibly other NEA applications), it is well known that the surfaces with the highest electron yields result from repetitive dosing with Cs and oxygen.[45] After each Cs and oxygen dose, the photoelectron (PE) yield from a visible light source is monitored. The process is stopped when a maximum PE yield is reached. In a variation of this technique that is used in our laboratory, a constant Cs flux on the semiconductor surface is maintained while an oxygen pressure of 1×10^{-9} Torr is turned sequentially on and off to increase the photoelectron yield steadily until it saturates. In these applications, it is clear that a slow Cs evaporation rate is necessary to maintain a reasonable level of process control. The desired evaporation rate of one monolayer in 10 min is achieved at a current between 4 and 4.2 A with a sample-dispenser distance of 25 mm. At these currents, the pressure rise in the chamber resulting from outgassing of the Cs dispenser is less than 2×10^{-10} Torr. Because the currents used are so low, an additional step has been added to the degassing procedure. At the end of the long degassing process, the current through the dispenser is increased to 5–5.2 A for 3 to 5 min and then reduced again to 4–4.2 A. Besides driving out yet more gas from the dispenser, the temperature reached at 5–5.2 A of over 600°C has the effect of reactivating the ZrAl reducing–gettering agent that had probably become saturated during the long degassing process. After the current is applied, the sample is shielded from the dispenser for 40 to 60 s before each evaporation in which the dispenser temperature is increased from room temperature. This procedure allows the dispenser to reach a steady-state temperature and any adsorbed gases to be pumped away. Operation in this mode has resulted in over 100 evaporation cycles with the same dispenser. In most cases, the actual lifetime of the dispenser is not measured nor is it important because the dispensers are usually replaced every time the system is vented. Experience has shown that cleaner films are obtained with dispensers that have not been exposed to air after degassing.

In surface science applications in which clean alkali overlayers are deposited onto clean surfaces, evaporation times shorter than those for the NEA surface preparation are desired. Sodium, K, Cs, and Rb have been used in our laboratory for adsorption studies on both semiconductor and metal oxide surfaces. Lithium will be discussed separately because it has not been used and is not yet part of the common database. Table 1 shows the alkali metal content in the different dispensers and the evaporation currents normally used in our laboratory to achieve one monolayer of coverage on Si in 1 min for a 10-mm sample-dispenser distance (Kendelwicz, pers. comm.; Refs. 45, 46). In those experiments where sample-heating from the 10-mm sample-dispenser distance could result unwanted surface diffusion, larger sample-dispenser distances have been used. In one study on the adsorption of K, Na, and Cs on Si(111), a 35-mm dispenser-sample distance was used resulting in evaporation rates of 0.25 mL/min at 5.7 A for K and Na and 0.14 mL/min at 5.8 A for Cs (Nelson, pers. comm.; Ref. 43). Because of the difficulties in measuring the fluence of the alkali metals and because their sticking coefficients are so low, it is more practical to relate the current through the dispenser to the adsorption of the alkali metal on a particular surface rather than to a pure evaporation rate. Note that in all cases except Cs the evaporation currents used in practice are lower than the manufacturer's stated current at the onset of evaporation (Table 1).[46] This is probably a result of the insensivity of the measurement technique for determining the "onset of evaporation," which relies on measuring the sudden increase in current flowing in a diode placed near the dispenser. In order to estimate the number of evaporations obtainable from a dispenser, consider Fig. 11, which shows the mass flow from a Cs dispenser as a function of current as measured by the manufacturer using a sensitive atomic absorption technique.[38] This figure shows that an evaporation rate of 1 µg/min is observed at a current of 3.5 A but increases to 50 µg/min at a current of 5 A. As in the previous application, assume that the sample is shielded from the dispenser for 60 s after the current listed in Table 1 is applied with the onset of evaporation being detected by an increase in

Table 1. Alkali Metal Content and Evaporation Currents for the Alkali Metal Dispensers

Alkali metal	Content (mg)			Evaporation current (A)	
	$A = 12$ mm	$A = 17$ mm	$A = 25$ mm	Data sheets	Experimental
Lithium	0.8	1.1	1.7	7.3	5–7.2
Sodium	1.7	2.4	3.5	6.2	5.5
Potassium	2.9	4.1	6.0	5.8	5.5
Rubidium	4.5	6.4	9.4	5.3	5.3
Cesium	5.2	7.3	10.8	4.8	5.2

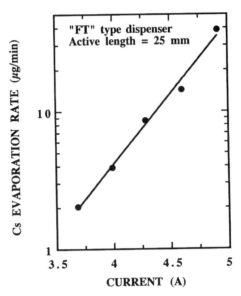

Figure 11. The dependence of the Cs evaporation rate with dispenser current for the dispenser in Fig. 9 with an active length of 25 mm.[38]

the chamber pressure. A simple calculation shows that the entire 5.2-mg content of a 12-mm-long dispenser can be consumed in approximately 50 one-minute evapo-rations (i.e., 2 min actual "on" time) when a 5-A current is used. This lifetime is typically more than adequate for most surface science applications.

Although Li dispensers are regularly used, there is less information available about their characteristics. As indicated in Table 1, the onset of evaporation from Li dispensers is at a current of 7.3 A, which is probably related to the lower vapor pressure of Li. This implies that degassing may be done at currents higher than the 4 A used for the other alkali atoms. The lowest evaporation current quoted in the literature is 5 A, which means that degassing currents of up to 4.75 A may be acceptable. The evaporation currents quoted in the literature are 5 to 7.2 A and yield an adsorption rate onto an Ag(111) surface of 3 monolayers/h at a 30-mm sample-dispenser distance.[42,47]

An alternative Li source has recently been described that consists of an Al–Li binary alloy disk containing Li at a concentration of 6.5%.[48,49] The disk is heated resistively to emit Li. Its temperature is measured directly with a K-type thermo-couple. The evaporator is outgassed for 30 min at a temperature below 680 K, which is the temperature at which Li evaporates from the surface of the alloy. At 695 K and a 10-mm sample-evaporator distance, a dosing rate onto an Al(111) crystal of 1 monolayer/min was observed.[48] The alternative Li source requires a minimum of outgassing and is reported to have comparable reproducibility and stability as the dispenser source. The sticking coefficient of Li to itself is much higher than that of the other alkali metals; for example, a 30-min evaporation resulted in 25 monolayers of Li adsorbed at room temperature on an Al(111) surface. The low deposition rate is also consistent with the much lower vapor pressure of Li relative to the other alkali metals.

In studies involving clean alkali metal films, the coverage is most easily determined by measuring a property of the alkali-covered sample. For example, one surface coverage method is based on the evolution of the line shapes of the alkali atom core levels as a function of coverage. At low coverage, the deposited alkali atoms are atomic-like and have a symmetric line shape consistent with a surface covered with small, isolated alkali clusters. As one monolayer is reached, the line shape of the core level develops a skewed Doniach–Sunjic line shape and plasmon losses appear.[50] In the case of room-temperature adsorption on silicon surfaces, such as K on Si(111), the work function decreases with increasing dosing time until one monolayer is reached where the work function saturates.[43,51] Concurrently, the intensity of the K $3p$ core level increases with dosing time and saturates at one monolayer, thus giving a good correlation with the work function measurement.[43] For low-temperature adsorption on metal surfaces, the work function reaches a minimum at one-half monolayer coverage and then increases to the work function of the bulk alkali metal at coverages greater than or equal to one monolayer.[42,52] Parker has also used the substrate and adsorbate AES to determine the coverage for Li on Ag(111); however, since the Li diffuses into the substrate, simple saturation of the signals does not occur and a more complicated analysis is necessary.[42] LEED can also be used when specific patterns for ordered adsorbates occur at the different coverages obtained and are known.[41,43]

4. Cleaving and Fracture

In situ cleavage has been, and continues to be, one of the simplest methods for preparing clean, well-ordered surfaces of many single-crystal materials.[3,53–64] Covalent and ionic materials, such as semiconductors, alkali halides, and metal oxides, are quite brittle and thus cleave readily. For these materials, cleavage is the method of choice to produce a clean, well-ordered surface quickly if, in fact, the surface of interest coincides with the cleavage plane. For example, in multicomponent materials, such as GaAs that cleaves along the (110) planes, cleavage can produce a close-to-ideal, stoichiometric surface in a matter of minutes, compared to many hours when the alternative technique of sputtering and annealing is used.[65] The well-defined surfaces created by cleavage in multicomponent systems can thus be used as a reference when other sample preparation techniques are used to prepare the same, or even the noncleavage, faces. The most reproducible surfaces are obtained when the sample is cleaved *in situ*;[66] however, for inert surfaces such as graphite, it is possible to cleave in air or an inert atmosphere and then insert the sample into the ultrahigh vacuum measurement chamber. Effects due to the emission of surface components[54] or included gases[67] during cleavage have been observed, and cleavage-induced defects in semiconductors have been extensively studied.[60,68–73]

In an excellent review of cleavage of brittle materials, Schultz[74] points out that the dynamic nature of the cleavage process makes it difficult to universally predict the preferred cleavage planes of crystals using criteria based solely on their static physical properties. However, it is often possible to develop individual criteria for crystal structures, thus obtaining a certain degree of physical insight. For example, crystals with the NaCl and CaF_2 structures cleave along the (100) and (111) planes, respectively, which correspond to their natural growth faces, implying that the growth and cleavage planes occur in directions that minimize the surface energies. In other examples, cleavage of homopolar semiconductors such as diamond, Si, and Ge occurs along the (111) planes where the minimum number of bonds per unit area is broken.[75] In these cases, cleavage occurs along planes that minimize the energy of creation of the resulting surfaces. Even this relatively straightforward argument tends to minimize the complexity of the cleavage process. For example, to obtain the best cleaves of Si that give a single-domain Si(111)–2 × 1 reconstruction, it has been found that the cleavage *direction* should be along the [211] rather than the [110] direction.[76] Cleaves made along the [110] direction failed to give reproducible single-domain surfaces and exhibited defects that extended several nanometers into the crystal.[77] Understanding this phenomenon requires detailed knowledge of the atomic bonding at the surface. Clearly, generalization of any of the criteria mentioned to all crystal structures inevitably results in exceptions.[74]

One criterion that does seem to apply universally is actually a negative one; i.e., crystals will not cleave along planes that result in non-neutral surfaces. For example, if a crystal in the ZnS structure such as GaAs were to cleave along the (100) or (111) planes, the resulting surfaces would have a sheet of positive charge on one surface and a sheet of negative charge on the other, which requires a very high total energy of formation. The cleavage planes for the ZnS structure are the (110) planes that have equal numbers of positively and negatively charged atoms within each plane. Similar arguments also apply to the NaCl and CaF_2 structures discussed. The cleavage planes are neutral, while the noncleavage planes, such as the (111) planes for NaCl or the (100) and (110) planes for CaF_2, would clearly result in two oppositely charged surfaces after cleavage.

Layered compounds, such as the high-temperature, superconducting materials (e.g., $YBa_2Cu_3O_x$), graphite, MoS_2, mica, etc., are characterized by strong covalent in-plane bonding and weak bonding between the layers resulting from van der

Waals forces acting between the planes. As a result, it is quite clear that preferred cleavage planes for these materials are between the layers along the basal planes.

Certain materials such as crystalline quartz (SiO_2) do not have strongly preferred cleavage planes. Therefore, fracture in these materials can result in a conchoidal surface, named for the similarity of its surface topography to a geometric conchoid. This is also the type of fracture observed in glasses, which by definition have no long-range crystalline order. Crystals without long-range order can be identified by examining the resistance to fracture, known as the fracture toughness, of the different planes in the crystal. The fracture toughness, a well-known parameter in fracture mechanics, has been determined experimentally for many materials and crystallographic orientations.[78] In the case of crystalline quartz, the fracture toughness for the ($10\bar{1}1$) and ($11\bar{2}0$) planes is nearly identical, resulting in conchoidal fractures or fracture along the cleavage planes depending on how the external stresses are applied.[74]

Although some metals can be cleaved, cleavage is not typically used for the preparation of single-crystal surfaces for surface science applications because most metals, including alloys, may readily be prepared by sputtering and annealing.[79] Cleavage studies using metal single crystals have been generally used to study fracture mechanisms rather than as surface preparation techniques. Many of these studies typically use specialized apparatus that is not easily adaptable to preparing clean surfaces. Because metals are typically ductile, it is necessary to cool them as low as 77 K before they will fracture instead of simply deforming.[78,80–82] Several different types of fracture stages have been used to cleave metals as well as expose internal surfaces of polycrystalline materials to investigate properties of grain boundaries for applications ranging from metallurgy to ceramics. With one such stage, notched bars are fractured by impact bending on a liquid-nitrogen-cooled blade.[83] Both sides of the sample are captured so that the complementary fracture faces may be studied. With another apparatus, a UHV version of a tensile tester allows quantitative stress–strain measurements to be made *in situ*.[82,84,85] Observation of the fracture with an electron microscope is permitted during the bending process in another apparatus.[86] Due to the specialized nature of such devices, they will not be treated in detail here. Examples of metals that can be cleaved include body-centered cubic metals, such as iron and tungsten in which cleavage occurs along the (100) planes, and hexagonal close-packed metals, such as zinc and beryllium in which cleavage occurs along the (0001) planes.

For brittle materials, a very simple and effective cleaver design uses a blade pressing against a sample that is supported by a flat anvil.[87–91] The sample is cleaved by either slowly driving the blade against the sample or by delivering a sharp blow to the blade. Although both methods have been reported to produce "good" cleavage, the author has found the former method to be highly reproducible. In the cleavers used in the author's laboratory, the blade is slowly driven against the sample by a screw until the sample cleaves. This type of cleaver has been used to cleave a variety of semiconductor and metal oxide samples with sizes ranging from 2 mm × 4 mm to 5 mm × 5 mm. Variations on this basic concept include driving wedges into a notched sample[68] or applying a bending force to a clamped bar.[53,92] Both require specially shaped samples and may not be applicable to a broad range of sample sizes and types of materials. A double-knife technique has been described to cleave relatively large 10 mm × 10 mm CdS crystals.[93] Special configurations have been described for low-temperature stages[66] and specific sample holders.[62] In addition, cleavers not readily adaptable for use in UHV have been described which can reproducibly cleave relatively large areas.[94]

An effective cleaver design includes having a rigid relative alignment between the blade and the anvil, a uniform distribution of force where the sample contacts the blade or anvil, the ability to increase the force gradually on the sample, a minimum transfer of force to the sample manipulator, and high reliability. In addition, the cleaver should allow mounting on a linear motion translator or port aligner to facilitate positioning. These criteria are readily met by the design in Fig.

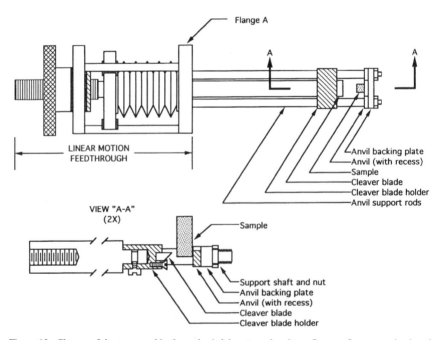

Figure 12. Cleaver of the type used in the author's laboratory showing a 5 mm × 5 mm sample placed in the cleaving position. The figure also shows how the cleaver blade holder slides on the anvil support rods ensuring a rigid relative alignment between the cleaver blade and anvil. The anvil support rods and linear motion feedthrough which drives the cleaver blade are both attached directly onto flange A. Note that the side view A–A is shown at double scale.

12.[4] In this design, the anvil, blade, and linear motion drive are all supported from a single 70-mm-diameter Conflat flange (labeled A in Fig. 12) forming a rigid, self-aligned unit. In this design, a custom linear motion feedthrough is attached directly onto flange A. A commercial feedthrough can be used with the simple addition of two tapped holes for the anvil support rods on the main flange. If the commercial feedthrough is to be used without modification, it may be necessary to use a double-sided flange in place of flange A. In the designs used in the author's laboratory, the anvil support rods are a pair of centerless ground 304 stainless steel rods (diameter of 6 mm) that are screwed into flange A. The cleaver blade holder consists of a bar of aluminum bronze, Ampco 18,[95] and slides on the anvil support rods by means of two close-tolerance, parallel holes drilled in the cleaver blade holder that act as precision bushings. The use of aluminum bronze eliminates the need for lubrication; using a single material for the entire part greatly simplifies fabrication. Ampco 18 is a copper–aluminum–iron alloy (with less than 0.3% each of Sn, Mn, and Si) that is recommended for heavy-duty applications in bearings, bushings, and gears.[95] In addition, the cleaver blade holder has a slot to accept the cleaver blade and a hole to accept the shaft from the linear motion feedthrough. The anvil consists of a 3-mm-thick, OFHC copper plate recessed to give only a 1-mm overlap with the cleaver blade supported by a 5-mm-thick, 304 stainless steel backing plate. This assembly slips onto the threaded ends of the anvil support rods and is secured by gold-plated nuts. The cleaver blade is a standard tungsten carbide tool blank singly beveled to a 45° angle. It is held onto the blade holder with a single screw, which allows some movement of the blade and easy replacement. The straight side of the blade faces toward the sample and allows the maximum number of cleaves to be obtained. A chip catcher of sheet metal with a glass slide window to view the sample can easily be mounted to the anvil support screws. For cleaving samples in air, a benchtop version of this cleaver can be easily constructed in which the elements required for use in UHV are eliminated.

Cleaving has always been an art, however, since the experimenter is working with, rather than against, the inherent properties of materials. Given a good cleaver and reasonable technique, it is relatively easy to repeatedly obtain good cleaves after a little practice. In many cases, material quality is critical. For example, in early studies of Fermi-level pinning in GaAs, it was difficult to consistently obtain unpinned surfaces.[96] Now after significant improvements have been made in growing more perfect GaAs crystals, it is difficult *not* to obtain an unpinned surface upon moderately careful cleavage.[97] The following describes the author's preferred method of cleaving. Once the cleaver is in position and the sample aligned so that it sets as squarely as possible against the anvil, the cleaver blade is brought gently in contact with the sample. Pressure is increased slowly until a snap is heard and the crystal is cleaved. The author prefers to turn the knob on the linear motion feedthrough in 20° increments and wait for a few seconds after each increment. In some cases, cleavage takes place during one of these waiting periods. Thicknesses of 1–2 mm are easily removed even in materials such as Si and Ge, which do not always give large areas that are perfectly flat. Thus a number of cleaves can be performed on samples which are typically 10 mm long.

This cleaver design is not compatible with layered materials, which are typically quite soft and thin. Cleaving these materials is accomplished by mechanically peeling or separating the weakly bound planes from each other. For example, materials such as graphite and GaSe have been cleaved in air prior to insertion into the vacuum system by simply peeling off a layer with adhesive tape or by using mechanical means such as manipulation with a razor blade.[98,99] The adhesive tape technique is commonly used to prepare graphite surfaces for scanning tunneling microscopy measurements. If necessary, adsorbed gases may be driven off the graphite by heating.[98] *In situ* cleavage may be accomplished by simply using a UHV wobble stick manipulator (VG Microtech, East Sussex, England; MDC Vacuum Products, Hayward, CA) to manipulate a razor blade or knife edge to peel off a part of the sample.[100] The advantage of the wobble stick manipulator is that several cleaves can be obtained from a single sample, but the number is clearly dependent on the experimenter's technique to achieve good cleave quality. Materials with sensitive surfaces, such as high-temperature superconductors that are typically grown as small crystals, require a more reproducible approach. In the most commonly used technique, the crystals are attached to the sample holder with epoxy and a post is attached to the top of the sample.[101–103] A fresh surface is exposed by knocking the post off the sample with a wobble stick manipulator or linear motion feedthrough. For ensuring electrical contact to the sample holder, the sample is bonded with a conductive epoxy, such as EpoTek H21D (Epoxy Technology, Billerica, MA), followed by baking in air for 30 min at 150°C. The top post can be bonded with a nonconducting epoxy, such as Torr Seal (Varian Associates), followed by baking in air for 30 min at 120°C. Excess glue on the side of the sample that could prevent cleavage can be removed by filing. If conductive epoxy is not used, it is possible to paint the side of the sample including part of the sample holder with colloidal graphite to ensure good electrical contact. For XPS applications, small samples are typically mounted on posts to eliminate the background signal that would be emitted from a large sample holder. This technique can also be applied to other layered materials such as WSe_2, GaSe, and InSe.

Table 2 lists the cleavage planes for a variety of crystal types with references to the original literature. The examples listed are only meant to be a representative sampling of crystals that have been used in surface science experiments. More detailed discussion on the subtleties related to cleavage of a particular material can be found in the references listed in Table 2. In addition, the reviews by Schultz *et al.*,[74] Henrich and Cox,[104] Gilman,[80] and Haneman[59] provide detailed information on a wider variety of materials.

Table 2. Cleavage Planes for Various Crystal Types

Crystal type	Cleavage planes	Examples	Comments
Rocksalt	(100)	LiF, NaCl, NaF, KCl[54]	Surface dissociation on cleavage
		MgO, CaO, NiO, CoO[104]	Excellent cleavage properties
		MnO, EuO[104]	Poor cleavage properties
		PbS	Excellent cleavage properties
Diamond	(111)	Si	Single domain 2×1 by cleaving along [211][51,76,77,105]
		Ge	2×1 LEED pattern
Zinc blende	(110)	GaAs, GaSb, GaP, InAs, InSb ZnTe, HgTe, CdTe[106-108]	1×1 LEED
Wurtzite	($10\bar{1}0$) (0001)	ZnO	(0001) cleavage gives the (0001)–Zn and ($000\bar{1}$)–O faces[104]
	($10\bar{1}0$) ($11\bar{2}0$)	CdS, CdSe, ZnS CdS, CdSe	CdS shown to have two cleavage planes[93]
Perovskite[104]	(100)	SrTiO$_3$	Conchoidal fracture, poor LEED
Rutile[104]	(110) (100)	TiO$_2$, SnO$_2$	Poor surface quality; shown to have two cleavage planes
Corundum[104]	($10\bar{1}2$)	Ti$_2$O$_3$	Excellent cleavage properties
		V$_2$O$_3$, Cr$_2$O$_3$	Poorer quality surfaces than Ti$_2$O$_3$
Layered materials	Basal plane	Graphite, mica, MoS$_2$ [74] WSe$_2$[109,110]	Unreactive surfaces
		TiSe$_2$,Bi$_2$Te$_3$,SnSe$_2$,SnS$_2$ GaSe, InSe[111]	
		YBa$_2$Cu$_3$O$_{7-x}$ [102,104] Bi$_2$Sr$_2$CaCu$_2$O$_{8-x}$ [102,104]	Cooling maintains surface quality[103]
Metals	(0001)	Be, Zn[80] Te[59]	Low temperatures required for cleavage

5. Conclusions

The sample preparation methods discussed in this chapter can be considered traditional since there has been little change in the basic techniques for over 30 years. The strength of these techniques lies in their simplicity, which has resulted in their continued use. In the case of gas dosing some of the recent calculations of doser geometries has resulted in significant improvements in emitter head geometries, while new applications with stringent uniformity requirements will further drive improvements. The alkali metal channels have also remained largely unchanged and are now used quite routinely with little detail given in the literature about their use. This chapter should serve to record some of the subtler aspects of their use. Finally, crystal cleaving is another area that does not receive much description in the recent literature due to its relatively routine use. It is the intent of this chapter to give those details necessary for the fabrication and operation of a reliable cleaver design.

Acknowledgments

The author acknowledges support from the Stanford Synchrotron Radiation Laboratory, which is supported by the Department of Energy, Office of Basic Energy Sciences. The author also gratefully acknowledges discussions with R. Cao, T. Kendelewicz, I. Lindau, E. J. Nelson, and S. Wanner. In addition, the author is indebted to W. E. Spicer, who brought many of the techniques discussed in this chapter to Stanford University and supported the author's early work in this field. Finally, special thanks goes to J. D. MacGowen, R. Morris, and P. McKernan from the Stanford University Tube Laboratory, whose technical ability and creativity brought many unique designs, including the cleaver described here, to our laboratory.

References

1. A. Roth, *Vacuum Technology*, North-Holland, New York (1983).
2. G. W. Rubloff, Photoemission studies of time-resolved surface reactions: Isothermal desorption of CO from Ni(111), *Surf. Sci.* **89**, 566 (1979).
3. P. Pianetta, I. Lindau, C. M. Garner, and W. E. Spicer, The oxidation properties of GaAs(110) surfaces, *Phys. Rev. Lett.* **37**, 1166 (1976).
4. P. Pianetta, I. Lindau, C. M. Garner, and W. E. Spicer, Chemisorption and oxidation studies of the (110) surfaces of GaAs, GaSb and InP, *Phys. Rev. B* **18**, 2792 (1978).
5. D. A. King and M. G. Wells, Molecular beam investigation of adsorption kinetics on bulk metal targets: Nitrogen on tungsten, *Surf. Sci.* **29**, 454 (1972).
6. T. E. Madey, Adsorption of Oxygen on W(100): Adsorption Kinetics and Electron Stimulated Desorption, *Surf. Sci.* **33**, 355 (1972).
7. J. Benziger and R. J. Madix, The Effects of carbon, oxygen, sulfur and potassium adlayers on CO and H_2 adsorption on Fe(100), *Surf. Sci.* **94**, 119 (1980).
8. C. T. Campbell and S. M. Valone, Design considerations for simple gas dosers in surface science applications, *J. Vac. Sci. Technol. A* **3**, 408 (1985).
9. D. C. Gray and H. H. Sawin, Design considerations for high-flux collisionally opaque molecular beams, *J. Vac. Sci. Technol. A* **10**, 3229 (1992).
10. B. B. Dayton, Gas flow patterns at entrance and exit of cylindrical tubes, in: *Proceedings of the 3rd AVS National Vacuum Symposium*, Pergamon Press, (1957), p. 5.
11. W. Steckelmacher, Knudsen flow 75 years on: The current state of the art for flow of rarefied gases in tubes and systems, *Rep. Prog. Phys.* **49**, 1083 (1986).
12. J. A. Giordmaine and T. C. Wang, Molecular beam formation by long parallel tubes, *J. Appl. Phys.* **31**, 463 (1960).
13. R. H. Jones, D. R. Olander, and V. R. Kruger, Molecular-beam sources fabricated from multichannel arrays. I. Angular distributions and peaking factors, *J. Appl. Phys.* **40**, 4641 (1969).
14. W. Steckelmacher, R. Strong, and M. W. Lucas, A simple atomic or molecular beam as target for ion-atom collision studies, *J. Phys. D: Appl. Phys.* **11**, 1553 (1978).

15. A. Winkler and J. T. Yates, Capillary array dosing and angular desorption distribution measurements: A general formalism, *J. Vac Sci. Technol. A* **6**, 2929 (1988).
16. S. Adamson and J. F. McGilp, Measurement of Gas Flux Distributions from Single Capillaries Using a Modified, UHV-compatible Ion Gauge, and Comparison with Theory, *Vacuum* **36**, 227 (1986).
17. S. Adamson, C. O'Carroll, and J. F. McGilp, The Spatial Distribution of Flux Produced by Single Capillary Gas Dosers, *Vacuum* **38**, 341 (1988).
18. S. Adamson, C. O'Carroll, and J. F. McGilp, The Angular Distribution of Thermal Molecular Beams Formed by Single Capillaries in the Molecular Flow Regime, *Vacuum* **38**, 463 (1988).
19. A. M. Glines, R. N. Carter, and A. B. Anton, An alternative for gas dosing in ultrahigh vacuum adsorption studies, *Rev. Sci. Instrum.* **63**, 1826 (1992).
20. D. R. Olander, Molecular-beam sources fabricated from multichannel arrays. II. Effect of source size and alignment, *J. Appl. Phys.* **40**, 4650 (1969).
21. D. R. Olander and V. R. Kruger, Molecular beam sources fabricated from multichannel arrays. III. The exit density problem, *J. Appl. Phys.* **41**, 2769 (1970).
22. H. C. W. Beijerinck and N. F. Verster, Velocity distribution and angular distribution of molecular beams from multichannel arrays, *J. Appl. Phys.* **46**, 2083 (1975).
23. M. J. Bozack, L. Muehlhoff, J. N. Russel, Jr., W. J. Choyke, and J. T. Yates, Jr., Methods in semiconductor surface chemistry, *J. Vac. Sci. Technol.* **5**, 1 (1987).
24. D. A. Scheinowitz, K. Werner, and S. Radelaar, Capillary arrays with variable channel density: An improved gas injection system, *J. Vac. Sci. Technol. B* **12**, 3228 (1994).
25. D. E. Ibbotson, T. S. Wittrig, and W. H. Weinberg, The chemisorption and decomposition of NO on the (110) surface of iridium, *Surf. Sci.* **110**, 294 (1981).
26. M. A. Henderson, R. D. Ramsier, and J. T. Yates, Minimizing ultrahigh vacuum wall reactions of $Fe(CO)_5$ by chemical pretreatment of the dosing system, *J. Vac. Sci. Technol. A* **9**, 2785 (1991).
27. P. K. Leavitt and P. A. Thiel, A warning concerning the use of glass capillary arrays in gas dosing: Potential chemical reactions, *J. Vac. Sci. Technol. A* **8**, 148 (1990).
28. E. Apen, R. Wentz, F. Pompei, and J. L. Gland, Glass and polymer based dosing system for the introduction of reactive gases into ultrahigh vacuum, *J. Vac. Sci. Technol. B* **12**, 2946 (1994).
29. M. A. Mendicino and E. G. Seebauer, Use of Teflon for minimizing spurious reactions in gas dosing and detection systems, *J. Vac. Sci. Technol. A* **10**, 3590 (1992).
30. F. C. Henn, M. E. Bussell, and C. T. Campbell, A simple means for reproducibly dosing low vapor pressure and/or reactive gases to surfaces in ultrahigh vacuum, *J. Vac. Sci. Technol. A* **9**, 10 (1991).
31. P. L. Hagans, B. M. DeKoven, and J. L. Womack, A laser drilled aperture for use in an ultrahigh vacuum gas doser, *J. Vac. Sci Technol. A* **7**, 3375 (1989).
32. S. M. Gates, J. N. Russell, Jr., and J. T. Yates, Jr., Scanning kinetic spectroscopy (SKS): A new method for investigation of surface reaction processes, *Surf. Sci.* **159**, 233 (1985).
33. R. C. Eden, Photoemission studies of the electronic band structures of gallium arsenide, gallium phosphide, and silicon, Stanford University, (1967), p. 303.
34. P. E. Gregory, P. Chye, H. Sunami, and W. E. Spicer, The oxidation of Cs-UV photoemission studies, *J. Appl. Phys.* **46**, 3525 (1975).
35. N. Szydlo, R. Poirier, and M. Kleefstra, Reverse I-V characteristics of the Na-Si Schottky barrier, *Appl. Phys. Lett.* **17**, 477 (1970).
36. R. D. Mathis Company, *Thin Film Evaporation Source Guide*, Long Beach, CA (1987).
37. P. della Porta, C. Emili, and S. J. Hellier, Alkali metal generation and gas evolution from alkali metal dispensers, SAES Getters, TR 18 (1968).
38. M. Succi, R. Canino, and B. Ferrario, Atomic absorption evaporation flow rate measurements of alkali metal dispensers, *Vacuum* **35**, 579 (1985).
39. T. A. Giorgi, B. Ferrario, and B. Storey, An updated review of getters and gettering, *J. Vac. Sci. Technol. A* **3**, 417 (1985).

40. R. A. dePaola, J. Hrbek, and F. M. Hoffmann, Potassium promoted C-O bond weakening on Ru(001). I. Through-metal interaction at low potassium coverage, *J. Chem. Phys.* **82**, 2484 (1985).
41. T. K. Sham, M.-L. Shek, G.-Q. Xu, and J. Hrbek, Coverage dependence of photoemission core levels of alkali-metal overlayers, *J. Vac. Sci. Technol. A* **7**, 2191 (1989).
42. S. D. Parker, Lithium adsorption on Ag(111): Characterization by AES and work function changes, *Surf. Sci.* **157**, 261 (1985).
43. E. J. Nelson, T. Kendelewicz, P. Liu, and P. Pianetta, New Surface Phases for Potassium Adatoms on Cleaved Si(111), *Surf. Sci.* **380**, 365 (1997).
44. R. Cao, H. Tang, and P. Pianetta, Negative Electron Affinity on the GaAs(110) with Cs and NF$_3$: A Surface Science Study, *SPIE Proc.* **2550**, 132 (1995).
45. R. L. Bell and W. E. Spicer, 3-5 compound photocathodes: A new family of photoemitters with greatly improved performance, *Proc. IEEE* **58**, 1788 (1970).
46. SAES Getters, *Alkali Metal Dispensers*, Milan, Italy (1983).
47. J. Lilja, M. Toivonen, P. Wysocki, and M. Pessa, Growth of Li-doped ZnSe by molecular beam epitaxy using an alkali metal dispenser, *Vacuum* **40**, 491 (1990).
48. F. J. Esposto, K. Griffiths, P. R. Norton, and R. S. Timsit, Simple source of Li metal for evaporators in ultrahigh vacuum (UHV) applications, *J. Vac. Sci. Technol. B* **12**, 3245 (1994).
49. A. Septier, F. Sabary, J. Dudek, and A. Boumiz, An alkali dispenser photocathode (Al-Li)-Ag-O-Li, *C. R. Acad. Sci. Paris* **314**, 569 (1992).
50. S. Doniach and M. Sunjic, Many-electron singularity in X-ray photoemission and X-ray line spectra from metals, *J. Phys. C (Sol. State Phys.)* **3**, 285 (1970).
51. B. Reihl and K. O. Magnusson, Surface electronic structure of K on Si(111)2X1 as a function of potassium coverage, *Phys. Rev. B* **42**, 11839 (1990).
52. R. L. Gerlach and T. N. Rhodin, Binding and charge transfer associated with alkali metal adsorption on single crystal nickel surface, *Surf. Sci.* **19**, 403 (1970).
53. G. W. Gobeli and F. G. Allen, Surface measurements on freshly cleaved silicon p-n junctions, *J. Phys. Chem. Solids* **14**, 23 (1960).
54. T. E. Gallon, I. G. Higginbotham, M. Prutton, and H. Tokutaka, The (100) surfaces of alkali halides, *Surf. Sci.* **21**, 224 (1970).
55. M. Henzler, Leed investigation of step arrays on cleaved germanium (111) surfaces, *Surf. Sci.* **19**, 159 (1970).
56. J. W. T. Ridgway and D. Haneman, Auger spectra and Leed patterns from vacuum cleaved silicon crystals with calibrated deposits of iron, *Surf. Sci.* **24**, 451 (1971).
57. R. Dorn, H. Luth, and G. J. Russell, Adsorption of oxygen on clean cleaved (110) gallium-arsenide surfaces, *Phys. Rev. B* **10**, 5049 (1974).
58. H. Froitzheim and H. Ibach, On the question of surface states on cleaved GaAs (110) surfaces, *Surf. Sci.* **47**, 713 (1975).
59. D. Haneman, Atomic Structure of Surfaces, in: *Surface Physics of Phosphorous and Semiconductors* (C. G. Scott and C. E. Reed, eds.), Academic Press, New York (1975), p. 1.
60. A. Huijser and J. van Laar, Work function variations of gallium arsenide cleaved single crystals, *Surf. Sci.* **52**, 202 (1975).
61. W. E. Spicer and P. E. Gregory, Surface and interface electronic structure of GaAs and other III–V compounds, *Crit. Rev. Solid State Sci.* **5**, 231 (1975).
62. E. E. Chaban and J. E. Rowe, Vacuum cleavage technique for the HP5950A ESCA spectrometer, *J. Electron Spectrosc.* **9**, 329 (1976).
63. J. D. V. Otterloo, Some Schottky barriers on clean-cleaved silicon, Thesis, Delft University of Technology (1977).
64. R. D. Bringans and R. Z. Bachrach, in: *Semiconductor Surface and Interface States, in Synchrotron Radiation Research* (R. Z. Bachrach, ed.), Plenum, New York (1992), p. 127.

65. P. Mark, P. Pianetta, I. Lindau, and W. E. Spicer, A comparison of LEED intensity data from chemically polished and cleaved GaAs(110) surfaces, *Surf. Sci.* **69**, 735 (1977).

66. T. Angot, J. Suzanne, and J. Y. Hoarau, An original in situ cleaver for low temperature surface experiments, *Rev. Sci. Instrum.* **62**, 1865 (1991).

67. J. T. Dickinson, L. C. Jensen, and M. R. McKay, Neutral molecule emission from the fracture of crystalline MgO, *J. Vac. Sci. Technol. A* **5**, 1162 (1987).

68. M. Henzler, The roughness of cleaved semiconductor surfaces, *Surf. Sci.* **36**, 109 (1973).

69. W. Gudat, D. E. Eastman, and J. J. Freeouf, Empty surface states on semiconductors: Their interactions with metal overlayers and their relation to Schottky barriers, *J. Vac. Sci. Technol.* **13**, 250 (1976).

70. W. E. Spicer, I. Lindau, P. E. Gregory, C. M. Garner, P. Pianetta, and P. W. Chye, Surface and interface electronic structure of GaAs and other III–V compounds, *J. Vac. Sci. Technol.* **13**, 780 (1976).

71. L. J. Whitman, J. A. Stroscio, R. A. Dragoset, and R. J. Celotta, Scanning-tunneling-microscopy study of InSb(110), *Phys. Rev. B* **42**, 7288 (1990).

72. B. M. Trafas, Y.-N. Yang, R. L. Siefert, and J. H. Weaver, Scanning tunneling microscopy of Ag growth on GaAs(110) at 300 K: From clusters to crystallites, *Phys. Rev. B* **43**, 14107 (1991).

73. W. Monch, in: *Semiconductor Surfaces and Interfaces*, Springer Series in Surface Sciences, Vol. 26 (G. Ertl, ed.), Springer, Berlin (1995).

74. R. A. Schultz, M. C. Jensen, and R. C. Bradt, Single crystal cleavage of brittle materials, *Int. J. Fracture* **65**, 291 (1994).

75. W. A. Harrison, *Electronic Structure and the Properties of Solids*, Freeman, San Francisco (1980).

76. D. Haneman, Surfaces of silicon, *Rep. Prog. Phys.* **50**, 1045 (1987).

77. R. M. Tromp, L. Smit, and J. F. van der Veen, Si(111)-(2X1) surface: buckling, chains, or molecules?, *Phys. Rev. Lett.* **51**, 1672 (1983).

78. R. E. Reed-Hill, *Physical Metallurgy Principles*, Van Nostrand, Princeton, NJ (1973).

79. M. Prutton, *Introduction to Surface Physics*, Oxford University Press, Oxford (1994).

80. J. J. Gilman, Direct measurement of the surface energies of crystals, *J. Appl. Phys.* **31**, 2208 (1960).

81. J. E. Cordwell and D. Hull, Observations of (100) cleavage in <110> axis tungsten single crystals, *Phil. Mag.* **26**, 215 (1972).

82. N. M. Giallourakis, D. K. Matlock, and G. Krauss, A cryogenic fracture technique for characterizing zinc-coated steels, *Metallography* **23**, 209 (1989).

83. P. Bruesch, W. Foditsch, and F. Stucki, An in situ fracture stage for x-ray photoelectron spectroscopy and Auger electron spectroscopy studies of internal surfaces in polycrystalline materials, *J. Vac. Sci. Technol. A* **5**, 3334 (1987).

84. R. D. Moorhead, Mechanical testing: In situ fracture device for Auger electron spectroscopy, *Rev. Sci. Instrum.* **47**, 455 (1976).

85. G. S. Was, H. H. Tischner, and J. R. Martin, An *in situ* slow extension rate fracture stage for Auger electron spectroscopy, *J. Vac. Sci. Technol. A* **1**, 1477 (1983).

86. C. E. Kalnas, J. F. Mansfield, G. S. Was, and J. W. Jones, In situ bend fixture for deformation and fracture studies in the environmental scanning electron microscope, *J. Vac. Sci. Technol. B* **12**, 883 (1994).

87. M. R. Barnes and L. D. Laude, A crystal cleaver and electron analyzer assembly for photoemission studies of vacuum cleaved crystals, *Rev. Sci. Instrum.* **42**, 1191 (1971).

88. A. P. Janssen and A. Chambers, A crystal cleavage device for use in Leed-Auger studies, *J. Phys. E: Sci. Instrum.* **7**, 425 (1974).

89. B. Dupoisson, P. Dumas, A. Steinbrunn, and J. C. Colson, Single crystal cleavage device adaptable to a UHV vessel, *J. Phys. E: Sci. Instrum.* **9**, 266 (1976).

90. R. Carr, New ultrahigh-vacuum cleaver for brittle materials, *Rev. Sci. Instrum.* **59**, 989 (1988).

91. C. Claeys, C. R. Henry, and C. Chapon, A crystal cleavage device for use with a standard UHV sample holder end, *Meas. Sci. Technol.* **2**, 81 (1991).
92. D. G. Li, N. S. McAlpine, and D. Haneman, Progression of cleavage in Si, Ge and GaAs, *Appl. Surf. Sci.* **65/66**, 553 (1993).
93. N. B. Kindig and W. E. Spicer, Vacuum system and cleaving mechanism for photoemission measurement of CdS single crystals in the vacuum ultraviolet, *Rev. Sci. Instrum.* **36**, 759 (1965).
94. F. C. Hallberg, B. E. Woodgate, and J. S. J. Benedicto, Cleaving machines for soft and hard crystals, *Rev. Sci. Instrum.* **52**, 759 (1981).
95. Copper and Brass Sales, CBS-003-06-96-Revised, Detroit, MI.
96. P. Pianetta, I. Lindau, P. E. Gregory, C. M. Garner, and W. E. Spicer, Valence band studies of clean and oxygen exposed GaAs(110) surfaces, *Surf. Sci.* **72**, 298 (1978).
97. C. Kim, R. Cao, and P. Pianetta, Fermi Level Variation on GaAs(110) Surface with Sb Overlayer Studied with a Photoelectron Microscope, *J. Vac. Sci. Technol B* **11**, 1575 (1993).
98. A. Humbert, M. Dayez, S. Granjeaud, P. Ricci, C. Chapon, and C. R. Henry, Ultrahigh vacuum and air observations of Pd clusters grown on clean graphite, *J. Vac. Sci. Technol. B* **9**, 804 (1991).
99. N. Nakayama, T. Kuramachi, T. Tanbo, H. Ueba, and C. Tatsuyama, AES, LEELS and XPS studies on the interface formation between layered semiconductors GaSe and InSe, *Surf. Sci.* **244**, 58 (1991).
100. R. H. Williams and A. J. McEvoy, Electron emission studies from GaSe surfaces, *J. Vac. Sci. Technol.* **9**, 867 (1972).
101. D. E. Fowler, C. R. Brundle, J. Lerczak, and F. Holtzberg, Core and valence XPS spectra of clean, cleaved single crystals of YBa$_2$Cu$_3$O$_7$, *J. Electron Spectrosc.* **52**, 323 (1990).
102. P. Akavoor, R. B. Phelps, and L. L. Kesmodel, Charging effects in measurements of high-temperature superconductors with high resolution electron energy loss spectroscopy, *J. Vac. Sci. Technol. B* **12**, 587 (1994).
103. H. L. Edwards, J. T. Markert, and A. L. de Lozanne, Surface structure of YBa$_3$Cu$_3$O$_{7-x}$ probed by reversed biased scanning tunneling microscopy, *J. Vac. Sci. Technol. B* **12**, 1886 (1994).
104. V. E. Henrich and P. A. Cox, *The Surface Science of Metal Oxides*, Cambridge University Press, Cambridge (1994).
105. W. Monch and P. P. Auer, On the geometrical structure of cleaved Si(111) surfaces, *J. Vac. Sci. Technol.* **15**, 1230 (1978).
106. K. O. Magnusson, U. O. Karlsson, D. Straub, S. A. Flodstrom, and F. J. Himpsel, Angle-resolved inverse photoelectron spectroscopy studies of CdTe(110), CdS(11$\bar{2}$0) and CdSe(11$\bar{2}$0), *Phys. Rev. B* **36**, 6566 (1987).
107. H. Qu, J. Kanski, P. O. Nilsson, and U. O. Karlsson, Angle resolved photoelectron spectroscopy of the surface electronic structure of ZnTe(110), *Phys. Rev. B* **43**, 9843 (1991).
108. X. Yu, L. Vanzetti, G. Haugstad, A. Raisanen, and A. Franciosi, Inequivalent sites for Hg at the HgTe(110) surface, *Surf. Sci.* **275**, 92 (1992).
109. C. A. Papageorgopoulos, M. Kamaratos, A. Papageorgopoulos, A. Schellenberger, E. Holub-Krappe, C. Pettenkofer, and W. Jaegermann, Adsorption of Cs on WSe$_2$ van der Waals surfaces: Temperature and sputter effects on growth properties, *Surf. Sci.* **275**, 311 (1992).
110. A. Klein, C. Pettenkofer, W. Jaegermann, M. Lux-Steiner, and E. Bucher, A photoemission study of barrier and transport properties of the interfaces of Au and Cu with WSe$_2$(0001) surfaces, *Surf. Sci.* **321**, 19 (1994).
111. J.-Y. Emery, L. Brahim-Ostmane, C. Hirlimann, and A. Chevy, Reflection high-energy electron diffraction studies of InSe and GaSe layered compounds grown by molecular beam epitaxy, *J. Appl. Phys.* **71**, 3256 (1992).

7

Chemical Modification of Surfaces

David G. Castner

1. Introduction

For many applications, the material selection criteria are based on the bulk properties of the material. For example, polyurethanes have been used in artificial hearts
because they can withstand the continual flexing required in this application.[1]
However, the surface composition and structure of a material can also play an
important role in the material's performance in a given application.[2] Typically, the
surface region is the interface or reaction zone of a material with the environment
in which the material is placed. For biomedical applications, the surface structure
and composition of the biomedical device determine the biological response of the
body to the implanted biomaterial.[3] It would be rather fortuitous if a polyurethane
prepared to meet the bulk mechanical demands of an artificial heart (e.g., continual
flexing without breaking or elongating) would also provide a surface structure that
would exhibit optimal biocompatibility. Therefore, it is necessary to be able to take
a material with the desired bulk properties and chemically modify its surface
structure to produce an optimal response of that material in the environment of its
intended application.[4]

David G. Castner • National ESCA and Surface Analysis Center for Biomedical Problems,
Departments of Chemical Engineering and Bioengineering, University of Washington, Seattle, WA
98195-1750

Specimen Handling, Preparation, and Treatments in Surface Characterization, edited by Czanderna *et
al.* Kluwer Academic / Plenum Publishers, New York, 1998.

The surface modification treatment must be done in such a manner that the bulk properties of the material are left unchanged. This can usually be accomplished by restricting the modification treatment to the outer few atomic layers of a material so that only a minute fraction of the material is altered. This places an important constraint on surface modification strategies: the modified surface region must be stable in its working environment. If this region is only a few atomic layers thick, then degradation or loss of material from this region can result in significant changes in the interactions between the material and its environment. Thus, the modified material (created either by deposition of new material or chemical reaction of the existing material) must be resistant to processes such as abrasion, chemical attack, delamination, and surface rearrangement. The degree of stability required will be determined by the properties of the environment in which the surface-modified material is placed. This does not mean that the surface-modified region must be nonreactive

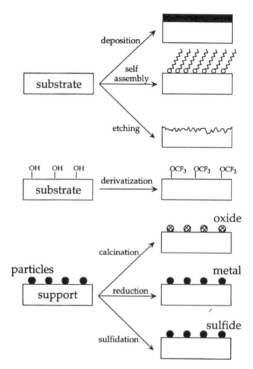

Figure 1. Schematic representation of the surface modification methods described in this chapter. The top three strategies depict methods of modifying the surface of a substrate by deposition (e.g., RF glow discharge with an organic monomer), self-assembly (e.g., alkyl thiol onto gold), and etching (e.g., by RF glow discharge with an inert gas). The middle strategy depicts derivatization of surface functional groups. The bottom three strategies depict the calcination (heating in air), reduction, and sulfidation processes used to convert the surface of the supported particles to oxide, metallic, and sulfide species.

with the environment, just that any interactions that occur must be beneficial. For example, the adsorption of a protein film followed by growth of a confluent layer of endothelial cells on a surface-modified vascular graft would be desirable.[5]

This chapter will discuss various methods that have been developed to chemically modify the surface of organic, metallic, and oxide materials. These methods include the derivatization or reaction of surface functional groups (hydroxyl, acid, etc.), self-assembly of molecules onto surfaces, thin-film deposition by radio-frequency glow discharge (RFGD), gas-phase treatments (oxidation, reduction, sulfidation, etc.), and liquid-phase treatments (e.g., acid etching). Schematics of these processes are shown in Fig. 1. In addition to presenting examples of these modification strategies, the benefits of *in situ* versus *ex situ* procedures will be discussed.

If surface modification is done, it is essential to characterize the resulting material to ensure that the desired surface composition and structure have been obtained. This is necessary since many of the surface modification methods can have undesirable side reactions or introduce surface contaminants. Hence, appropriate techniques must be used to assess the extent of surface modification. Since the objective of surface modification is just to change the composition and structure of the outer few atomic layers of a material, characterization techniques that can detect changes in this small amount of the material are required. Surface-sensitive techniques such as X-ray photoelectron spectroscopy (XPS), secondary-ion mass spectrometry (SIMS), and Auger electron spectroscopy (AES) are commonly used to characterize modified surfaces.[2,6–12] Results from these techniques will be presented along with examples of surface modification strategies to demonstrate the effectiveness of each method. Surface modification applications in adhesion, biomedical, catalytic, microelectronics, wettability, and sensor processes will be discussed.

2. Chemical Reactions of Organic Surfaces

The chemical modification of organic surfaces can be divided into two categories: specific and nonspecific. Nonspecific reactions introduce several types of surface functional groups, typically in a rather random distribution across the surface. For example, the oxidation of polymer surfaces by chromic acid or corona discharge treatments creates a range of carbon functional groups on the polymer surfaces. Specific reactions introduce one surface functional group or molecular species, often in a well-defined arrangement across the surface. Examples of specific reactions include derivatization of surface functional groups and immobilization of biomolecules.

2.1. Nonspecific Surface Modification

Nonspecific surface modification is typically employed when the general properties of the surface require modification. For example, the wettability of hydrocarbon polymers such as polystyrene (PS) and polyethylene (PE) can be significantly improved by adding hydrophilic functional groups (hydroxyl, acid, etc.) to their surfaces. One method of introducing these functional groups is acid etching. For example, a few minutes exposure of a low-density PE film to concentrated chromic acid at temperatures up to 75°C followed by treatment with aqueous nitric acid at 50°C generates a wettable surface on the PE film.[13,14] A distribution of oxidized carbon species can also be introduced by exposure of a hydrocarbon polymer to an air discharge.[15-17] Alternatively, an inert glow discharge can be used to create reactive sites at the surface (defects, radicals, unsaturation, etc.) that upon exposure to air will form oxidized carbon species.[17] One application for nonspecific surface modification is the modification of polymer surfaces for use in tissue cell culture.[15,18,19] Another biomedical application where wettability is important is contact lenses.[20] In addition, increasing wettability can improve subsequent processing steps where the material is exposed to aqueous-based solutions (e.g., improved wettability for inks used in the printing process). Also, incorporation of oxidized carbon species can improve the interfacial adhesion of material by adding functional groups that can chemically react with the material being bonded to it (e.g., carboxylic acid groups on the substrate reacting with amine groups in the overlayer).

2.2. Surface Derivatization

In contrast to the nonspecific surface modification processes, there are many applications which demand a specific sequence of reactions to be executed to produce the desired surface properties. For organic surfaces, many of the techniques used to functionalize or derivatize organic molecules in the liquid or gas phase have been adapted for surface modifications.[21] These strategies have been developed not only to change the surface properties of a material, but also to enhance the ability of surface analytical techniques to characterize organic surfaces.[21-23] For example, hydroxyl (C–OH) and ether (C–O–C) carbon species have indistinguishable C $1s$ binding energies in XPS analysis, but using trifluoroacetic anhydride (TFAA) to derivatize the surface it is straightforward to differentiate between the two groups.[24] This is because TFAA reacts with OH groups and not with C–O–C groups. Therefore the concentration of hydroxyl groups in the surface region can be determined by measuring either the fluorine concentration (assuming the surface did not contain any fluorine prior to derivatization with TFAA) or the CF_3 contribution to the C $1s$ spectrum. Using the F $1s$ peak to measure the fluorine concentration has a double advantage for XPS analysis. First, the F $1s$ photoionization cross section is a factor of 4 larger than the C $1s$ photoionization cross section. Second, three fluorine atoms are introduced for each hydroxyl groups. These two advantages significantly increase the XPS sensitivity for detection of OH groups. However, when using derivatization reactions to enhance XPS or other surface analysis techniques, care must be taken to ensure that no experimental artifacts are introduced by the derivatization reaction. Points to consider when selecting a derivation strategy include cross reactions with other functional groups, the extent of reaction, diffusion limitations, and the stability of the derivatized surface.[23]

An example of how subtle changes in the surface chemistry can produce large changes in the effectiveness of a derivatization reaction is the trifluoroethanol derivatization of carboxylic acid groups. This reaction goes to completion with poly(acrylic acid) (PAA), but proceeds to ~60% of stoichiometry with poly(methacrylic acid) (PMAA).[25] The only difference between the structure of the two polymers is an additional methyl group on the backbone of PMAA.

In addition to the TFAA and trifluoroethanol reactions, many other derivatizations have been proposed to enhance surface characterization techniques such as XPS and static SIMS. A few of the more widely used reactions are shown in Fig. 2. Additional reactions have been discussed elsewhere.[21–23] In addition, the field of organic chemistry serves as a source for a vast array of possible derivatization strategies.[26]

Although the previous examples were developed to enhance surface characterization abilities, they also find use as methods of changing surface chemistry. For example, if an application requires a surface consisting of CF_3 groups, the TFAA reaction can be used to convert a hydroxyl surface into a CF_3 surface. Organic surfaces prepared by self-assembly are good candidates for derivatization reactions. Molecules used to prepare self-assembled monolayers (SAMs) contain a functional group that will form a bond to the surface (thiol–gold, chlorosilane–hydroxyl, etc.).[27–29] This bond serves to anchor the molecules to the surface. At the opposite end of the molecule from the anchor another functional group can be present. The latter group can be the actual functional group desired in the final surface, or it can be a functional group which can be derivatized to produce the final surface. The derivatization method is especially useful if the functional group targeted for the final surface is highly reactive such as an amine. In this case, if the reactive functional group is incorporated directly into the starting molecule, it will compete with the anchor group for surface sites, thereby limiting the ability of the self-assembly process to produce a highly ordered, well-defined surface.[30] Again, the experience of organic chemistry can be used to produce a wide range of functionalized surfaces through derivatization reactions. For example, SAMs on silica surfaces can be prepared using Br-terminated hexadecyltrichlorosilane.[31] Then, via a nucleophilic substitution reaction, the Br surface is converted to a range of different surface chemistries.[31] Another way of introducing reactive functional groups into an organic film is the use of blocking agents (e.g., see Ref. 32). The

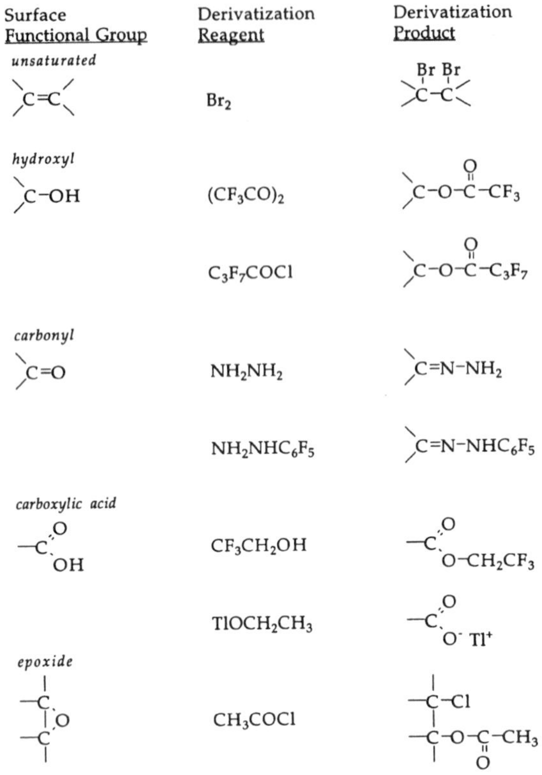

Surface Functional Group	Derivatization Reagent	Derivatization Product
unsaturated $\diagup C=C \diagdown$	Br_2	$\diagup \overset{Br}{C}-\overset{Br}{C} \diagdown$
hydroxyl $\diagup C-OH$	$(CF_3CO)_2$	$\diagup C-O-\overset{O}{\overset{\|}{C}}-CF_3$
	C_3F_7COCl	$\diagup C-O-\overset{O}{\overset{\|}{C}}-C_3F_7$
carbonyl $\diagup C=O$	NH_2NH_2	$\diagup C=N-NH_2$
	$NH_2NHC_6F_5$	$\diagup C=N-NHC_6F_5$
carboxylic acid $-\overset{O}{\overset{}{C}}\diagdown_{OH}$	CF_3CH_2OH	$-\overset{O}{\overset{}{C}}\diagdown_{O-CH_2CF_3}$
	$TlOCH_2CH_3$	$-\overset{O}{\overset{}{C}}\diagdown_{O^- \; Tl^+}$
epoxide $-\overset{\|}{\underset{\|}{C}}\diagdown_O-\overset{\|}{\underset{\|}{C}}$	CH_3COCl	$-\overset{\|}{\underset{\|}{C}}-Cl$ $-\overset{\|}{\underset{\|}{C}}-O-\overset{O}{\overset{\|}{C}}-CH_3$

Figure 2. Selected examples of derivatization reagents and the products formed from reaction of these reagents with unsaturated, hydroxyl, carbonyl, carboxylic acid, or epoxide functional groups present at the surface of organic substrates. Before using a given derivatization reaction, the possibility of cross reactions with other functional groups, the extent of reaction, diffusion limitations, and the stability of the derivatized surface must be addressed.

reactive group is first protected with a suitable blocking agent, then a film with the protected groups at the surface is assembled, and finally the reactive groups are deprotected to generate the desired surface.

2.3. Biomolecule Immobilization

The previous two sections discussed methods for introducing functional groups onto a surface. It is also possible to attach entire biomolecules onto a surface.[33–38] Examples of biomolecules that can be immobilized on a surface include proteins, peptides, saccharides, lipids, DNA, and drugs. Even living cells can be immobilized on a surface. A variety of methods can be used to accomplish this surface modification, including physical adsorption, ionic adsorption, cross-linking, hydrophobic adsorption, and covalent binding with or without a spacer arm. Sometimes two or more methods are combined (e.g., physical adsorption followed by cross-linking). Criteria for evaluating the success of an immobilization strategy are the number of biomolecules immobilized, the amount of biological activity retained by the immobilized biomolecules, and the stability of the immobilized biomolecules.

The immobilization methods that involve adsorption of the biomolecule onto a surface from solution are usually the easiest and cheapest. For example, a glass slide can be coated with albumin simply by placing it in a solution of albumin. Coating a surface with albumin is often done to prevent nonspecific adsorption of other biomolecules. However, because covalent bonds between the biomolecules and the surface are not typically formed upon adsorption, the stability of the biomolecule coating can be a problem. The stability of adsorbed biomolecules can be improved by cross-linking them to each other and the surface. One method of accomplishing this cross-linking is to briefly expose the adsorbed biomolecules to an Ar glow discharge.[39] However, the length and power of Ar glow discharge treatment must be carefully controlled. If the biomolecules are too lightly cross-linked, their stability will still be a problem. Conversely, if the biomolecules are too highly cross-linked, they will lose activity.

Compared to adsorption methods, covalent attachment of biomolecules to a surface involves more complex chemistry. The primary benefit of covalent attachment is the improved stability of the immobilized biomolecules. The particular sequence of chemical reactions used to covalently immobilize biomolecules depends on the structure of the support and biomolecule. If a polymer such as poly(methacrylic acid) is used as the support, then carboxylic acid groups of the polymer can be used to form the covalent bond to the biomolecule. Polymers that do not contain a reactive functional group must be activated by introduction of the desired functional group. Glow discharge (see next section) is a common method used to functionalize the surface of polymers. For example, an ammonia discharge can be used to introduce amine groups onto polyethylene substrates.[40] Fluorocarbon substrates can also be activated for attachment of peptides by glow discharge treatment.[41,42]

Biomolecules can also be immobilized onto metal and oxide surfaces. Self-assembly is a common method used for this purpose. On gold surfaces, alkanethiols with the appropriate terminal functional groups have been used to bind proteins.[43,44] One such terminal group is biotin, which selectively binds streptavidin. Since streptavidin contains four biotin binding sites and only one site is required to bind it to the biotinylated monolayer, the remaining sites can then be used to bind other biotinylated molecules.[43] Another strategy uses a nitrilotriacetic acid terminal group on the alkanethiol.[44] A Ni(II) ion is then complexed with the nitrilotriacetic acid group. This complexed Ni(II) ion can then be used to bind proteins with histidine tags. Often a mixture of alkanethiols is used to functionalize gold surfaces.[44] For example, an alkanethiol terminated with a tri(ethylene glycol) group that resists protein adsorption is often mixed with the alkanethiol containing the terminal group that is used for binding protein. If an appropriate silane is substituted for the alkanethiol, similar strategies to the alkanethiol–gold system can be used to immobilize biomolecules onto metal oxides.[45]

3. Radio-Frequency Glow Discharge

In radio-frequency glow discharge (RFGD) processes, substrates to be treated are placed in a vacuum chamber, pumped down to a modest vacuum (typically 10^{-3} Torr or lower), a gas precursor is flowed over the samples, and then a glow discharge in the flowing gas is initiated (typically by using a 13 MHz RF generator).[46] The discharge contains a wide variety of species, including ions, electrons, photons, molecular fragments, etc.[46] Varying the RFGD experimental conditions (RF power, gas type, gas flow rate, substrate position, etc.) allows a range of different sample treatments to be obtained. A schematic for a capacitively coupled, tubular RFGD reactor system is shown in Fig. 3. Also, discharges can be powered by different excitation sources (microwave, DC, etc.) and at atmospheric pressure (corona). This section will cover only the two classes of RFGD surface modifica-

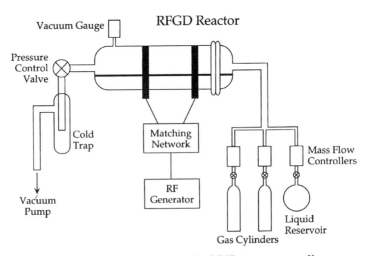

Figure 3. A schematic of a capacitively coupled, tubular RFGD reactor system. Key components are the RF generator with matching network, the gas–liquid inlet system with mass flow control, pressure control valve, and pumping system with cold trap. Variation of the operating conditions for these components along with the type of monomer and the location of the substrate produce a wide range of surface modifications.

tion, etching and deposition. After a brief overview of etching, the remainder of this section is devoted to deposition treatments.

3.1. RFGD Etching Processes

The etching treatment can be accomplished either by ablating the surface with high-energy ions and particles or by chemically reactive species produced in the RFGD. A couple of examples of etching processes (e.g., Ar discharges) were discussed in the previous section. Another example of etching treatment is fluoro-carbon RFGD processes using molecules such as CF_4 and C_2F_6.[47–49] These molecules, when ionized in a RFGD, produce a high concentration of fluorine atoms, which are known to promote substrate etching. This process is widely used in the microelectronics industry to etch features into silicon wafers. The fluorine atoms react with silicon atoms in the wafer to form volatile silicon fluorides which are then pumped away, resulting in loss of material from the wafers.[50] With appropriate masks, various patterns and shapes can be etched in the silicon wafer substrates. Another reactive gas that can be used in RFGDs for etching substrates is oxygen. Typically, oxygen discharges are used to remove organic material from a substrate or roughen the surface of the substrate to improve the adhesion of a subsequent coating.[51,52] Etching substrates with inert-gas discharges can also be used for substrate cleaning and surface roughening.

3.2. RFGD Deposition Processes

Reducing the concentration of etching components in a RFGD allows the process to be switched from etching to deposition. For fluorocarbon discharges, this means reducing the F atom concentration relative to the concentration of film-forming species (e.g., CF_x).[47–49] This can be accomplished by two means. First, H_2 can be mixed with the fluorocarbon gas being used.[47,53,54] The hydrogen will then combine with the reactive F atoms, thereby reducing the F atom concen-tration. Hydrocarbon molecules (e.g., CH_4) can also be used for this purpose. The second method is to reduce the F concentration in the fluorocarbon gas molecules themselves (e.g., switch the feed gas from C_2F_6 to C_2F_4).[47–49] Both methods have been successfully used to deposit fluorocarbon films onto substrates.

RFGD deposition of thin organic films has several advantages.[55] It is a mature technology that allows for easy surface modification of materials. A virtually unlimited range of substrate types and shapes can be coated. This is because the low-pressure discharge does not rely on line-of-sight paths to reach the substrate, and it can penetrate the pores and convoluted shapes of materials and devices. Since the deposition process does not require specific functional groups on the substrate for reaction, substrates with a wide range of surface compositions can be coated. For substrates located directly in the glow, the deposited films are typically highly cross-linked and free of pinholes and voids. This makes them durable barrier films with low levels of leachable components. Also, because of the energetic nature of the discharge which creates these cross-linked films, they typically exhibit good adhesion to the substrate.

Deposition onto substrates located in the glow produces films that contain a range of chemical functional groups. For example, overlayers deposited onto substrates placed in the glow region of a C_2F_4 RFGD typically contain C–C, CF, CF_2, and CF_3 species.[47,56] This can result in the creation of unique ensembles of surface functional groups which can provide surface properties not available through conventional organic chemical methods. For example, carbonyl groups present on the surface of RFGD films deposited from discharges of oxygen containing hydrocarbons exhibit excellent cell growth properties.[57] In contrast, a conventional polymer containing carbonyl groups, poly(vinyl methyl ketone), does not support cell growth. This result implies that a surface ensemble including the carbonyl group is required for good cell growth. Another advantage of RFGD is that films can be deposited from precursors such as acetone which cannot be polymerized by conventional techniques.[58] A final advantage of RFGD for biomedical applications is that the devices removed from the reactor after a RFGD treatment are sterile.[55]

If retention of the gas-phase precursor structure is desired, suitable changes in the RFGD experimental conditions must be made. Two important conditions to consider are the position of the substrate and the power of the RF discharge. Since fragmentation of the molecules fed to the discharge typically increases as the power level of RF excitation increases, low-power discharges are used to deposit films that retain the structural features of the gas-phase precursor. For tetrafluoroethylene (TFE), placing the substrates downstream from such a low-power discharge (remote RFGD) results in films that contain vertical, CF_3-terminated chains of CF_2 groups.[56] Although these chains are nominally similar to the CF_2 chains formed from conventional polymerization of TFE (PTFE), the orientation differences of the CF_2 chains result in significantly different surface properties.[56,59] The downstream TFE film has a surface of CF_3 groups, while PTFE has a surface of CF_2 groups, resulting in different biological behavior of these two materials. The downstream TFE films bind proteins more strongly than PTFE films, probably because of the more hydrophobic nature of the downstream TFE surface.

The substrate position for deposition of RFGD films from glyme precursors also has a noticeable effect on the retention of precursor structure and protein adsorption.[60] In this case, substrates placed upstream from the discharge exhibit the best retention of precursor structure and minimize protein adsorption, an important consideration for nonfouling applications such as biological sensors.

Another method of retaining precursor structure is to cool the substrate below room temperature.[58,61,62] This will increase the condensation rate of the gas-phase precursor molecules onto the substrate. However, condensation by itself is not sufficient to deposit a stable film, since some cross-linking and reaction of the deposited species must occur so that the film will remain intact at room temperature. Light cross-linking can be accomplished by the photons and ions produced in the discharge. Also, the co-condensation of reactive fragments from the discharge can contribute to the stability of the deposited film. Films that retain a substantial amount of the precursor structure have been successfully deposited from several gas-phase molecules, including hydroxyethyl methacrylate, ethylene oxide, hexafluorobutadiene, tetrahydrofuran, acetone, hexafluoro-2-propanol, and ethylene glycol oligomers using substrate cooling.

A recent method developed for retaining precursor structure involves pulsing the RFGD. Varying the duty cycle (the time the power is on relative to the sum of the time the power is on and off), optimizes the retention of precursor structure. This approach has been successfully applied to several monomers.[53,63,64] One advantage of the pulsed method compared to placing the substrates downstream from the RFGD is the higher deposition rates obtained with pulsed plasmas. In fact, it appears there is significant film growth occurring between RF pulses (i.e., when no RF power is being supplied to the reactor) when ions, electrons, etc., are not present.[53] This observation may explain why precursor structure is retained in the pulsed RFGD experiments.

4. Self-Assembled Monolayers (SAMs)

As discussed in the following section, the use of metallic single-crystal substrates has allowed researchers to uncover important relationships between catalytic activity and catalyst surface structure. For processes that rely on the use of organic substrates (e.g., biomedical applications), the lack of similar organic single-crystal substrates has limited the ability of researchers to develop relationships between material performance and surface structure. With the advent of self-assembly processes, researchers studying processes that occur on organic surfaces now have a technique for producing structurally well-defined surfaces.[27,65] The initial work on SAMs has heavily emphasized alkyl thiols on Au surfaces, but other sulfur-containing molecules such as disulfides and thiol ethers self-assemble onto Au surfaces.[27,29,65–70] In fact, films prepared from symmetric alkyl disulfides exhibit properties that are indistinguishable from those prepared from the corresponding alkyl thiol.[67,70] Alkyl thiol systems are discussed, and then examples of other systems are described.

4.1. Alkyl Thiol SAMs

The preparation of alkyl thiol SAMs is straightforward.[27,29,65–70] First, a film of Au is evaporated onto a substrate (typically a glass slide or a silicon wafer). The thickness of the Au film is usually 10 to 300 nm. If extensive treatments in solutions will be carried out, then a thin (2 nm or less) layer of Cr or Ti is deposited onto the substrate before the Au deposition to increase the adhesion of the Au layer to the substrate. When the freshly deposited Au films are removed from the evaporator, they are immediately immersed into a solution containing the molecule to be self-assembled. Concentrations of 1 mM or less are typically used, and the adsorption is allowed to occur for a day or more at room temperature. After removal from solution, it is important to thoroughly rinse the monolayer with pure solvent to remove any physisorbed molecules.[67] The main driving force for self-assembly of sulfur-containing molecules on Au surfaces is the formation of the Au–S thiolate bond. The strength of this bond is estimated to be 40 kcal/mole,[27] which is strong enough to displace weakly bound hydrocarbon contaminants that adsorb onto the Au surface when it is removed from the evaporator and briefly exposed to the ambient atmosphere before immersion into the self-assembly solution. A second driving force for self-assembly of molecules with hydrocarbon tails is van der Waals interactions between the hydrocarbon chains:[27] the longer the hydrocarbon chain, the higher degree of van der Waals stabilization. The van der Waals stabilization has been attributed to the fact that C_{12} and longer hydrocarbon chains assemble more readily than the shorter hydrocarbon chains. Also, hydrocarbon chains are tilted by 20° to 30° from the surface normal because this angle maximizes the van der Waals interactions of the hydrocarbon chains.[27] The initial adsorption of sulfur-containing molecules on the Au surface is fairly rapid, typically occurring within a few minutes of immersion.[71] However, to obtain well-ordered, defect-free SAMs, the adsorption step is typically extended to periods of one to four days. This time is sufficient to anneal out any defects in the initially formed film and to allow the hydrocarbon chains to become fully extended and aligned with each other.[71]

Assembly of long-chain alkyl thiols with a functionalized head group (hydroxyl, acid, etc.) at the opposite end of the alkyl chain from the thiol group will result in the formation of a surface layer of that functionalized group.[27] This result allows the researcher to vary systematically the chemistry at the surface of these organic films to examine its effect on material performance. For example, the attachment and growth of endothelial cells were studied on a range of functionalized SAMs.[72] It was observed that carboxylic-acid-terminated SAMs exhibited the highest levels of cell growth.[72] These studies were done on homogeneous surfaces where the same functional group was present on the end of each alkyl thiol chain. It is also possible to form patterned surfaces containing different functional groups. A variety of methods have been developed for forming patterned surfaces, including photolithography, selective chemisorption, stamping, etc.[73–79] Also, mixed monolayers of different alkyl thiolates can be assembled from solutions containing two or more alkyl thiols.[68,70,80–83] Depending on the alkyl thiols used and the adsorption conditions, either phase-mixed or phase-separated SAMs can be formed. Also, the surface concentrations of the components in the SAM can be significantly different from the solution concentrations of the components used to assemble the monolayer.

A major problem with the alkyl thiolate SAMs is the single-point attachment to the surface through the sulfur group. Although bound by ca. 40 kcal/mole, this bond is not strong enough to prevent displacement or desorption of the individual thiol molecules. It has been shown that when a SAM formed from one alkyl thiol is placed in a solution containing a different alkyl thiol, some of the previously adsorbed thiolate molecules in the SAM are displaced by the thiols in solution.[84] A second pathway for removal of thiolate SAMs is by oxidation of the thiolate to a sulfone.[85–88] This process occurs over time as SAMs are exposed to ambient laboratory conditions. Once the thiol has been oxidized to a sulfone, its interaction with the Au surface becomes significantly weaker and the oxidized thiol molecules can readily be rinsed off the surface at room temperature with water or other solvents. A third method for breaking the Au–S bond is to heat the sample.[69,89] Complete removal of the SAM is accomplished, typically through desorption from the Au surface. This removal can either be done in a solvent (e.g., hexadecane) or in air. When heating in a solvent, temperatures of only ca. 80°C are required to remove the SAMs.

4.2. Polymeric Monolayers

One method that has been developed to improve the stability of SAMs is to form the organic monolayers with polymers.[73,89–93] Polymeric materials offer several attractive properties for the formation of organic monolayers (Fig. 4). First, multifunctional polymers can be prepared using conventional polymer and organic chemistry techniques with a portion of the monomer units along the polymer backbone functionalized with a long-chain alkyl thiol or disulfides.[94] When a freshly deposited Au surface is placed in a solution of this polymer, the sulfur atoms will chemisorb to the Au. Since each polymer unit contains many sulfur-bearing side chains, multiple Au–S bonds will be formed between each polymer molecule and the Au surface.[73,89–93] This process results in a significant increase in the stability of the organic molecules.[73,89] In the case of multiple-point attachment, one or more of the Au–S bonds can be broken without the molecule becoming detached because at any time several Au–S bonds will still be intact. The multipoint attachment also results in increased stability to thiol oxidation, because even if a portion of the bound thiols have been oxidized there will still be unoxidized thiols bound to the Au surface. These benefits of multipoint attachment result in enhanced thermal stability of polymeric monolayers compared to their monomeric counterparts.

An additional advantage of polymeric monolayers is that different functional groups can be attached to the polymer molecule.[73,92] For a polymer containing both perfluoro and hydrocarbon side chains, phase separation of the polymer will

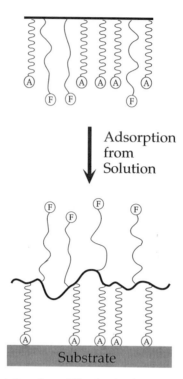

Figure 4. Schematic representation of a multifunctional polymer assembling from solution onto a substrate to form a monolayer. By using a flexible polymer backbone (e.g., siloxane) grafted with side chains containing anchor groups (A) that selectively adsorb onto the substrate (e.g., alkyl thiol/gold), multipoint attachment of the polymer is obtained. Further variation in the surface chemistry of these polymeric monolayers can be obtained by grafting side chains containing different functional groups (F) onto the polymer backbone. Functional groups commonly used include hydrocarbon, perfluoroalkyl, perfluoroether, and ethylene glycol oligomers. Biomolecules can also be attached to the polymer chain.

occur as it unfolds from solution to adsorb on the Au surface. The hydrocarbon chains, which contain the sulfur-anchoring group, will be closest to the surface, next will be a layer containing the polymer backbone, and finally at the outer surface will be the perfluoro chains.[73,92] To optimize the formation of such layers it is best to use a flexible polymer backbone such as a siloxane. In this example, it has been assumed that the polymer was functionalized with both types of side chains prior to assembly onto the Au surface. Functionalization of the polymer backbone with a second side chain can also be done after the polymer has been assembled onto the Au surface, providing the polymer backbone still contains some attachment sites.[91] Thus, polymeric monolayers provide increased stability and derivatizability compared to their monomeric counterparts. However, these advantages come with a price: the polymeric monolayers are not as well ordered as the monomeric systems.

Studies are currently underway to optimize the ordering in polymeric monolayers and assess the trade-off in stability versus order.[95]

4.3. Silane Monolayers

Another route to forming organic monolayers is the attachment of functionalized silane molecules to a hydroxylated surface.[28–31,96–104] Alkyltrichlorosilanes are the most commonly used, although other types of silanes such as dichloro, monochloro, methoxy, ethoxy, etc., have also been employed. Like self-assembly of thiol molecules, the assembly of silanes is commonly carried out in dilute organic solutions. For silanes with sufficient vapor pressure, assembly from the gas phase is possible.[102] A two-step mechanism was initially proposed for attachment of organic chlorosilanes to silica surfaces. The first step was the hydrolysis of a Si–Cl bond to form a Si–OH bond. The second step was condensation of this organic silanol with a surface silanol to form a covalent bond between the silane and the silica surface.[28,97] It has been shown that hydrolysis of the silane bonds does occur, but that few, if any, Si–O–Si covalent bonds are formed at room temperature between hydrolyzed alkylchlorosilanes and silanol groups on the silica surface.[96,102] However, direct coupling of hydrolyzed fluoroalkylsilanes and surface silanol groups can occur at room temperature.[103] Hydrolysis of the trifunctional silanes to produce trisilanol species can occur with water present in the solvent or water adsorbed on the silica surface.[102,104] The organic trisilanol species can then react via condensation of silanol groups on adjacent molecules to form a cross-linked network. Obviously, the rate of hydrolysis and cross-linking depends on the surface and solution concentrations of the various species. To prepare monolayer films from trifunctional silanes, it is important to minimize or eliminate water in the solvent so that the hydrolysis and cross-linking of the silanes occur at the silica surface and the cross-linked or polymeric layer becomes adsorbed onto the silica surface.[96,102]

In spite of the above drawbacks, formation of organic monolayers using silanization is being actively pursued by several research groups because the cross-linked monolayer formed is significantly more stable than the Au-thiolate SAMs.[97] Just as with thiol SAMs, surfaces with a variety of different functional groups can be prepared by selecting a silane with the desired functional group.[31,97] In many cases, the silane containing this desired end functional group can be directly assembled onto a surface. Sometimes, the reactivity of a functional group is too high, and both the silane group and functional group (e.g., amine) will react with the surface. In these cases an intermediate functional group that can be converted or derivatized to the final functional group after formation of the monolayer must be used.[31] Various examples of creating different surface chemistries using silane molecules were discussed in Sec. 2.2.

Like the thiol SAMs, patterned surfaces of silane monolayers can be fabricated. For example, lines of hydrophilic, amide-terminated silane monolayers surrounded by hydrophobic, perfluoro-terminated silane regions have been used to direct the growth of neuron cells on SAMs.[101] By using a mixture of hydrocarbon and fluorocarbon silanes, monolayers with phase-separated domains can be formed due to the immiscibility of the hydrocarbon and fluorocarbon chains of the two silanes.[98] Also, polymeric molecules with silane anchor groups have been assembled onto hydroxylated surfaces.[105] In fact, anchor groups can be used to form patterned organic monolayers on surfaces. The surface is first patterned by fabricating areas of different materials (Au, hydroxyl groups, oxides). Then molecules with the appropriate anchor groups are preferentially attached to different regions of the surface (e.g., alkyl thiol to gold). This strategy works for assembly of monomeric or polymeric monolayers.[73,106]

5. Catalytic Materials

In industrial processes such as petroleum refining, catalytic materials are used at high pressures (>1 atm) to produce a wide range of products (gasoline, jet and diesel fuel, lubricants, etc.) from crude oil. The active component in many of these catalytic materials is a transition metal (Pt, Ni, Co, etc.).[107–111] Since the cost of metals such as Pt can be high, they are usually prepared as small particles supported on a high-area oxide support such as alumina.[109] This approach allows a high metal surface area to be loaded into a reactor at a reasonable cost. (Catalytic processes in petroleum refining occur on the surface of the supported particles, so the high surface area of the supported particles allows high conversions to be obtained.) To obtain optimum catalyst activity and selectivity for a given reaction, the catalyst must be activated through a series of treatment steps. Some of the treatments involved in catalyst activation are drying, calcination, reduction, and sulfidation.[111] These treatments and the instrumentation required to characterize the catalyst surface structure (dispersion or particle size and oxidation state) produced by these treatments will be discussed in this section. In addition, many reaction studies have been done using metallic single-crystal surfaces as models of the supported catalysts. The instrumentation and issues involved in these model studies have been reviewed elsewhere.[112]

A common method for preparing supported catalysts is pore-fill impregnation.[109] A volume of solution containing metal ions that will just fill the pore volume of the oxide support is mixed with the support. For example, $Co(NO_3)_2$ can be used to produce a solution of Co ions. Solutions containing two metal ions are used to prepare bimetallic catalysts. The impregnated catalyst is then dried at low temperatures (ca. 100°C) under vacuum, nitrogen, or air to remove the solvent (typically water). The dried catalyst now contains both the positive metal ions and the corresponding anions dispersed on the oxide support, and is ready for conversion into the catalytically active state.[113] Usually this involves removal of the anion by calcination (heating in air) followed by reduction and sulfidation of the resulting supported metal-oxide particles. Sometimes it is desirable to retain the anion (e.g., chloride from chloroplatonic acid impregnation of alumina for reforming catalytic applications[114]).

A wide range of techniques are used to characterize catalyst surface structure and composition during the conversion from the starting dried state to the final active state.[115] Two important quantities to track during this conversion are the size of the supported particles and their oxidation state. One technique well suited to this task is XPS. From the intensity ratio of a peak from the supported particle to a peak from the support, the particle size can be calculated.[116–118] The benefit of using XPS to determine particle size is that it can be directly applied to all forms of supported particles (oxide, metallic, sulfide, etc.). From the XPS binding energy (BE) of a characteristic peak from the supported particle, the oxidation state can be determined.[119] Also, for treatments that result in only a partial reduction or sulfidation of the supported particles, XPS can provide a direct measure of the extent of reduction and sulfidation from the relative areas of the oxide, metallic, and sulfide peaks of a given element.[120,121] This information is particularly useful in bimetallic catalysts (e.g., CoMo) since the extent of reduction and sulfidation can be determined for each metal.

5.1. UHV Catalyst Treatment Instrumentation

The working environment of most catalytic materials used in petroleum refining is reducing or sulfiding, so it is essential to be able to examine catalysts in their reduced and sulfided states. On the other hand, XPS analysis is done under ultrahigh vacuum (UHV) conditions. To bridge this pressure gap, methods must be developed to remove the catalyst from the high-pressure and high-temperature environment of the reactor and insert it into the UHV analysis chamber of the XPS system.[112,120–133] In addition, the chemical composition and structure of the catalysts must be maintained during this transfer. Reduced or sulfided catalysts cannot be exposed, even for a fraction of second, to air since they readily oxidize. Several methods have been developed to avoid air exposure.[112] One method uses an isolation cell inside the UHV chamber.[122,123] This approach allows the sample to remain stationary while being cycled between UHV and high pressures. Thus, the standard methods of mounting metallic single crystals (resistive heating, LN_2 cooling, etc.) can be used with this setup. After the initial preparation and surface characterization of the sample under UHV, the isolation cell is closed and then filled with a gas mixture for either further treatment (e.g., reduction) or reaction. During the reaction, the gas mixture is circulated through the cell and the composition of the gas mixture can be sampled periodically using gas chromatography. At the end of the treatment or reaction, the gas is pumped away and the isolation cell is reopened to the UHV system for surface characterization. The major advantage of this system is that catalytic reactions can be run at pressures as high as 100 atm. The major drawback is that it is difficult to completely pump the reactor down to UHV conditions before the isolation cell is opened. As a result the UHV environment can be degraded.

To overcome the above limitation, a variety of methods that involve transferring the sample between the UHV analysis chamber and the high-pressure reaction cell were developed. Some just involved a probe that could be sealed around a sample after high-pressure treatment or reaction, then inserted and reopened inside the UHV chamber once the gas had been pumped away.[124–126] Other systems were developed where the sample was mounted on a rod with sliding seals and differential pumping so that the sample could be rapidly transferred between the UHV analysis chamber and the high-pressure reactor.[127–130] Depending on the sliding-seal design, pressures of up to 30 atm could be maintained in the reactor cell.[127] The major advantage of the sliding-seal design is that the catalyst can be transferred from the reactor to the UHV chamber within seconds without pumping down the reactor.[112] The rapid sample transfer minimizes the chance of contamination or changes in the catalyst structure. Since the reactor can be maintained at pressure, the sample can be easily shuttled back and forth to follow changes in catalyst composition that are occurring as a function of treatment or reaction time.[112] Like the isolation-cell system, the typical sample mounting techniques for metallic single crystals can be used with sliding-seal systems.

Many of the advantages of the sliding-seal and isolation-cell designs are not important when studying supported catalysts used in industrial processes. For most industrial applications, the steady-state, not transient, behavior of the catalyst is important. Thus, rapid sample transfer back and forth between the UHV analysis chamber and the high-pressure reactor is usually not essential. Also, since supported catalysts cannot be heated resistively, the reactor, gas mixture, and catalyst must all be heated. This situation means that the reactors can now be separated from the UHV analysis chamber by one or more valves, and a combination of transfer rods and sample trolleys can be used to transport the samples.[120,131–133] The catalyst is typically ground into a fine powder and then pressed into a pellet or disk and mounted on a holder that can be readily transferred by the rod and trolley systems.[120] In addition, if several different types of supported catalysts are to be examined, the issue of cross-contamination must be addressed. For example, catalysts used in hydrodesulfurization applications must be sulfided,[110] while other catalysts (e.g., catalysts used for CO hydrogenation) are readily poisoned by sulfur.[134] The easiest method to avoid sulfur cross-contamination is to use multiple reactors.[120] One reactor can be dedicated to sulfur-free treatments (calcination, reduction), while a second reactor can be used for sulfidation. An additional advantage of separating the UHV analysis chamber from the high-pressure reactors with valves and a movable sample holder is that the surface analysis equipment and sample treatment facilities can be operated independently. That is, the acquisition of XPS data can be done on one sample while others are being treated. This approach allows the expensive surface analysis equipment (e.g., a modern XPS system can cost over $500,000) to be in continuous use. In contrast, the surface analysis equipment in the isolation-cell and sliding-seal systems typically sit idle while the sample is in the high-pressure reactor.

5.2. UHV System for Supported Catalysts

A UHV catalyst treatment system developed by the author to meet the design criterion for supported catalysts discussed in the previous section is shown in Fig. 5.[120,135] This was a modular system that contained separate stations for XPS analysis, AES analysis, sample treatment at 1 atm, metal evaporation, sample storage, and sample introduction. These stations were connected by a 3.5-m, 10-cm-diameter transfer tube. Samples were moved through the transfer tube by a chain-driven trolley riding on a 1.25-cm shaft. Every 27.5 cm along the tube a set of four ports allowed access to the various stations. Sample transfer between the trolley and a given station was accomplished using a magnetically coupled manipulator.

The system used for XPS analysis was a modified Hewlett–Packard 5950A ESCA spectrometer. This spectrometer uses a monochromatized Al K_α X-ray source, a hemispherical analyzer, and a multichannel plate–resistive strip detector. A low-energy electron flood gun is also available for charge control during analysis of supported catalysts. A Physical Electronics 10-155 single-pass cylindrical mirror

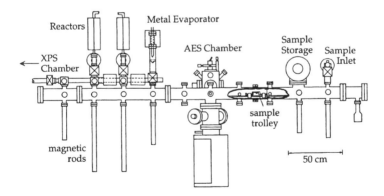

Figure 5. Schematic of a UHV system used to modify the surface structure of supported catalysts by calcination, reduction, and sulfidation treatments. The modular system is designed so that samples can be transferred between the surface analysis stations (XPS, AES) and the reactors under UHV conditions. See the text for a detailed description of this system.

analyzer with a coaxially mounted high-energy electron gun was used for AES analysis.

The catalyst treatment system in Fig. 5 had two quartz reactors for sample treatments at pressures of 1 atm. One reactor was used exclusively for calcination and hydrogen reduction treatments.[113,120,135] The second reactor was used for sulfidation treatments with 1% H_2S in hydrogen.[121] Either static or flowing gas treatments could be carried out. The quartz reactors were heated by focused quartz–halogen lamps, and the temperature profiles were monitored and controlled with a microprocessor. Each reactor was evacuated by a separate turbomolecular pump and was isolated from the rest of the UHV system with all-metal valves and differentially pumped conduction barriers.

To prepare supported catalysts for analysis in the system in Fig. 5, they are first ground to a fine powder and then a few milligrams of powder are pressed into a 200-mesh gold grid. The gold grid, which is completely covered by the pressed catalyst powder, adds strength to the pressed wafer, thereby minimizing the amount of powdered material required to make a wafer that can be transferred around the UHV system without breaking apart. For ease of transfer and treatment, the pressed wafer is mounted on a gold holder that can be easily manipulated by the transfer rods and trolley of the system. The combination of minimizing the amount of catalytic material in the pressed wafer and using a gold sample holder reduces outgassing when the sample and holder are transferred back into the UHV chamber after a high-pressure treatment.[120,121]

The typical experimental procedure after introduction of a pressed wafer into the UHV system is first to characterize the sample by XPS, then transfer it into one of the quartz reactors for treatment. The sample is transferred into the reactor under vacuum; then the transfer rod is retracted and the reactor isolated from the pumps and the UHV transfer tube. The gas flow (approximately 1 atm and >50 mL/min) is commenced, and the temperature of the flowing gas and sample are linearly ramped up to the desired treatment temperature. A high gas flow is used to ensure that any reaction products (e.g., water from reduction of metal-oxide particles) are rapidly swept out of the reactor.[120] After holding at the treatment temperature for a specified time, the reactor furnace is turned off and the sample is cooled to room temperature in the presence of the flowing gas. If a sulfidation treatment in H_2S/H_2 is being done, the gas flow is switched from H_2S/H_2 to pure H_2 when the temperature drops below 100°C to avoid adsorption of H_2S onto the catalyst.[121] After cooling, the gas flow is stopped, the reactor is evacuated, and the sample is returned to the XPS station for analysis. Typically, the reactor pumpdown and sample transfer can be accomplished in less than 5 min. The process of treatment and characterization is then repeated as required to document the change in catalyst structure at each step of the activation process.

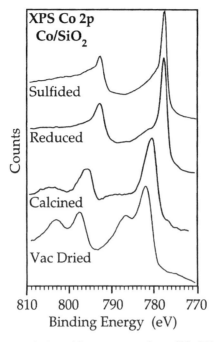

Figure 6. XPS Co $2p$ spectra acquired at ambient temperature from a SiO_2-923 support that was pore-fill impregnated with a $Co(NO_3)_2$ solution and then vacuum-dried at 100°C, calcined at 450°C, reduced in H_2 at 500°C, and sulfided in 1% H_2S/H_2 at 400°C.

5.3. Treatment and Characterization of Supported Catalysts

The Co $2p$ spectra obtained after sequential vacuum drying, calcination, reduction, and sulfidation of a silica-supported cobalt catalyst are shown in Fig. 6. This catalyst was prepared by pore-fill impregnation with a cobalt nitrate solution to produce a 9 wt% metal loading.[113] The binding-energy (BE) scales of all spectra were referenced by setting the Si $2p$ peak from the silica support to 103.0 eV. After calcination in 21% O_2/N_2 for 30 min at 450°C, only Si, O, and Co were detected. From the line shape and BEs of the Co $2p$ peaks it was determined that the cobalt species was Co_3O_4.[113] After treatment for 30 min at 500°C in hydrogen, the supported Co_3O_4 particles are completely reduced to metallic cobalt.[120] After the final treatment for 30 min at 400°C in 1% H_2S/H_2, the metallic cobalt is converted into cobalt sulfide.[121] Co sulfide and Co metal have similar Co $2p$ line shapes and BEs, but the presence of the S $2p$ peak at 162.5 eV after heating in 1% H_2S/H_2 confirms that the supported Co sulfide particles have been formed. Since after each treatment only one type of Co species is present, the particle size of the supported

Table 1. Average Sizes of Cobalt Particles Supported on Silica Determined from XPS Measurements[a]

		Average Co particle size (nm)	
Treatment	Co species	Co/SiO$_2$-62	Co/SiO$_2$-923
Vacuum dried	Co(NO$_3$)$_x$	2	2.5
Calcined	Co$_3$O$_4$	18	5.0
Reduced	Co	9	3.5
Sulfided	Co$_9$S$_8$	15	4.0

[a]The mean pore sizes of the two silica supports were 14.3 nm (SiO$_2$-62) and 3.3 nm (SiO$_2$-923).[(113)]

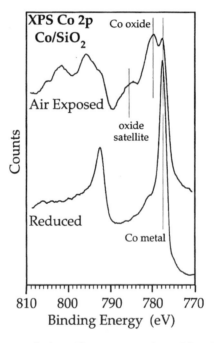

Figure 7. XPS Co 2p spectra acquired at ambient temperature from a fully reduced Co/SiO$_2$-62 catalyst analyzed after direct UHV transfer from the reactor to the XPS chamber (lower spectrum) and after exposure to air at ambient temperature for less than 1 min during the transfer from the reactor to the XPS chamber (upper spectrum). The brief air exposure results in significant oxidation of the metallic cobalt particles.

Co particles can readily be determined by the method of Kerkoff and Moulijn.[118] This has been done for two different silica-supported cobalt catalysts after drying, calcination, reduction, and sulfidation.[113,120,121] The results are shown in Table 1. Where Co particle sizes on the catalysts have been determined by other techniques, the agreement with the XPS results is reasonable.[113,120] For supported catalysts that contain multiple supported phases, more detailed calculations are required to determine the particle size of each phase.[116]

Several conclusions can be drawn from the results in Table 1. First, the dried catalysts have the highest dispersion (smallest particle size). Second, a large decrease in dispersion is observed when the catalysts are calcined. Third, smaller cobalt particles are produced on the silica support with the smaller pore size (SiO$_2$-923). Fourth, the cobalt particle size decreases upon reduction and then increases again after sulfidation. These two changes are probably due to removal of oxygen from the particles during reduction and to addition of sulfur to the particles during sulfidation.

To document the reasons for transferring the samples between the reactors and XPS analysis chamber under UHV conditions, Fig. 7 shows the Co $2p$ spectrum from a silica-supported cobalt catalyst that had been first reduced to metallic cobalt and then exposed to air for less than 1 min before acquiring the XPS data. Instead of just detecting metallic cobalt, both cobalt oxide and metallic cobalt species were detected. This result is due to the high reactivity of metallic cobalt particles with oxygen.

Acknowledgment

The author gratefully acknowledges the support of NIH grant RR-01296 during the preparation of this manuscript.

References

1. M. D. Lelah and S. L. Cooper, *Polyurethanes in Medicine*, CRC Press, Boca Raton, FL (1986).
2. B. D. Ratner, Advances in the Analysis of Surfaces of Biomedical Interest, *Surf. Interface Anal.* **23**, 521–528 (1995).
3. B. D. Ratner and D. G. Castner (eds.), *Surface Modification of Polymeric Biomaterials*, Plenum Press, New York (1997).
4. B. D. Ratner, Surface Modification of Polymers: Chemical, Biological, and Surface Analytical challenges, *Biosensors Bioelectron.* **10**, 797–804 (1995).
5. S. Shindo, A. Takagi, and A. D. Whitmore, Improved Patency of Collagen-impregnated Grafts after *in vitro* Autogenous Endothelial Cell Seeding, *J. Vasc. Surg.* **6**, 325–332 (1987).
6. B. D. Ratner, A. Chilkoti, and D. G. Castner, Contemporary Methods for Characterizing Complex Biomaterial Surfaces, *Clin. Mater.* **11**, 25–36 (1992).
7. D. Briggs and M. P. Seah (eds.), *Practical Surface Analysis,* 2nd ed., Vol. 1. Wiley, Chichester (1990).

8. E. E. Johnston and B. D. Ratner, Surface Characterization of Plasma Deposited Organic Thin Films, *J. Electron. Spectrosc. Rel. Phenom.* **81**, 303–317 (1996).

9. J. C. Vickerman (ed.), *Surface Analysis—Techniques and Applications*, Wiley, Chichester (1997).

10. C. Linsmeier, Auger Electron Spectroscopy, *Vacuum* **45**(6/7), 673–690 (1994).

11. A. Benninghoven, Surface Analysis by Secondary Ion Mass Spectrometry (SIMS), *Surf. Sci.* **299/300**, 246–260 (1994).

12. P. C. Zalm, Secondary Ion Mass Spectrometry, *Vacuum* **45**(6/7), 753–772 (1994).

13. J. R. Rasmussen, E. R. Stedronsky, and G. M. Whitesides, Introduction, Modification, and Characterization of Functional Groups on the Surface of Low-Density Polyethylene Film, *J. Am. Chem. Soc.* **99**(14), 4736–4745 (1977).

14. R. G. Nuzzo and G. Smolinsky, Preparation and Characterization of Functionalized Polyethylene Surfaces, *Macromolecules* **17**, 1013–1019 (1984).

15. C. F. Amstein and P. A. Hartman, Adaptation of Plastic Surfaces for Tissue Culture by Glow Discharge, *J. Clin. Microbiol.* **2**(1), 46–54 (1975).

16. L. J. Gerenser, J. F. Elman, M. G. Mason, and J. M. Pochan, ESCA Studies of Corona-Discharge-Treated Polyethylene Surfaces by Use of Gas-Phase Derivatization, *Polymer* **26**(8), 1162–1166 (1985).

17. J. L. Grant, D. S. Dunn, and D. J. McClure, Argon and Oxygen Sputter Etching of Polystyrene, Polypropylene, and Poly(ethylene terephthalate) Thin Films, *J. Vac. Sci. Technol.* A **6**(4), 2213–2220 (1988).

18. T. A. Horbett and L. A. Klumb, Cell Culturing: Surface Aspects and Considerations, in: *Interfacial Phenomena and Bioproducts* (J. L. Brash and P. W. Wojciechowski, eds.), Marcel Dekker, New York (1996), pp. 351–445.

19. Y. Suzuki, M. Kusakabe, and M. Iwaki, Wettability Control of Polystyrene by Ion Implantation, *Nucl. Inst. Meth. Phys. Res.* B **91**, 584–587 (1994).

20. P. L. Valint, Jr., D. M. Ammon, Jr., George L. Grobe III, and J. A. McGee, In-Situ Surface Modification of Contact Lens Polymers, in: *Surface Modification of Polymeric Biomaterials* (B. D. Ratner and D. G. Castner, eds.), Plenum Press, New York (1997), pp. 21–26.

21. J. D. Andrade, X-ray Photoelectron Spectroscopy (XPS), in: *Surface and Interfacial Aspects of Biomedical Polymers: Surface Chemistry and Physics*, Vol. 1 (J. D. Andrade, ed.), Plenum Press, New York (1985), pp. 105–195.

22. C. D. Batich, Chemical Derivatization and Surface Analysis, *Appl. Surf. Sci.* **32**, 57–73 (1988).

23. A. Chilkoti and B. D. Ratner, Chemical Derivatization Methods for Enhancing the Analytical Capabilities of X-ray Photoelectron Spectroscopy and Static Secondary Ion Mass Spectrometry, in: *Surface Characterization of Advanced Polymers* (L. Sabbatini and P. G. Zambonin, eds.), VCH, Weinheim, Germany (1993), pp. 221–256.

24. A. Chilkoti and B. D. Ratner, An X-ray Photoelectron Spectroscopic Investigation of the Selectivity of Hydroxyl Derivatization Reactions, *Surf. Interface Anal.* **17**, 567–574 (1991).

25. M. R. Alexander, P. V. Wright, and B. D. Ratner, Trifluoroethanol Derivatization of Carboxylic Acid-Containing Polymers for Quantitative XPS analysis, *Surf. Interface Anal.* **24**, 217–220 (1996).

26. J. D. Roberts and M. C. Caserio, *Basic Principles of Organic Chemistry*, Benjamin, New York (1965).

27. L. H. Dubois and R. G. Nuzzo, Synthesis, Structure, and Properties of Model Organic Surfaces, *Ann. Rev. Phys. Chem.* **43**, 437–463 (1992).

28. J. Gun, R. Iscovici, and J. Sagiv, On the Formation and Structure of Self-Assembling Monolayers, *J. Coll. Interface Sci.* **101**(1), 201–213 (1984).

29. A. Ulman, *An Introduction to Ultrathin Organic Films*, Academic Press, Boston (1991).

30. P. Harder, K. Bierbaum, Ch. Woell, and M. Grunze, Induced Orientational Order in Long Alkyl Chain Aminosilane Molecules by Preadsorbed Octadecyltrichlorosilane on Hydroxylated Si(100), *Langmuir* **13**, 445–454 (1997).
31. N. Balachander and C. N. Sukenik, Monolayer Transformation by Nucleophilic Substitution: Applications to the Creation of New Monolayer Assemblies, *Langmuir* **6**, 1621–1627 (1990).
32. H. Mori, A. Hirao, and S. Nakahama, Protection and Polymerization of Functional Monomers. 21. Anionic Living Polymerization of (2,2-dimethyl-1,3-dioxolan-4-yl)methyl methacrylate, *Macromolecules* **27**(1), 35–39 (1994).
33. G. H. Engbers and J. Feijen, Current Techniques to Improve the Blood Compatibility of Biomaterial Surfaces, *Int. J. Artif. Org.* **14**(4), 199–215 (1991).
34. A. S. Hoffman, Bioreactions in Medicine, *Biomat., Art. Cells, Art. Org.* **18**(4), 523–528 (1990).
35. A. C. Jen, C. Wake, and A. G. Mikos, Review: Hydrogels for Cell Immobilization, *Biotechnol. Bioeng.* **50**(4), 357–364 (1996).
36. J. F. Kennedy, E. H. M. Melo, and K. Jumel, Immobilized Enzymes and Cells, *Chem. Eng. Prog.* **86**(7), 81–90 (1990).
37. W. Marconi, Immobilized Enzymes: Their Catalytic Behaviour and Their Industrial and Analytical Applications, *Reactive Polym.* **11**, 1–19 (1989).
38. A. S. Hoffman, Surface Modification of Polymers: Physical, Chemical, Mechanical, and Biological Methods, *Macromol. Symp.* **101**, 443–454 (1996).
39. M.-S. Sheu, A. S. Hoffman, B. D. Ratner, J. Feijen, and J. M. Harris, Immobilization of Polyethylene Oxide Surfactants for Non-Fouling Biomaterial Surfaces Using an Argon Glow Discharge Treatment, *J. Adhesion Sci. Tech.* **7**(10), 1065–1076 (1993).
40. P. Favia, M. Stendardo, and R. d'Agostino, Selective Grafting of Amine Groups on Polyethylene by Means of NH_3-H_2 RF Glow Discharges, *Plasmas Polym.* **1**(2), 91–112 (1996).
41. T. G. Vargo, E. J. Bekos, Y. S. Kim, J. P. Ranieri, R. Bellamkonda, P. Aebischer, D. E. Margevich, P. M. Thompson, F. V. Bright, and J. A. Gardella, Jr., Synthesis and Characterization of Fluoropolymeric Substrata with Immobilized Minimal Peptide Sequences for Cell Adhesion Studies. I, *J. Biomed. Mater. Res.* **29**, 767–778 (1995).
42. T. G. Vargo, J. A. Gardella, Jr., A. E. Meyer, and R. E. Baier, Hydrogen/Liquid Vapor Radio Frequency Glow Discharge Plasma Oxidation/Hydrolysis of Expanded Poly(tetrafluoroethylene) (ePTFE) and Poly(vinylidene fluoride) (PVDF) Surfaces, *J. Polym. Sci. A* **29**, 555–570 (1991).
43. J. Spinke, M. Liley, H. J. Guder, L. Angermaier, and W. Knoll, Molecular Recognition at Self-Assembled Monolayers: The Construction of Multicomponent Multilayers, *Langmuir* **9**, 1821–1825 (1993).
44. G. B. Sigal, C. Bamdad, A. Barbaris, J. Strominger, and G. M. Whitesides, A Self-assembled Monolayer for the Binding and Study of Histidine-tagged Proteins by Surface Plasmon Resonance, *Anal. Chem.* **68**, 490–497 (1996).
45. Y. W. Lee, J. Reed-Mundell, C. N. Sukenik, and J. E. Zull, Electrophilic Siloxane-based Self-assembled Monolayers for Thiol-mediated Anchoring of Peptides and Proteins, *Langmuir* **9**, 3009–3014 (1993).
46. H. Yasuda, *Plasma Polymerization,* Academic Press, New York (1985).
47. R. d'Agostino, F. Cramarossa, F. Fracassi, and F. Illuzzi, Plasma Polymerization of Fluorocarbons, in: *Plasma Deposition, Treatment and Etching of Polymers* (R. d'Agostino, ed.), Academic Press, Boston (1990), pp. 95–162.
48. E. Kay, J. W. Coburn, and A. Dilks, Plasma Chemistry of Fluorocarbons as Related to Plasma Etching and Plasma Polymerization, in: *Plasma Chemistry III*, Vol. 94 (S. Veprek and M. Venugopalan, eds.), Springer-Verlag, Berlin (1980), pp. 1–41.
49. J. W. Coburn and H. F. Winters, Plasma Etching—A Discussion of Mechanisms, *J. Vac. Sci. Technol. B* **7**, 391 (1979).

50. H. F. Winters and I. C. Plumb, Etching Reactions for Silicon with F Atoms: Product Distributions and Ion Enhancement Mechanisms, *J. Vac. Sci. Technol. B* **9**(2), 197–207 (1991).

51. M. A. Hartney, D. W. Hess, and D. S. Soane, Oxygen Plasma Etching for Resist Stripping and Multilayer Lithography, *J. Vac. Sci. Technol. B* **7**, 1–13 (1989).

52. F. D. Egitto, V. Vukanovic, and G. N. Taylor, Plasma Etching of Organic Polymers, in: *Plasma Deposition, Treatment and Etching of Polymers* (R. d'Agostino, ed.), Academic Press, Boston (1990), pp. 321–422.

53. N. M. Mackie, N. F. Dalleska, D. G. Castner, and E. R. Fisher, Comparison of Pulsed and Continuous Wave Deposition of Thin Films from Saturated Fluorocarbon/H_2 Inductively Coupled RF Plasmas, *Chem. Mater.* **9**, 349–362 (1997).

54. P. Favia, V. H. Perez-Luna, T. Boland, D. G. Castner, and B. D. Ratner, Surface Chemical Composition and Fibrinogen Adsorption-Retention Properties of Fluoropolymer Films Deposited from an RF Glow Discharge, *Plasmas Polym.* **1**, 297–324 (1996).

55. B. D. Ratner, A. Chilkoti, and G. P. Lopez, Plasma Deposition and Treatment for Biomaterial Applications, in: *Plasma Deposition, Treatment and Etching of Polymers* (R. d'Agostino, ed.), Academic Press, Boston (1990), pp. 463–516.

56. D. G. Castner, K. B. Lewis, D. A. Fischer, B. D. Ratner, and J. L. Gland, Determination of Surface Structure and Orientation of Polymerized Tetrafluoroethylene Films by Near Edge X-ray Absorption Fine Structure, X-ray Photoelectron Spectroscopy, and Static Secondary Ion Mass Spectrometry, *Langmuir* **9**, 537–542 (1993).

57. S. I. Ertel, A. Chilkoti, T. A. Horbett, and B. D. Ratner, Endothelial Cell Growth on Oxygen-containing Films Deposited by Radio-Frequency Plasmas: The Role of Surface Carbonyl Groups, *J. Biomater. Sci. Polym. Ed.* **3**, 163–183 (1991).

58. G. P. Lopez and B. D. Ratner, Substrate Temperature Effects of Film Chemistry in Plasma Deposition of Organics. I: Nonpolymerizable Precursors, *Langmuir* **7**, 766–773 (1991).

59. D. Kiaei, A. S. Hoffman, and T. A. Horbett, Tight Binding of Albumin to Glow Discharge Treated Polymers, *J. Biomater. Sci. Polym. Ed.* **4**(1), 35–44 (1992).

60. E. E. Johnston and B. D. Ratner, XPS and SSIMS Characterization of Surfaces Modified by Plasma-Deposited Oligo(glyme) Films, in: *Surface Modification of Polymeric Biomaterials* (B. D. Ratner and D. G. Castner, eds.), Plenum Press, New York (1997), pp. 35–44.

61. G. P. Lopez and B. D. Ratner, Substrate Temperature Effects on Film Chemistry in Plasma Deposition of Organics. II: Polymerizable Precursors, *J. Polym. Sci., Polym. Chem. Ed.* **30**, 2415–2425 (1992).

62. G. P. Lopez and B. D. Ratner, Molecular Adsorption and the Chemistry of Plasma-deposited Thin Organic Films: Deposition of Oligomers of Ethylene Glycol, *Plasmas Polym.* **1**(2), 127–151 (1996).

63. C. R. Savage, R. B. Timmons, and J. W. Lin, Molecular Control of Surface Film Compositions via Pulsed Radio Frequency Plasma Deposition of Perfluoropropylene, *Chem. Mater.* **3**, 575–577 (1991).

64. V. Panchalingam, B. Poon, H. H. Huo, C. R. Savage, R. B. Timmons, and R. C. Eberhart, Molecular Surface Tailoring of Biomaterials via Pulsed RF Plasma Discharges, *J. Biomater. Sci. Polymer. Ed.* **5**(1/2), 131–145 (1993).

65. R. G. Nuzzo and D. L. Allara, Adsorption of bifunctional organic disulfides on gold surfaces, *J. Am. Chem. Soc.* **105**, 4481–4483 (1983).

66. G. M. Whitesides, G. S. Ferguson, D. L. Allara, D. Scherson, L. Speaker, and A. Ulman, Organized Molecular Assemblies, *Crit. Rev. Surf. Chem.* **3**, 49–65 (1993).

67. D. G. Castner, K. Hinds, and D. W. Grainger, X-ray Photoelectron Spectroscopy Sulfur 2p Study of Organic Thiol and Disulfide Binding Interactions with Gold Surfaces, *Langmuir* **12**(21), 5083–5086 (1996).

68. E. B. Troughton, C. D. Bain, G. M. Whitesides, R. G. Nuzzo, D. L. Allara, and M. D. Porter, Monolayer Films Prepared by the Spontaneous Self-assembly of Symmetrical and Unsymmetrical Dialkyl Sulfides from Solution onto Gold Substrates: Structure, Properties, and Reactivity of Constituent Functional Groups, *Langmuir* **4**, 365–385 (1988).

69. C. D. Bain, E. B. Troughton, Y. T. Tao, J. Evall, G. M. Whitesides, and R. G. Nuzzo, Formation of Monolayer Films by the Spontaneous Assembly of Organic Thiols from Solution onto Gold, *J. Am. Chem. Soc.* **111**(1), 321–335 (1989).

70. C. D. Bain, H. A. Biebuyck, and G. M. Whitesides, Comparison of Self-Assembled Monolayers on Gold: Coadsorption of Thiols and Disulfides, *Langmuir* **5**, 723–727 (1989).

71. G. Hahner, C. Woll, M. Buck, and M. Grunze, Investigation of Intermediate Steps in the Self-Assembly of n-Alkanethiols on Gold Surfaces by Soft X-ray Spectroscopy, *Langmuir* **9**(8), 1955–1958 (1993).

72. C. D. Tidwell, S. I. Ertel, B. D. Ratner, B. Tarasevich, S. Atre, and D. L. Allara, Endothelial Cell Growth and Protein Adsorption on Terminally Functionalized, Self-Assembled Monolayers of Alkanethiolates on Gold, *Langmuir*, **13**(13), 3404–3413 (1997).

73. F. Sun, D. G. Castner, G. Mao, W. Wang, P. J. McKeown, and D. W. Grainger, Spontaneous Polymer Thin Film Assembly and Organization Using Mutually Immiscible Side Chains, *J. Am. Chem. Soc.* **118**(8), 1856–1866 (1996).

74. C. D. Frisbie, E. W. Wollman, and M. S. Wrighton, High Lateral Resolution Imaging by Secondary Ion Mass Spectrometry of Photopatterned Self-Assembled Monolayers Containing Aryl Azide, *Langmuir* **11**, 2563–2571 (1995).

75. J. J. Hickman, D. Ofer, C. Zou, M. S. Wrighton, P. E. Laibinis, and G. M. Whitesides, Selective Functionalization of Gold Microstructures with Ferrocenyl Derivatives via Reaction with Thiols or Disulfides: Characterization by Electrochemistry and Auger Electron Spectroscopy, *J. Am. Chem. Soc.* **113**(4), 1128–1132 (1991).

76. J. M. Behm, K. R. Lykke, M. J. Pellin, and J. C. Hemminger, Projection Photolithography Utilizing a Schwarzschild Microscope and Self-Assembled Alkane Thiol Monolayers as Simple Photoresists, *Langmuir* **12**, 2121–2124 (1996).

77. A. Kumar, N. L. Abbott, E. Kim, H. A. Biebuyck, and G. M. Whitesides, Patterned Self-Assembled Monolayers and Meso-Scale Phenomena, *Acc. Chem. Res.* **28**, 219–226 (1995).

78. J. M. Calvert, Lithographic Patterning of Self-Assembled Films. *J. Vac. Sci. Technol.* **11**(6), 2155–2163 (1993).

79. A. Kumar, H. A. Biebuyck, and G. M. Whitesides, Patterning Self-Assembled Monolayers: Applications in Materials Science, *Langmuir* **10**(5), 1498–1511 (1994).

80. K. L. Prime and G. M. Whitesides, Self-Assembled Organic Monolayers: Model Systems for Studying Adsorption of Proteins at Surfaces, *Science* **252**, 1164–1167 (1991).

81. D. A. Offord, C. M. John, M. R. Lindford, and J. H. Griffin, Contact Angle Goniometry, Ellipsometry, and Time-of-Flight Secondary Ion Mass Spectrometry of Gold Supported, Mixed Self-Assembled Monolayers Formed From Alkyl Mercaptans, *Langmuir* **10**, 883–889 (1994).

82. M.-W. Tsao, C. L. Hoffman, J. F. Rabolt, H. E. Johnson, D. G. Castner, C. Erdelen, and H. Ringsdorf, Studies of Molecular Orientation and Order in Self-Assembled Semifluorinated n-Alkylthiols: Single and Dual Component Mixtures, *Langmuir*, **13**(16), 4317–4322 (1997).

83. S. J. Stranick, A. N. Parikh, Y. T. Tal, D. L. Allara, and P. S. Weiss, Phase Separation of Mixed-Composition Self-Assembled Monolayers into Nanometer Scale Molecular Domains, *J. Phys. Chem.* **98**, 7636–7646 (1994).

84. H. A. Biebuyck and G. M. Whitesides, Interchange Between Monolayers on Gold Formed from Unsymmetrical Disulfides and Solutions of Thiols: Evidence for Sulfur-Sulfur Bond Cleavage by Gold Metal, *Langmuir* **9**, 1766–1770 (1993).

85. J. Huang and J. C. Hemminger, Photooxidation of Thiols in Self-Assembled Monolayers on Gold, *J. Am. Chem. Soc.* **115**, 3342–3343 (1993).

86. D. A. Hutt and G. J. Leggett, Influence of Adsorbate Ordering on Rates of UV Photo-Oxidation of Self-Assembled Monolayers, *J. Phys. Chem.* **100**, 6657–6662 (1996).

87. M. J. Tarlov, D. R. F. Burgess, Jr., and G. Gillen, UV Photopatterning of Alkanethiolate Monolayers Self-Assembled on Gold and Silver, *J. Am. Chem. Soc.* **115**, 5305–5306 (1993).

88. C. G. Worley and R. N. Linton, Removing Sulfur from Gold Using Ultraviolet/Ozone Cleaning, *J. Vac. Sci. Technol. A* **13**, 2281–2284 (1995).

89. F. Sun, D. G. Castner, and D. W. Grainger, Ultrathin Self-Assembled Polymeric Films on Solid Surfaces. II. Formation of 11-(n-pentyldithio)undecylate-bearing Polyacrylate Monolayers on Gold, *Langmuir* **9**(11), 3200–3207 (1993).

90. F. Sun, D. W. Grainger, and D. G. Castner, Ultrathin Self-Assembled Polymeric Films on Solid Surfaces. III. Influence of Acrylate Dithioalkyl Side Chain Length on Polymeric Monolayer Formation on Gold, *J. Vac. Sci. Technol. A* **12**(4), 2499–2506 (1994).

91. F. Sun, D. W. Grainger, D. G. Castner, and D. K. Leach-Scampavia, Adsorption of Ultrathin Films of Sulfur-Containing Siloxane Oligomers on Gold Surfaces and Their In Situ Modification, *Macromolecules* **27**, 3053–3062 (1994).

92. W. Wang, D. G. Castner, and D. W. Grainger, Ultrathin Films of Perfluoropolyether-Grafted Polysiloxanes Chemisorbed via Alkylthiolate Anchors to Gold Surfaces, *Supramol. Sci.* **4**, 83–99 (1997).

93. T. J. Lenk, V. M. Hallmark, J. F. Rabolt, L. Haussling, and H. Ringsdorf, Formation and Characterization of Self-Assembled Films of Sulfur-Derivatized Poly(methyl methacrylates), *Macromolecules* **26**, 1230–1237 (1993).

94. F. Sun and D. W. Grainger, Ultrathin Self-Assembled Polymeric Films on Solid Surfaces. I. Synthesis and Characterization of Acrylate Copolymers Containing Alkyl Disulfide Side Chains, *J. Polym. Sci. Part A: Polym. Chem.* **31**, 1729–1740 (1993).

95. D. G. Castner, W. Wang, and D. W. Grainger, Structural Anisotropy in Perfluoroalkyl-Grafted Polysiloxane Monolayer Films Bound to Gold Surfaces, manuscript in preparation (1997).

96. C. P. Tripp and M. L. Hair, Direct Observation of the Surface Bonds Between Self-Assembled Monolayers of Octadecyltrichlorosilane and Silica Surfaces: A Low-Frequency IR Study at the Solid/Liquid Interface, *Langmuir* **11**(4), 1215–1219 (1995).

97. R. Maoz, L. Netzer, J. Gun, and J. Sagiv, Self-Assembling Monolayers in the Construction of Planned Supramolecular Structures and as Modifiers of Surface Properties, *J. Chim. Phys.* **85**(11/12), 1059–1064 (1988).

98. S. Ge, A. Takahara, and T. Kajiyama, Phase Separated Morphology of an Immobilized Organosilane Monolayer Studied by a Scanning Probe Microscope, *Langmuir* **11**(4), 1341–1346 (1995).

99. S. R. Wasserman, G. M. Whitesides, I. M. Tidswell, B. M. Ocko, P. S. Pershan, and J. D. Axe, The Structure of Self-Assembled Monolayers of Alkylsiloxanes on Silicon: A Comparison of Results from Ellipsometry and Low-Angle X-ray Reflectivity, *J. Am. Chem. Soc.* **111**, 5852–5861 (1989).

100. H. Hoffman, U. Mayer, and A. Krischanitz, Structure of Alkylsiloxane Monolayers on Silicon Surfaces Investigated by External Reflection Infrared Spectroscopy, *Langmuir* **11**, 1304–1312 (1995).

101. D. A. Stenger, J. H. Georger, C. S. Dulcey, J. J. Hickman, A. S. Rudolph, T. B. Nielsen, S. M. McCort, and J. M. Calvert, Coplanar Molecular Assemblies of Amino- and Perfluorinated Alkylsilanes: Characterization and Geometric Definition of Mammalian Cell Adhesion and Growth, *J. Am. Chem. Soc.* **114**(22), 8435–8442 (1992).

102. C. P. Tripp and M. L. Hair, Reaction of Alkylchlorosilanes with Silica at the Solid/Gas and Solid/Liquid Interface, *Langmuir* **8**, 1961–1967 (1992).

103. C. P. Tripp, R. P. N. Veregin, and M. L. Hair, Effect of Fluoroalkyl Substituents on the Reaction of Alkylchorosilanes with Silica Surfaces, *Langmuir* **9**, 3518–1522 (1993).

104. C. R. Kessel and S. Granick, Formation and Characterization of a Highly Ordered and Well-Anchored Alkylsilane Monolayer on Mica by Self-Assembly, *Langmuir* **7**, 532–538 (1991).
105. G. Mao, D. G. Castner, and D. W. Grainger, Polymer Immobilization to Alkylchorosilane Organic Monolayer Films Using Sequential Derivatization Reactions, *Chem. Mater.,* **9**(8), 1741–1750 (1997).
106. P. E. Laibinis, J. J. Hickman, M. S. Wrighton, and G. M. Whitesides, Orthogonal Self-Assembled Monolayers: Alkanethiols on Gold and Alkane Carboxylic Acids on Alumina, *Science* **245**, 845–847 (1989).
107. G. A. Somorjai, *Introduction to Surface Chemistry and Catalysis*, Wiley, New York (1994).
108. E. Christoffel, *Laboratory Studies of Heterogeneous Catalytic Processes*, Elsevier, Amsterdam (1989).
109. G. C. Bond, *Heterogeneous Catalysis: Principles and Applications*, Clarendon Press, Oxford (1987).
110. B. C. Gates, J. R. Katzer, and G. C. A. Schuit, *Chemistry of Catalytic Processes*, McGraw-Hill, New York (1979).
111. V. Ponec and G. C. Bond, *Catalysis by Metals and Alloys*, Elsevier, Amsterdam (1995).
112. C. T. Campbell, Studies of Model Catalysts with Well-Defined Surfaces Combining Ultrahigh Vacuum Surface Characterization with Medium- and High-Pressure Kinetics, *Adv. Catal.* **36**, 1–49 (1989).
113. D. G. Castner, P. R. Watson, and I. Y. Chan, X-ray Absorption Spectroscopy, X-ray Photoelectron Spectroscopy and Analytical Electron Microscopy Studies of Cobalt Catalysts. I. Characterization of the Calcined Catalysts, *J. Phys. Chem.* **93**, 3188–3194 (1989).
114. S. R. Tennison, Recent Advances in Catalytic Reforming, *Chem. Britain* **17**, 536–540 (1981).
115. J. W. Niemantsverdriet, *Spectroscopy in Catalysis*, VCH, Weinheim (1993).
116. A. Frydman, D. G. Castner, M. Schmal, and C. T. Campbell, A Method for Accurate Quantitative XPS Analysis of Multimetallic and Multiphase Catalysts on Support Particles, *J. Catal.* **157**, 133–144 (1995).
117. H. P. C. E. Kuipers, H. C. E. van Leuven, and W. M. Visser, The Characterization of Heterogeneous Catalysts by XPS Based on Geometrical Probability. 1. Monometallic Catalysts, *Surf. Interface Anal.* **8**, 235–242 (1986).
118. F. P. J. M. Kerkhof and J. A. Moulijn, Quantitative Analysis of XPS Intensities for Supported Catalysts, *J. Phys. Chem.* **83**, 1612–1619 (1979).
119. B. D. Ratner and D. G. Castner, Electron Spectroscopy for Chemical Analysis, in: *Surface Analysis—Techniques and Applications* (J. C. Vickerman, ed.), Wiley, Chichester (1997), pp. 43–98.
120. D. G. Castner, P. R. Watson, and I. Y. Chan, X-ray Absorption Spectroscopy, X-ray Photoelectron Spectroscopy, and Analytical Electron Microscopy Studies of Cobalt Catalysts. II. Hydrogen Reduction Properties, *J. Phys. Chem.* **94**, 819–828 (1990).
121. D. G. Castner and P. R. Watson, X-ray Absorption Spectroscopy and X-ray Photoelectron Spectroscopy Studies of Cobalt Catalysts. III. Sulfidation Properties in H_2S/H_2, *J. Phys. Chem.* **95**, 6617–6623 (1991).
122. D. W. Blakely, E. Kozak, B. A. Sexton, and G. A. Somorjai, New Instrumentation and Techniques to Monitor Chemical Surface Reactions on Single Crystals over a Wide Pressure Range in the Same Apparatus, *J. Vac. Sci. Technol.* **13**, 1091–1096 (1976).
123. A. L. Cabrera, N. D. Spencer, E. Kozak, P. W. Davies, and G. A. Somorjai, Improved Instrumentation to Carry Out Surface Analysis and to Monitor Chemical Surface Reactions *In Situ* on Small Area Catalysts over a Wide Pressure Range (10^{-8}–10^5 Torr), *Rev. Sci. Instrum.* **53**, 1888–1893 (1982).
124. T. P. Debies and D. M. Schrall, An Economical Transfer Vessel with Detachable Reactor for XPS Studies, *J. Electron Spec. Rel. Phenom.* **31**, 191–196 (1983).

125. T. A. Patterson, J. C. Carver, D. E. Leyden, and D. M. Hercules, A Surface Study of Cobalt-Molybdena-Alumina Catalysts Using X-ray Photoelectron Spectroscopy, *J. Phys. Chem.* **80**, 1700–1708 (1976).

126. W. Grunert, R. Feldhaus, K. Anders, E. S. Shapiro, G. V. Anoshin, and Kh. M. Minachev, A New Facility for Inert Transfer of Reactive Samples to XPS Equipment, *J. Electron Spec. Rel. Phenom.* **40**, 187–192 (1986).

127. H. D. Polaschegg, A. Jungel, and E. Schirk, A High Pressure Preparation Lock for Surface Analysis, *Surf. Interface Anal.* **1**, 100–103 (1979).

128. H. J. Krebs, H. P. Bonzel, and G. Gafner, A Model Study of the Hydrogenation of CO over Polycrystalline Iron, *Surf. Sci.* **88**, 269–283 (1979).

129. C. T. Campbell and M. T. Paffett, Model Studies of Ethylene Epoxidation Catalyzed by the Ag(110) Surface, *Surf. Sci.* **139**, 396–416 (1987).

130. O. Ganschow, R. Jede, L. D. An, E. Manske, J. Neelsen, L. Wiedmann, and A. Benninghoven, A Combined Instrument for the Investigation of Catalytic Reactions by Means of Gas Chromatography, Secondary Ion and Gas Phase Mass Spectrometry, Auger and Photoelectron Spectroscopy, and Ion Scattering Spectroscopy, *J. Vac. Sci. Technol. A* **1**, 1491–1506 (1983).

131. D. W. Goodman, R. D. Kelley, T. E. Madey, and J. T. Yates, Kinetics of the Hydrogenation of CO over a Single Crystal Nickel Catalyst, *J. Catal.* **63**, 226–234 (1980).

132. C. W. Miller, J. R. Fagan, D. H. Karweik, and T. Kuwana, A Versatile Sample Isolation, Chemical Modification, and Introduction System Designed for a Physical Electronics Model 548 Electron Spectrometer, *Appl. Surf. Sci.* **9**, 214–226 (1981).

133. D. W. Dwyer and J. H. Hardenbergh, The Catalytic Reduction of Carbon Monoxide over Iron Surfaces: A Surface Science Investigation, *J. Catal.* **87**, 66–76 (1984).

134. C. H. Bartholomew and J. R. Katzer, Sulfur Poisoning of Nickel in CO Hydrogenation, in *Catalyst Deactivation* (B. Delmon and G. F. Froment, eds.), Elsevier, Amsterdam (1980), pp. 375–396.

135. D. G. Castner and D. S. Santilli, X-ray Photoelectron Spectroscopy of Cobalt Catalysts: Correlation with Carbon Monoxide Hydrogenation Activities, in: *Catalytic Materials: Relationship Between Structure and Reactivity* (T. E. Whyte, Jr., R. A. Dalla Betta, E. G. Derouane, and R. T. Baker, eds.), ACS Symposium Series 248, Washington, DC (1984), pp. 39–56.

8

Physical and Chemical Methods for Thin-Film Deposition and Epitaxial Growth

John R. Arthur

1. INTRODUCTION

Thin-film deposition is the process of adding a thin layer of one material on another in order to modify the properties of the initial underlying material, i.e., to increase hardness, to change electrical conduction, or to alter the optical properties of the initial material referred to as the substrate. In this chapter we consider thin-film deposition primarily as a method for improving the structural and electronic properties of surfaces in order to prepare these surfaces for other experiments. For many experiments the desired surface will be that of a highly perfect single crystal. This chapter is intended for the surface scientist who may have little experience with film deposition and, especially, with epitaxial film growth.

We begin with a summary of the most common methods for deposition of clean, ordered films on single-crystal substrates. We then discuss in more detail those processes leading to epitaxial films, with particular emphasis on those ultrahigh vacuum (UHV) methods that allow the use of *in situ* surface tools to provide real-time feedback on the growth process, namely, molecular beam epitaxy (MBE) and chemical beam epitaxy. We conclude with some discussion of the nature of

John R. Arthur • Department of Electrical and Computer Engineering, Oregon State University, Corvallis, Oregon 97331.

Specimen Handling, Preparation, and Treatments in Surface Characterization, edited by Czanderna *et al.* Kluwer Academic / Plenum Publishers, New York, 1998.

defects in epitaxial films, and the problems associated with mismatch between the lattice constant of the film and that of the substrate.

Thin films have become a significant part of microelectronic circuitry because of their small thickness, i.e., their nearly two-dimensional nature, which permits the fabrication of very small device structures, or microcircuits. The economic value of these electronic applications continues to be a tremendous driving force resulting in the development of a large arsenal of ever more sophisticated methods for deposition and characterization of very high purity films. Since contamination of integrated circuits is of major concern, many deposition methods utilize a UHV environment, which allows high purity to be maintained at the interface between film and substrate as well as on the surface of the growing film because contamination from the residual gases during deposition is essentially eliminated. In some very important cases, the high purity of the deposition environment combined with prior cleaning of the substrate permits the films to be deposited with a single-crystalline structure that "fits" on the crystal structure of the substrate, a process called *epitaxial growth*. Methods for achieving epitaxial growth will be the principal focus of our discussion.

Thin films have become important in recent years both from a scientific viewpoint and from their technological value to the semiconductor industry. In semiconductor devices, thin films are used to form the conducting lines connecting individual devices, as well as the contact pads to which are bonded the wires that connect the circuit to the encapsulating structure. They are also used to form insulating layers to isolate conducting films from the underlying device structure. In many cases, part or all of the active device may be composed of a single crystal of semiconductor material deposited epitaxially on the underlying supporting substrate, which provides a semiconductor material with properties not available from the bulk substrate crystal. In brief, thin-film technology is an essential component of microcircuit fabrication, and advances in circuit complexity have kept pace with advances in our understanding of how to control the deposition and properties of thin films. More thorough discussions of the complexities of thin film deposition are presented in Refs. 1–4; here we simply give an overview of the most common approaches.

Deposited thin films have also been extremely important in developing an understanding of surfaces and interfaces. Because of the essentially two-dimensional nature of films, many of the properties of films are dominated by the proximity of interfaces. Perhaps the most dramatic demonstration of thin-film science has resulted from the precisely controllable electronic properties of quantum wells,[5] i.e., the properties of thin films that have one or more dimensions smaller than the wavelength of the electrons in the film. Observations of the transport properties of films as a function of the thickness show clearly the transition from a continuum of energy states to discrete, quantized electronic states.[5]

Much of the recent scientific interest in films has come from the increasingly sophisticated and powerful array of analytical techniques for examining the atomic and electronic structures of thin films. Electron, ion, and photon spectroscopic techniques are now used to identify the chemical composition and bonding state of atoms in films with ever greater sensitivity and greater spatial resolution, while the new tunneling probe microscopies display the atomic microstructure. These techniques have provided a much better understanding of the nature of thin films and have played a crucial role in optimizing the methods for their deposition.[6,7]

Thus, while the main thrust of this chapter is on the application of thin films to modify and improve substrates, it is important to recognize that the modern tools of surface science have in turn provided the direct feedback about the structure and composition of films that has led to the modern improvements in deposition technology.

2. Techniques for the Deposition of Thin Films

Methods for the deposition of thin films [1-3] can be conveniently divided into two classes: (1) *physical deposition*, in which atoms or molecules of the source material are vaporized by physical processes such as thermal evaporation or bombardment of the source by an energetic beam of electrons, photons, or ions, and (2) *chemical deposition*, which rely on a chemical reaction with vapor species containing the film constituents that are incident on a substrate to produce a film of the desired composition. Physical deposition is typically carried out in high vacuum or UHV to avoid contamination of the film by the ambient atmosphere, while chemical deposition often utilizes an inert carrier gas, and may even be carried out at atmospheric pressure.

The steps common to all methods of film deposition are

1. Generating a supply of atoms or molecules from a source that may be solid, liquid, or vapor. This is accomplished by supplying energy to the source by heating, by adding kinetic energy from incident energetic particles, or by using a chemical reaction to produce volatile species.

2. Transport of the constituent atoms or molecules to the substrate. From a technological point of view, this is a crucial part of the deposition, because the manner of transport will determine the uniformity of coverage of the substrate. Obviously, large substrates are more difficult to cover uniformly than are small substrates; however, the mode of transport plays a considerable role in this.

3. Deposition of atoms or molecules on the substrate. The condition of the substrate surface is the key issue in this part of the process. If the substrate surface is contaminated with a fraction of a monolayer of impurity atoms, the surface mobility of the arriving film atoms may be reduced to an extent

that the crystal orientation of the film bears little relationship to that of the substrate, and the structure is likely to be polycrystalline. Conversely, if the substrate surface is essentially clean, the arriving film atoms are likely to adsorb in sites related to the substrate orientation, and the film may be oriented or even epitaxial.

4. Postgrowth treatment. In this part of the process, a prolonged annealing of the film at elevated temperatures may yield a film with superior properties as a result of crystal grain growth in the film or the changes in composition produced by diffusion of constituents. Focused energy beams such as lasers can be used to recrystallize an amorphous or polycrystalline film.

5. Process analysis. Modern analytical techniques[6,7] have been essential in improving the deposition parameters to increase the quality of the film in terms of structure or purity, or both. These procedures can involve surface cleaning and characterization of the substrate prior to growth, *in situ* analysis during growth using, e.g., electron diffraction to monitor film structure or optical measurements of growth rate, and postgrowth analysis, which might consist of compositional depth profiling by various surface spectroscopies coupled with controlled ion etching of the film (see Chapter 3, Volume 5 in this series), or optical or electrical measurements of film properties, or structural analysis by electron or X-ray diffraction. Regardless of the specific measurement, the feedback from the results of these techniques has been crucial in improving deposition processes.

3. Physical Deposition

Physical deposition is the transfer of atoms and molecules from a source to a substrate by a process that relies on physical methods to produce the vapor species. We now consider some typical physical deposition processes along with their advantages and disadvantages that relate to material transport to the substrate. Specific substrate treatments necessary to obtain epitaxial growth are discussed in Sec. 5.

3.1. Vacuum Evaporation

Except in a few notable instances that will be discussed, high-purity films are deposited in vacuum, most generally in UHV, which we define as a pressure of 10^{-8} Torr or less. The reasons for requiring UHV are evident from a discussion of elementary gas kinetics, as follows.

Surfaces have an atomic density of ~3 × 10^{14} to 1.2 × 10^{15} atoms/cm^2, depending on the atomic spacing and the exposed crystal plane. From fundamental gas kinetic theory, the number n of gas atoms impinging on a unit area of surface in unit time is

$$\frac{dn}{dt} = \frac{P}{\sqrt{2\pi m k T}} \tag{1}$$

where P is the gas pressure, m is the atomic mass, k is the Boltzmann constant, and T is the absolute temperature. Equation (1), the Knudsen equation, is extremely important in surface science because it permits a comparison of the rate of deposition with the rate of contamination from the residual gas in a vacuum system. For example, if the vacuum pressure P is measured in Torr and if the atomic mass m is converted to molecular weight M in g, then Eq. (1) becomes

$$\frac{dn}{dt} = 3.5 \times 10^{22} \frac{P}{\sqrt{MT}} \text{ cm}^{-2}\text{ s}^{-1} \tag{1a}$$

For a typical residual gas molecular weight of 40 g and a temperature of 298 K, the rate of gas arrival is 3.2 × 10^{20}P (Torr), and at a pressure of 10^{-6} Torr the arrival rate is 3.2 × 10^{14} cm^{-2} s^{-1}. In other words, the arrival rate in just 1 s at 10^{-6} Torr is nearly equal to the number of atoms in 1 cm^2 of surface. Thus, it takes only seconds to completely cover an initially clean surface with residual gas, assuming all of the arriving atoms or molecules stick to and remain on the surface. A pressure of 10^{-6} Torr is a fairly representative ultimate vacuum attained in older diffusion-pumped evaporation systems that are cycled frequently between high vacuum and atmosphere. The cautionary point is that the rate of deposition must be orders of magnitude faster than the contamination rate to avoid impure films. This requires either a lower pressure or higher rates of supply from the film source.

The need for better vacuum in thin-film technology has been a major driving force in the development of improved pumping systems. Pumping specifics and, more importantly, vacuum technology in general are discussed elsewhere.[7–13] Today there are many choices of pumping schemes, including diffusion pumps, ion pumps, turbomolecular pumps, and cryogenic pumps, for achieving vacuum less than 10^{-6} Torr. Each type of pumping technology has advantages and disadvantages, and the possible choices depend on the specific details of the system to be pumped. In many cases, it is desirable to include multiple types of pumping simultaneously to take advantage of the strengths of each.

Vacuum evaporation is possible for a large range of materials, including all but the highly refractory metals with very low vapor pressures. Even with these elements it is usually possible to obtain low deposition rates using a heated filament source. By heating the film source, such as a crucible containing an Al charge, the vapor pressure of the charge is raised to the point at which sublimation or evaporation occurs, depending on whether the charge is solid or molten. The relevant physical constraint is that the deposition rate should be large compared to the rate of contamination from the residual gases; i.e., the pressure of the source material at the substrate must be much larger than the residual pressure in the system. (Normally there is no equilibrium pressure of depositing atoms at the substrate but rather a unidirectional flux; however, for simplicity we can describe the flux in terms of an effective pressure that would produce the same arrival rate at the surface.) It can easily be shown by a geometrical argument, assuming that the background pressure is low enough that scattering in the vapor flux can be neglected, that the effective pressure at a substrate distance L from a source of area A is related to the vapor pressure P at the source by

$$P_{\text{eff}} = \frac{(PA)}{4\pi L^2} \tag{2}$$

Consequently, for a 1-cm^2 source 10 cm from a substrate, the pressure at the substrate is reduced by a factor of more than 1000.

The rate of deposition can be obtained from Eqs. (1) and (2), and by using the fact that the rate of increase in thickness per unit area, R_t, is just the rate of mass increase divided by the density ρ:

$$R_t = \sqrt{\frac{M}{2\pi k \rho^2}} \left(\frac{P_e}{\sqrt{T}}\right) \frac{A}{4\pi L^2} \tag{3}$$

where P_e is the equilibrium vapor pressure at the source temperature T, and M is the molecular weight. Written in this way, it is evident that the first term is material dependent, the second term is temperature dependent, and the third term is dependent on chamber geometry and sample size. For P in dyne/cm^2, Eq. (3) becomes

$$R_t = 4.62 \times 10^{-3} \frac{\sqrt{M}}{\rho} \frac{P_e}{\sqrt{T}} \frac{A}{L^2} \text{ cm/s} \tag{3a}$$

For Al at 1200°C with a vapor pressure[14] of 10^{-2} Torr and based on the geometry described earlier, a deposition rate of 50 nm/s, or about 3 μm/min, is obtained.

Uniformity of film thickness is an important issue in the production of integrated circuits. There are two physical limitations to consider. The first limitation is indicated in Eq. (2); i.e., for a point source of atoms, the flux (and growth rate) decreases as the square of the distance between substrate and source; the edges of a large substrate positioned perpendicular to a line from the source will be farther away from the source than the center and will thus receive a lower flux. The second limitation is that for a source with a finite area, the flux decreases as the cosine of the angle from the normal to the source; i.e., the substrate "sees" a smaller area of source. In electronic technology in which many semiconductor wafers are coated simultaneously, the configuration of source and substrates must be arranged so that all the wafers receive the same average deposition flux. This can be accomplished by placing the substrates and the source crucible at positions corresponding to points on the circumference of a sphere, often described as a "planetary configuration." In this configuration substrates closer to the source are also located at an increasing angle from the normal to the source, so the distance and angle effects exactly compensate. Rotation of each substrate then ensures that the average flux across the wafer is uniform. [15,16]

With large-diameter substrates, such as the 200–300-mm Si wafers increasingly used in microelectronic applications, the thickness uniformity of films across the substrate becomes a very important issue. It may be improved by rotating the wafer during deposition, using multiple deposition sources, or doing both. However, the exact angular distribution of the atom flux from the sources must be determined empirically. We discuss film thickness uniformity in more detail in Sec. 5.

A further issue that may be of considerable importance is the extent of nonuniform coverage at steps in the surface topography. An integrated circuit is typically etched into mesas and steps, and the metal interconnects must pass over these features without significant reduction in thickness. Without wafer rotation during deposition, shadowing at step edges will cause thinning. The coating of holes with high aspect ratio (defined as the depth-to-width ratio) is even more of a problem. Shadowing is unavoidable in these geometries, and the only way to achieve sufficiently uniform coverage is to rely on surface migration of the

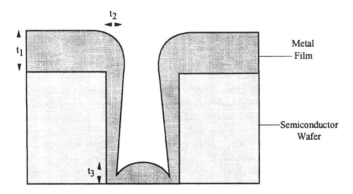

Figure 1. Schematic cross-sectional view of a trench or via hole in a semiconductor wafer showing the variation of metal film thickness at the edge of the trench and across the floor. The aspect ratio is the ratio of the depth of the trench to the diameter. t_1, t_2, and t_3 represent the thicknesses of the metal film on the wafer, on the sidewall, and in the bottom of the trench, respectively. The effect of shadowing on the wall and bottom depositions is indicated.

deposited atoms to the regions of lower direct coverage. As a result, heating the substrate to increase surface mobility during deposition improves the coverage; however, to cover high-aspect-ratio features it is necessary to use other deposition techniques, as discussed in other sections. Figure 1 shows a schematic view of the cross section of a via hole or a trench in a semiconductor wafer with a metal film deposited on it. It is evident that the film thickness uniformity is reduced within the hole. The nonuniformity is increased as the aspect ratio is increased. One of the advantages of Al for semiconductor metallization is that substrate heating during deposition produces significant flow to even the thickness. In fact, as circuit dimensions decrease, the electrical conductivity of Al becomes a serious limitation; however, an important current research problem is to find a more conductive replacement metal that has sufficient surface mobility during deposition.

Some metals have such low vapor pressures that they are difficult to evaporate from a resistively heated crucible. Electron beam evaporators utilize the energy of a beam of high-energy electrons to heat a small portion of a metallic charge held in a water-cooled hearth to a high temperature. In effect, the molten portion of the charge is contained in an unmolten container of the same material, a phenomenon sometimes referred to as "skull melting." This containment solves the crucible problem for these metals, but the thermal energy radiated from the hot center of the charge can be excessive. Furthermore, the electron beam used to heat the charge excites the emission of secondary electrons that may impinge on the substrate with sufficient energy to produce heating or bombardment damage.

The self-crucible formed during electron beam evaporation is obviously a good solution for avoiding contamination from containers used to hold the evaporating charge. While refractory metals are used as resistively heated boats to contain evaporating materials with high vapor pressure, at very high temperatures there may be some solution of the crucible material into the charge, leading to contamination of the evaporating flux. For semiconductor materials, trace amounts of transition metals in a film are not acceptable, so, consequently, it is necessary to use inert crucibles such as graphite or, preferably, pyrolytic BN to achieve the needed film purity.

3.2. Sputter Deposition

For thermal evaporation of a material, it is necessary to heat it to a temperature where it has a significant vapor pressure. Many materials of interest for electronic films have vapor pressures too low for the practical evaporation of them due to either the difficulty in finding a suitable crucible or, more often, because of the problems caused by thermal radiation from a source that must be operated at a very high temperature. For example, the radiant power emitted by a hot source can cause decomposition of organic films such as the photoresist films used in lithography. Furthermore, increasing the distance between source and substrate is not useful because both the emitted radiant thermal flux and the deposition flux decrease as the inverse square of the distance.

An alternative method of deposition that avoids these problems is the "sputtering" or ejection of atoms or molecules from a target bombarded by energetic ions.[17–22] In the simplest form, depicted in Fig. 2a, sputtering is obtained by generating a beam of inert-gas ions, accelerating the beam to energies of a few kV, and directing the beam at a target composed of the material to be deposited. Atoms or molecules from the target are ejected by momentum transfer from the incident ion flux and are deposited onto the substrate. More detail about the energy transfer and ejection mechanisms is given in Chaps. 3 and 5 of Volume 5 in this series. Sputtering can also be accomplished with a diode configuration where the target is biased negatively to attract the positive ions from a glow discharge, as shown in Fig. 2b. A magnetic field is used to increase the plasma density and, thus, the sputtering rate in this magnetron configuration.

Sputtering has a number of advantages for the controlled deposition of films. First, nearly any thermally stable conducting material can be sputtered, including compounds or alloys where one component may have a much larger vapor pressure than the other. While the evaporated flux from such a material will have a composition very different from the source if the vapor pressures of the constituents differ widely, the *sputtered flux* from compound materials, after a brief time to reach surface equilibrium, will have the same composition as the source target.[23] The sputtering efficiency for different materials can vary widely, but conservation of matter requires that eventually the surface concentration on the sputtering target of the more efficiently sputtered atoms is reduced to exactly compensate for the greater sputtering rate in order that the loss of all species becomes equal.[6] Second, the power radiated to the substrate is much less than from a hot evaporating source, because the target temperature is only slightly increased by the impact of the incident ion beam. Third, the arriving atoms or molecules typically have higher kinetic energies than evaporated atoms, which improves step coverage by enhancing surface migration; furthermore, in some methods of sputtering, there may be resputtering of deposited material on the substrate that can also improve step coverage. Fourth, sputtering can produce much more useful deposition rates, particularly for refractory metals or oxides, for which evaporation rates are quite low. Fifth, sputtered films tend to have a finer grain size than do evaporated films, which may be either advantageous or undesirable. Sixth, sputtering targets are readily fabricated by hot pressed-powder sintering; consequently the composition can be easily controlled. Finally, ion bombardment of the substrate before deposition often enhances adhesion of the film by removing surface impurities and by increasing the surface topography.

Compared to evaporation, sputter deposition of films has five disadvantages: (1) sputtering requires an inert gas that must be highly purified to avoid contamination of the film; (2) sputtered films nearly always include some of the sputtering

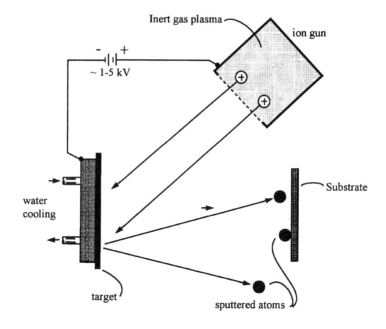

(a) Ion Beam Sputtering

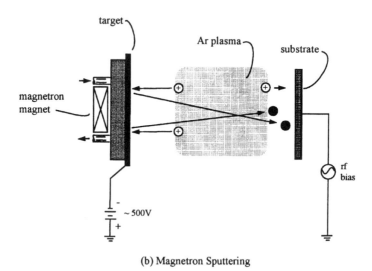

(b) Magnetron Sputtering

Figure 2. Configuration for (a) direct ion beam sputtering and (b) magnetron sputtering.

gas; (3) there may be ion bombardment damage of the substrate at ion energies above about 50 eV (see Chapter 3 in Volume 5 of this series); (4) impurities from the fixtures within the system may also be sputtered and incorporated into the film; (5) sputtered films usually do not grow epitaxially; however, with appropriate substrate temperature and cleanliness, epitaxy can sometimes be obtained.

A variety of magnetron sputtering configurations are possible, depending on the application. Planar magnetron sputtering, shown in Fig. 2b, is most common in manufacturing because of the greater ion density and consequent faster sputtering rate; the plasma is confined by a magnetic field parallel to the target.

Reactive-ion sputtering is another option used to deposit insulator films.[24,25] For example, if a Ta target is sputtered in a mixture of inert gas and N_2, a film of Ta and/or Ta_2N and TaN is produced. The exact composition of the film depends reproducibly on the gas composition; thus, this technique can be used to produce films of varying electrical resisitivity. Reactive-ion sputtering is commonly used to deposit oxides, carbides, and nitrides. It is particularly useful for the deposition of the oxynitrides of Ti, Ta, and Si, where the controlled addition of both oxygen and nitrogen can be used to minimize the tensile stress due to the mismatch between the lattice constant of the film and that of the Si substrate.

3.3. Ion Beam Deposition

While less commonly used in manufacturing of integrated circuits, the direct deposition of ion beams onto a substrate has some very useful applications.[26] By ionizing material to be deposited, focusing the ions into a fine beam, and then directing the beam electrostatically to a desired location, it is possible to deposit small amounts of film material at very precise locations, i.e., to carry out direct "writing" of features. A useful characteristic of this technique is that by increasing the ion beam energy, the operational mode changes from deposition to ion beam etching. At still higher energies, implantation of the ions into the substrate takes place. Focused ion beams, which are obtained by using field-enhanced ionization from sharpened tips, combine the versatility of localized etching and deposition with the optical simplicity of a high-brightness point source of ions. These systems are most generally used for the repair of photomasks or in prototype integrated circuit failure analysis studies. In some cases the ion beam is used to decompose a precursor vapor containing the metal to be deposited.[27]

4. Chemical Deposition

Chemical methods of deposition take advantage of a chemical reaction between a compound vapor containing the element or elements comprising the film and the substrate. In many cases the substrate acts primarily as a catalyst to mediate the energy involved in the reaction between two vapor species. If energy is needed to initiate the reaction, it may be supplied thermally or from an electron or photon beam incident on the substrate. In some cases, the reaction is stimulated by RF or optical excitation of the reactant gases.

4.1. Chemical Vapor Deposition

Chemical vapor deposition (CVD) of thin films occurs when the reacting gas species come in contact with a heated substrate that catalyzes the reaction to produce the solid film.[28] The substrate may supply the heat used to pyrolyze the reactant(s); alternatively the entire reaction chamber may be heated. For example, the pyrolysis of silane can be used to deposit Si films:

$$SiH_4(g) \Leftrightarrow Si(s) + 2H_2(g) \tag{6}$$

In fact this reaction is commonly used in LPCVD (low-pressure CVD) systems operating at ~600°C and 1–10 Torr to deposit polycrystalline Si used for IC interconnects.[29] Deposition rates of 10–20 nm/min are typical. To make the polysilicon conducting, gases such as diborane, phosphine, or arsine are added to dope the film with B, P, or As, respectively.

Other important CVD reactions for the deposition of dielectric films are

$$SiH_4(g) + O_2 \rightarrow SiO_2(s) + 2H_2 \tag{7}$$

and

$$3SiH_4(g) + 4NH_3(g) \rightarrow Si_3N_4(s) + 12H_2(g) \tag{8}$$

Refractory metals are difficult to evaporate because of their low vapor pressure, and are more readily deposited by CVD:[30]

$$WF_6(g) + 3H_2(g) \rightarrow W(s) + 6HF(g) \tag{9}$$

Since CVD reactions are sometimes carried out at pressures in the viscous flow regime, gas flow dynamics can play a significant role in determining reaction rates and uniformity of the resulting films. A large variety of chamber designs have been used to minimize turbulence and to facilitate the removal of reaction products;[31,32] however, these are principally problems of production systems where throughput is a foremost consideration. In this chapter we are most concerned with CVD reactions that occur in UHV, where the interest is primarily directed toward the crystalline perfection of the epitaxial layer.

4.2. Plasma-Enhanced Chemical Vapor Deposition

Plasma-enhanced chemical vapor deposition (PECVD) [33] is used extensively to deposit a large variety of films, usually compounds such as oxides or nitrides. In the electronics industry PECVD is particularly used for depositing insulating films on integrated circuits. Silicon nitride is an important dielectric material sometimes used for passivating IC wafers after the completion of fabrication. At this stage of processing, the circuits are not able to withstand high temperatures. By using a RF plasma to aid in the decomposition, reactions such as that in Eq. (8) can take place with little additional heating of the substrate. Heating by the plasma alone will typically keep the substrate below 300°C. With the addition of small amounts of N_2O or H_2 the composition of the dielectric SiN can be modified to an oxynitride to minimize the stress in the film.

In a recent improvement, energy is supplied to the gas mixture by electron cyclotron resonance (ECR) plasmas that have the desirable effect of gaining efficient ionization of the source gas at lower pressures, typically 10^{-5}–10^{-3} Torr. This allows efficient deposition with even lower power input to the substrate.[34]

Additionally, decomposition of the reactant gases can be accelerated by use of a focused laser beam by a pyrolytic process, in which the substrate is heated by the laser beam, or by a photolytic process, in which the reactants are directly dissociated by energetic photons. This technique can be used for direct "writing" of deposits, e.g., metal lines on prototype circuits, because the deposition occurs only where the laser beam is incident. Direct writing is not suitable for production processes because it is slow compared with photolithographic processes on full wafers.

5. Epitaxial Growth

The discussion so far has been a brief overview of very general methods for supplying film constituents to a substrate without regard for the structure of the film that is deposited. These techniques are quite important for many technological applications where the film properties are not critically dependent on the crystallography of the film. For example, optically reflective films, interconnect metallization on integrated circuits, and passivating or very hard oxides all provide functionality without requiring epitaxial deposition. However, most active semiconductor devices require the longer carrier lifetimes available from single-crystal materials. In the following sections, we deal specifically with the additional steps in film deposition needed to obtain single-crystal films.

Epitaxy describes the extended single-crystal growth of a film on a crystalline substrate. The term comes from the Greek *epi*, placed upon, and *taxis*, arrangement, and was first used by the French mineralogist L. Royer.[35] In early work, the formation and ripening of crystalline nuclei on various substrates such as rocksalt were observed in the electron microscope and were described as epitaxial growth; however, such films were highly imperfect due to defects such as twins formed when the nuclei coalesced into a continuous film. With high-quality substrate surfaces and deposition in UHV, modern epitaxial films can be nearly perfect. For example, growth on whole semiconductor wafers can be carried out such that the lateral dimensions of the single-crystal film are 100–200 mm while the film is only a few monolayers thick! Epitaxial layers are extremely important in semiconductor devices because they can provide unique carrier or potential profiles that are not possible to obtain by other deposition technologies, and because in many cases it is possible to produce films that are purer or more defect free than the substrate material.

The use of *in situ* epitaxial growth for the purpose of substrate preparation for surface science studies has a long history. For example, Saltsburg and coworkers carried out some elegant studies of atom diffraction from epitaxial metal films grown *in situ* on mica substrates.[36,37] Foxon and Joyce[38] and Arthur and Brown[39,40] studied the reaction kinetics of molecular beams with GaAs surfaces grown by MBE. Tsao et al[41] expanded on their approach in further studies of Group V element beam interactions with MBE surfaces, which clearly showed the power of this method. More recently the increasing availability of STM technology has led to an explosive increase in the use of epitaxial layers for studies with controlled surface structures.[42–44]

Two types of epitaxial growth can be distinguished: *homoepitaxy* in which the film is grown upon a crystalline substrate of the same material; and *heteroepitaxy* in which the film and substrate are different materials. Homoepitaxy has the obvious advantage that the substrate and film have the same lattice constant. If the substrate surface is initially free of contamination, it should not be difficult to find conditions in which the deposited film will grow as an extension of the single crystal substrate. Homoepitaxy is most useful for semiconductor-on-semiconductor growth, in which the film has different properties from the substrate. For example, the film might have a much lower doping concentration than the substrate. This is not possible to achieve in a diffusion process where the diffusion flux into the film is driven by a higher concentration at the surface. Heteroepitaxy is obviously the method of choice to use for preparing large area crystal surfaces of materials that are not available as bulk crystals.

Heteroepitaxy almost invariably involves mismatch between the lattice constants of the film and substrate. However a clean, ordered substrate provides a strongly orienting template for deposited atoms. Various film structures occur depending on the amount of lattice mismatch and the resulting interfacial strain. If the mismatch is < 0.1%, epitaxial growth will probably occur with the formation of a lightly strained layer. For a larger lattice mismatch, the growth will often initially be "pseudomorphic" in which the film grows with the lattice constant of the substrate, resulting in strain in the plane of the interface.[45] As a pseudomorphic film increases in thickness, the strain energy increases to a point at which misfit dislocations can be generated, leading to a relaxation of strain, and many defects are introduced into the film. Strain relaxation typically causes the flat, planar epitaxial growth to become rough on an atomic scale.[45]

The structure of the clean surface frequently supplies an energetic template sufficient to force the growing film into a pseudomorphic configuration. The energetic bonding between the surface and the growing film can produce a mode of growth, the Frank–van der Merwe mode, in which each layer of the growing film becomes complete before the next begins. This mode produces a desirable planar geometry for the epitaxial layer. If the bonding between substrate and film is less strong, the growth may be of the Volmer–Weber type with the formation and growth of islands of epitaxial material on the otherwise bare substrate; this mode is typical for metal films on ionic crystals (e.g., rocksalt). In the intermediate Stranski–Krastanov mode, the first few layers may grow in a layer-by-layer mode, but as strain energy increases the subsequent film forms islands to relieve the strain.[46] After a discussion of the methods for obtaining epitaxy, we consider growth morphology and strain in more detail.

5.1. Molecular Beam Epitaxy

Molecular beam epitaxy is conceptually the simplest epitaxial process from an atomistic viewpoint, and is also the best understood.[47–51] A good background in the early work in MBE is given by the references in the exhaustive bibliography of MBE literature by Ploog.[52] MBE occurs by the interaction of beams of atoms or molecules with a clean substrate surface contained in UHV. Atoms arriving on the surface can migrate to lattice positions to extend the crystalline interface, can desorb back into the vapor, or may form three-dimensional nuclei, which leads to nonplanar growth. In elemental source MBE (ESMBE), the beams are generated by the thermal evaporation of the constituent elements contained in refractory crucibles so arranged that the beams can be turned on or off by opening or closing shutters in the line of sight to the substrate. This allows the growth of a specific material to be initiated or terminated extremely rapidly by the mechanical action of the shutter for that material. As a result, on very flat substrates, interfaces between different materials can be produced that are nearly atomically abrupt. While the first use of MBE was for the growth of III–V compound materials,[47] in recent years a wide range of materials have been grown by MBE ranging from insulators to semiconductors to metals. There is increasing evidence that the surface processes involved in growth are very similar, regardless of material. Because of intense commercial interest in semiconducting films, most of the results of MBE experiments have been obtained from the growth of semiconductors, and the system description below is based primarily on semiconducting film growth. This is not necessarily a limitation, because the preparation and processing of semiconductors are, in general, more demanding than that of other materials.

Vacuum systems for MBE studies are usually constructed with large areas of internal cryopanels on which the unused beam flux (that which is not incorporated into the growing film) condenses with high efficiency. This feature, coupled with high-speed vacuum pumping, maintains a nearly UHV environment around the substrate, where the principal surrounding gas species are constituents of the film being grown. The UHV environment prevents contamination of the surface when growth is not taking place and facilitates using a variety of analytical tools in the vacuum chamber for surface analysis, beam flux measurement, and film structure studies. This *in situ* analysis capability has been a major factor for developing an unusually detailed understanding of the process. Figure 3 shows the arrangement of sources and substrate in a typical MBE system.

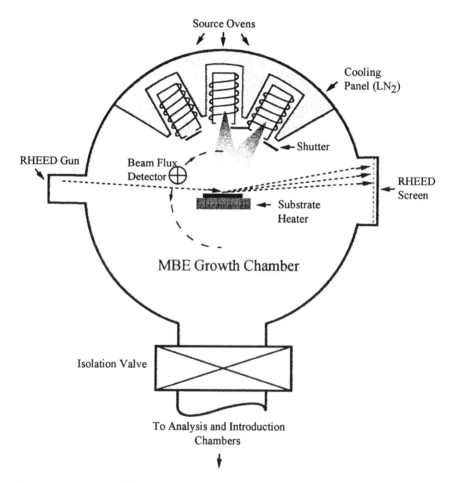

Figure 3. Top view of a MBE growth chamber showing the reflection high-energy electron diffraction (RHEED) system for monitoring structure during growth.

Historically, the beam sources of atoms or molecules first used in MBE were the elemental evaporation sources described before. However, in recent years many of the film deposition sources described in the previous sections have also been used. Particularly significant in the development of MBE as a production process for microelectronic circuits are various forms of gas sources in which gaseous compounds containing the constituent elements are leaked into the growth chamber through specially heated ovens that decompose the gases to form elemental beams. This type of process is usually referred to as gas source MBE (GSMBE).[53] More accurately, GSMBE has two subdivisions. The first is hydride source MBE (HSMBE), which utilizes beams of nonmetal hydrides (e.g., AsH_3 or PH_3) which are thermally cracked in the beam source to provide the nonmetal components. HSMBE has the advantage that films containing more than one Group V species can be readily grown by using mass flow controllers to determine the desired beam composition. It has the disadvantage that the Group V hydrides are extremely toxic. The second subdivision is metal–organic MBE (MOMBE), which uses not only the nonmetal hydrides but also beams of volatile organometallic species such as trimethylgallium. These Group III metal alkyls dissociate on the substrate at growth temperature. MOMBE has the very important advantage of providing a virtually unlimited supply of source material, whereas ESMBE requires that the source evaporation crucibles be refilled frequently, with subsequent pumping and recleaning of the growth chamber. The disadvantages of MOMBE are that toxic vapors are involved and that dissociation of Group III metal alkyls has a strong dependence on substrate temperature. A feature of MOMBE, which has both good and bad implications, is the presence of considerable pressures of reaction products, e.g., various hydrocarbon species, in the growth chamber. These species can be used advantageously to incorporate carbon dopants into the semiconductor films, but growth conditions must be controlled to minimize unwanted carbon incorporation.

The generic MBE growth process consists of the arrival of atoms or molecules at the heated substrate surface, surface migration to step edges or kinks, and incorporation into the crystal surface at these active sites. At steady state, after a transient period during which the substrate becomes leveled by the growth,[47] the surface consists of monatomic steps progressing across the crystal. Impurities can interfere with this process by creating nucleation sites or pinning sites that inhibit the flow of steps. If the substrate is too cool, the mean surface diffusion length for arriving atoms is not long enough for them to reach step edges, and random nucleation on the terraces is likely, which may roughen the interface. If the substrate is too hot, the arriving atoms simply reevaporate before they can reach the site of stronger bonding at step edges. Thus the beam flux, the substrate temperature, and, most importantly, the beam and vacuum purity must be precisely controlled for film growth to form the desired flat surface with only monatomic steps.

The III–V materials, GaAs and GaP, were used as source materials in the earliest MBE experiments. When these compounds are heated, the nonmetallic elements As and P both have high vapor pressures, while the metal Ga has a much lower pressure; thus, the beam fluxes of metal and nonmetal are significantly different. The growth of a stoichiometric compound by evaporation of the constituents would seem to require very precise control of the flux (and thus the temperature) from each constituent source; however, it was found that the adsorption of the Group V element was much stronger on a layer of Ga atoms adsorbed on the surface than on the surface of the GaAs substrate. Thus, stoichiometry could be achieved in this instance by maintaining an overpressure of As or P. The Ga atoms arriving at the surface all adsorbed, while only an equal number of nonmetal atoms reacted; the excess As or P reevaporated from the growing film.[54] Thus, stoichiometry is attained almost automatically by maintaining an excess of the Group V element and keeping the substrate temperature high enough that the excess is lost from the growing crystal.

The MBE systems used in early surface kinetic studies[54-57] were single UHV chambers with one or more source ovens, a substrate holder that could be heated to temperatures above 600°C, and a minimum of surface analysis capability, typically Auger electron spectroscopy (AES). A major advance was made when reflection high-energy electron diffraction (RHEED) apparatus was included in the growth chamber to allow real-time observations of the film structure.[58] This relatively simple type of system is still useful for experiments where small-area substrates are used and where a high throughput of semiconducting device material is not needed. For example, an extensive series of elegant STM studies of Si growth mechanisms have been carried out using simple Si sublimation sources to produce MBE surfaces at slow growth rate.[59-61]

5.1.1. Molecular Beam Epitaxy Systems

Modern MBE systems were developed with the interest in electronic devices in mind. Material purity and throughput are both important, and present systems normally consist of several vacuum chambers, each with a suitable pumping system. A typical system (Fig. 4) consists of at least three independently pumped chambers, although there can be many variations. A vacuum interlock or introduction chamber is used to allow substrates to be installed on a transport device in a small volume that is pumped prior to opening the valve to the rest of the system to reduce the amount of gas load introduced into the main vacuum. The introduction chamber normally has a provision for heating the substrates prior to introduction into the main vacuum system to outgas the substrates in the holders. A second chamber is often used for additional substrate preparation and surface characterization, using tools such as AES or X-ray photoelectron spectroscopy (XPS), or both. The third chamber is used for the actual growth and usually can be isolated from the rest of the system during both growth and substrate introduction and offers the two advantages of avoiding contamination of the analysis chamber and the surface

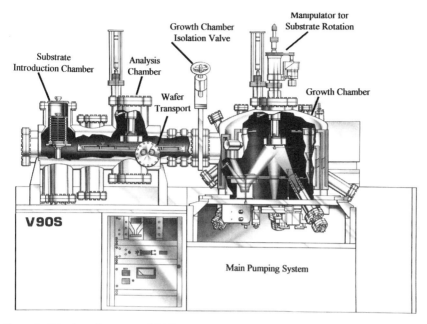

Figure 4. Side view of a commercial MBE system with a substrate introduction chamber on the left, analysis chamber, and growth chamber on right configured for upward deposition to minimize particulates on the substrate. (Courtesy of Vacuum Generators, Ltd.)

probes with vapors produced during growth, and minimizing vacuum contamination of the growth chamber during pressure bursts when the intro chamber is opened to the main vacuum. Additional growth chambers are sometimes added for the epitaxy of additional layers where the constituents of the different layers may be incompatible. For example, it is not desirable to grow II–VI and III–V semiconductors in a single chamber because components of one are dopants in the other, and there are usually significant residual pressures remaining after growth from the more volatile species.

Sample transport from chamber to chamber has been accomplished in a variety of innovative ways. The requirements are to have a means for selecting a wafer from a multiple wafer stack in the introduction chamber, then to move that wafer into another chamber and to dock it securely onto a heater–manipulator, all without mishap or contamination of the wafer by dust particles in the system. The bearings providing smooth rotation must not seize in the UHV environment, and there must be positive control over the sample position, particularly when docking and undocking from the chamber manipulators. These are demanding requirements, which have been met by using magnetically coupled transport rods, by trolley systems with sealed rotary linkage to the exterior, and by large bellows-sealed rods providing an extended linear motion. There has been a great deal of reliability testing by MBE researchers, to an extent that present systems work remarkably well, regardless of the particular method used. The success of these techniques has made it possible to construct large systems with many growth and analysis chambers that can be used by several operators simultaneously. Of course, these techniques have added to the cost and complexity of MBE systems, but the improvements in vacuum quality brought by relatively infrequent exposures to atmosphere have been quite significant.

Substrate mounting for epitaxy is important because quite often the precise control of temperature during growth can be critical. It is quite difficult to make good thermal contact between a semiconductor wafer and an underlying metal heater. Even when both matching surfaces are well polished and when clamps are used to hold the semiconductor substrate in place, temperature differences of over 100°C can be measured between the substrate and holder. One successful approach that has been used since the early MBE work is to bond the semiconductor wafer to a metal heater plate using a low melting metal such as In, which provides a liquid thermal contact at the growth temperature. Indium is particularly useful because it is usually relatively insoluble in either the substrate or the metal heater plate, and because its vapor pressure is low up to about 600°C. Another advantage of liquid metal bonding is that the surface tension of the liquid provides a fairly strain free adhesion of the substrate to the heater so that clamping can be kept to a minimum; as a result, the epitaxial layers after growth do not show the strain-induced slip lines that appear near the points of attachment on substrates that are firmly clamped to the heater unit. Finally, metal bonding is especially useful for odd-shaped substrates, where no special provision need be taken for the geometry. There are problems with metal bonding, however. For higher growth temperatures, the In will vaporize and may contaminate the growth surface; In-induced defects can frequently be observed near the edges of In-mounted wafers. Also, if the wafers are to be further processed in a semiconductor fabrication facility, the In metal must be thoroughly removed from the back of the wafer to avoid contamination of the process line.

Sample holders providing radiative coupling between substrate and heater have become quite popular in recent years, in order to eliminate the need for In backing. The sample wafers are held by the rim in front of a heated ribbon filament.[62] Radiation from the heater that is more energetic than the band gap of the semiconductor wafer is absorbed by the wafer and heats it rapidly because of the low thermal mass of the wafer. To avoid nonuniformity of temperature due to hot spots directly over the radiant elements, a thermal diffusing plate made from sapphire or BN is usually placed between the heater and the substrate to even out the radiation pattern. This also prevents evaporation products from the wafer impinging on the hot heater ribbon. (In the case of GaAs, some evaporation of As_2 occurs at growth temperatures, and this vapor is corrosive to heated Ta or Mo foil.) Advantages of this method include a clean, metal-free substrate and the ability to make faster changes in temperature due to the smaller thermal mass of heater and substrate. A disadvantage is that a particular holder will fit only a single size of substrate.

5.1.2. Temperature Measurement

Measurement of the substrate temperature is more difficult than it might seem. Accurate temperature measurement with a thermocouple requires a very good thermal contact between the thermocouple and the substrate material. Most semiconductor MBE systems are arranged to allow rotation of the substrate during growth to produce a more uniform film thickness. Thus, it is not feasible to bond a thermocouple directly to either the substrate or the heater. A noncontacting stationary thermocouple can be mounted directly behind the substrate to measure the radiant flux from the substrate, but this, of course, does not provide a direct measurement of the temperature. It does provide a relative measure of temperature changes, but accurate temperature calibration is necessary.

To calibrate substrate temperatures, many investigators use an infrared optical pyrometer to measure temperatures of 400–600°C. If semiconducting substrates are used, the bandpass of the pyrometer must be chosen to be an energy window centered above the band gap of the semiconductor; otherwise, the pyrometer simply detects the radiation from the substrate heater element that is transmitted by the substrate. Further complicating matters are the changes in total emissivity that can be produced by the growth of an epitaxial layer. Coating of the optical window through which the pyrometer detects the substrate radiation can also cause problems. The reader contemplating the use of this technique would be well advised to consult the papers on the subject that discuss the pitfalls.[63–68] Single-temperature-point calibration is possible using the melting of In or Sn or other metal eutectic structures attached to the surface of the substrate, or by RHEED observations of surface phase transitions. However, these methods are obviously not as accurate as is possible with a plot of thermocouple reading versus substrate temperature covering the entire temperature range of interest.

5.1.3. Substrate Preparation

The key to successful epitaxial growth by MBE has been the ability to obtain and characterize clean, ordered substrates on which the film is grown. We define "cleaning" as the removal of damage by polishing and etching, followed by removal of all impurities that may influence the surface ordering of the epitaxial layer. The most suitable cleaning process is, of course, strongly material dependent. Substrate preparation is discussed in three other chapters in this volume; however, we describe three quite different approaches to sample cleaning that have been shown to produce the necessary surface purity for epitaxy on semiconductor surfaces that are particularly difficult to clean.

The first approach is based on the fairly standard method of using a wet, oxidizing etch to produce a passivating oxide layer, followed by a subsequent oxide desorption in UHV. InSb is a particularly difficult material to clean because the melting point of the crystal is low (~527°C) and because the vapor pressure of the major oxide (In_2O_3) is low. A detailed series of experiments have been reported using a variety of etches and microscopic analysis of post-etch morphology as well as XPS, RHEED, and AES studies of the surface before, during, and after oxide desorption.[69,70] They found that the etch CP4A ($HNO_3:CH_3COOH:HF:H_2O$ in the ratio 2:1:1:10) produces the flattest surfaces. The surfaces are also protected by a passivating oxide produced by the oxidizing etch. XPS studies showed that the oxide consists predominantly of Sb_2O_3 but contains some In_2O_3 as well. Heating in vacuum produces a loss of Sb_2O_3 abruptly above 380°C, and AES shows that all oxygen is lost after heating at 425°C for 25 min. Other etchants required heating to temperatures much closer to the melting point to remove the oxide. Heating near the melting point causes problems because it is above the congruent evaporation temperature for InSb, i.e., above the temperature at which InSb begins to lose an excess of Sb. Consequently, it is preferable to direct a beam of Sb_4 at the surface during the In oxide desorption to prevent the loss of Sb and the formation of droplets of excess In on the surface. With a flux of Sb directed at the surface, the remaining surface oxide is lost in a very narrow temperature range of 410–420°C. SEM studies after cleaning showed that In droplets initially are formed, but these subsequently react with the Sb_4 to form smooth layers of InSb. Following these treatments, epitaxial InSb could be readily deposited. The point emphasized here is that solving this difficult cleaning problem required information provided by several surface characterization tools, namely, optical interference microscopy, XPS, RHEED, AES, and electron microscopy.

The second approach is to etch and passivate the surface using dry processing techniques rather than wet etching. GaAs surfaces are typically etched in a wet oxidizing etch such as $H_2SO_4:H_2O_2$, which leaves the surface with a passivating oxide layer to prevent the chemisorption from the laboratory air of carbon-containing compounds that are difficult to remove by heating. A study of two B(111) surfaces was carried out using both the conventional wet etch and a UV–ozone treatment, which also produces a passivating oxide.[71] Both surfaces were given identical pre-etch degreasing in trichloroethane, acetone, and H_2O. After oxidation, the crystals were introduced into an MBE system where the surface structure could be studied by RHEED during the thermal desorption of the oxide. The emergence of a structured RHEED pattern was taken as an indication of the removal of oxide. The wet-etched surface became clean at 580°C (which is similar to the cleaning temperature for the (100) plane), while the apparently thicker oxide grown by UV–ozone did not desorb until 610°C. Neither surface had carbon impurity after the thermal desorption when examined by AES. The biggest difference between the two methods of oxidation were found in microscopic studies of the topography. The ozone-etched surface showed noticeable roughness with asperities of ~1 µm clearly visible. An additional finding in these studies was that the growth of some 60–100 monolayers of GaAs on the rougher, ozone-etched surface was sufficient to smooth the surface to an extent that the RHEED measurements were optimized; i.e., the epitaxial growth flattened the surface so that after some 100 monolayers of growth it was comparable to the smoother wet-etched surface.

Finally, a third approach in substrate preparation involves passivating the surface in the laboratory with hydrogen rather than oxygen. This has the advantage that the hydrogen is readily desorbed at lower temperatures than are required to remove the passivating oxide. Semiconductor surfaces such as Si and GaAs can be hydrogenated by immersion in HF solutions. In some cases, exposing the surface to a H-plasma prior to the HF treatment improves hydrogen coverage and passivation.[72] In our experience, we have occasionally been able to produce a clean GaAs surface by HF treatment prior to insertion in the vacuum system; however, the process is not very reproducible. Clearly some further work is necessary to establish the best conditions. Nevertheless, all of these approaches are fundamentally similar in that they attempt to produce a clean, smooth surface chemically and then protect the surface with a layer that can subsequently be removed by thermal evaporation in vacuum. The oxide passivation, particularly on GaAs, is highly reproducible—so much so that the temperature for oxide desorption serves as a convenient calibration for the substrate heating.

The combination of substrate processing under UHV conditions and the inclusion of surface analysis tools in the MBE apparatus has made it possible to prepare and characterize atomically clean substrates by thermal treatment after these etch-passivation methods and to preserve the pristine state before and during growth. However, the analysis chamber, which began by playing a vital role in optimizing surface preparation, has become increasingly important as the center of the experiments to be carried out on well-characterized and nearly perfect surfaces. A number of systems are now designed and built specifically for highly specialized experiments, such as scanning tunneling microscopy on growth surfaces.[73] MBE and atomic layer epitaxy (ALE) systems have been incorporated into synchrotron beam lines to prepare surfaces *in situ* for high-energy photoemission experiments.[74,75]

5.1.4. Growth Chamber

The growth chamber and its components remain the keys to good films since the vacuum requirements are stringent, and the presence of heated source crucibles can generate considerable contamination by outgassing themselves or by radiatively heating other parts of the system. A large amount of research has been done on the designing of source ovens that can be thoroughly outgassed ahead of time, have extensive radiative shielding, have a low thermal mass so that rapid temperature changes can be made to vary the growth rate, and have reproducible and uniform temperatures. For growth on large semiconductor wafers, it is also important to achieve a uniform flux distribution over a large area. As a result of this research and the interest in a variety of material systems and the increase in industrial interest in high-performance devices, there are now available commercially a broad spectrum of source ovens optimized for specific purposes.

In the studies of GaAs MBE in the 1970s, it became apparent that the evaporation of Ga and Al presented problems because of their reactivity, and both would alloy with nearly any refractory metal at high temperatures. Pyrolytic boron nitride (PBN) was just becoming available in custom configurations at that time and was found to provide an excellent container material. Now there are a large variety of PBN crucibles with a range of shapes and sizes to fit any of the commercially manufactured source ovens. Pyrolytic graphite (PG) and tungsten crucibles are also available for very high temperature applications (PBN begins to lose nitrogen above 1500°C).

Great improvements in source oven design have been made recently. The function of the source oven is to heat the source material to a constant temperature and to hold that temperature constant to a small fraction of a degree without contributing contamination to the substrate. The temperature control of the source should be highly reproducible so that growth rates, which are determined by the temperature setting, are constant and reproducible to within 1% or less. To indicate the difficulty posed by this requirement, we note that because of the exponential dependence of vapor pressure on temperature, extreme precision in temperature control of the source is needed. For example, a Ga source at 1100 K has a vapor pressure of 10^{-4} Torr; this pressure changes 2% for a 1° change in temperature. To achieve the needed temperature control therefore requires extremely stable thermocouples that make very good thermal contact to the source crucibles. One way to accomplish this has been described by Miller and Arthur in which a metal foil belt forms the hot junction of the thermocouple and is wrapped tightly around part of the crucible.[76] Figure 5 shows a typical oven configuration, with crucible, thermocouple, heater winding, and radiation shielding.

Another requirement for a source oven is that little or no contamination be produced by it even at very high temperatures. This means that the oven itself must be constructed from clean, refractory materials, and the radiation from the hot oven must be adequately shielded from the rest of the growth chamber to avoid radiatively heating chamber walls that may be covered with material deposited from previous growth runs. Alloy materials that contain trace amounts of high-vapor-pressure elements cannot be allowed in the hot zone. For example, many stainless steels contain small amounts of Mn that is somewhat volatile and acts as a doping impurity in semiconductors. Thus, extensive radiation shielding is needed to limit the heated region, and only refractory materials such as pure tantalum and molybdenum can be included in the heated zone.

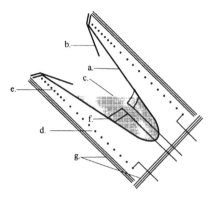

Figure 5. Side view of MBE effusion furnace; (a) pyrolytic BN crucible; (b) PBN crucible insert; (c) molten charge; (d) main resistive heater winding; (e) extra heater winding at crucible lip; (f) strap thermocouple; (g) metal foil radiation shields.

More subtle requirements have been identified as a result of the intense effort to fabricate high-performance electronic devices by MBE. For example, mechanical shutters are normally placed in front of the opening in each of the ovens to intercept the molecular beams until such time as they are needed for growth. When it is desired to use a particular beam to provide a component of the film, the shutter can be withdrawn from blocking the beam in much less time than is required to grow a single monolayer. However, it has been observed by several investigators that the beam flux from an oven shows an initial transient burst after opening a shutter, which subsequently decays to the steady value, even though the oven temperature is constant. This transient results from a change in the temperature distribution in the crucible caused by the shutter operation. That is, with the shutter closed, the radiation is reflected back into the crucible and the temperature becomes more uniform throughout, but with the shutter open there is radiation loss from the outer end of the crucible, reducing the temperature of the surface of the charge and thus decreasing the vapor flux. New oven designs minimize the "shutter overshoot" by the use of dual heater windings to maintain the outer, more open portion of the crucible at the same or higher temperature than the inner part. Another modification that effectively reduces shutter beam flux transients involves placing a conical insert in the crucible. Figure 5 shows an extra heating element in the crucible lip area as well as a PBN crucible insert. The insert ensures that within a limited cone angle the area of the source charge viewed by the substrate is constant, and this also provides a more reproducible flux as the volume of charge varies.[77]

Another nonobvious but very important problem resulting from variation in crucible temperatures is that droplets of the evaporant can condense on the cooler portions of the crucible, eventually reaching a size where they fall back into the molten charge. This can cause very small liquid droplets of evaporant to be ejected from the crucible onto the substrate. These liquid metal droplets on the surface of the growing film can nucleate defects in the film. In the case of III–V compound films, this process is one source of the pernicious "oval defects" that occur on III–V epilayers.[78] The dual-heater ovens greatly reduce the numbers of such defects by preventing the condensation of droplets.[79,80] Evidence that temperature nonuniformities are a major source of defects is provided by Schlom et al.,[81] who used a sapphire crucible with a much greater thermal conductivity than the usual PBN crucible. Defect densities in GaAs epilayers decreased to the extremely low value of < 50 cm^{-2}.

While the source ovens discussed, consisting of a crucible and thermocouple contained in a filament heater and surrounded by foil heat shielding, are relatively simple, ovens used with high-vapor-pressure materials, such as some nonmetals, are considerably more complicated. For example, the element P is used in the growth of III–V phosphides. While the vapor pressure of red phosphorus is not excessive, the P vapor condenses as white phosphorus, which has a vapor pressure at room temperature that is higher than the UHV limit. Furthermore, when exposed to air, white phosphorus ignites spontaneously. I well remember the early dramatic demonstration of this problem as I opened the main flange on a vacuum system that had been used for growing GaP and found a blue flame filling the interior with white clouds of P_2O_5 pouring out the opened flange into the laboratory atmosphere. Needless to say, this is not a desirable outcome from a health and safety standpoint! The deposition of quantities of white phosphorus can be greatly reduced by providing a valve to isolate the source when the system is opened to the air. In addition, the nonmetal elemental beam pressures are typically maintained higher than those of the metals to stabilize the growth surface in a nonmetal-rich surface structure, which means that considerably more nonmetal constituent is used than metal constituent. Thus, the nonmetal sources become exhausted more quickly unless a large reservoir of material is provided. However, a large mass of source material means that beam flux response to changes in control temperature will be slow, which may be inadequate for growth of layered structures. By using a valved source, the beam flux of nonmetal can be varied quite rapidly with the valve.

Various investigators have contributed to the development of valved nonmetal vapor sources that have a low-temperature heated reservoir containing the solid source material, which then vaporizes through a heated tube into the main chamber to create a beam of the nonmetal molecules incident on the substrate. (The Group V elements typically vaporize as tetramers, e.g., P_4.) The source flux is regulated by a valve in the tube, which can be completely closed off when the growth is terminated, thus preventing excessive amounts of material from entering the growth chamber. Usually the sources also contain a heated "cracker" filament or ribbon, creating a hot zone at the end of the tube in the main chamber to allow the molecules

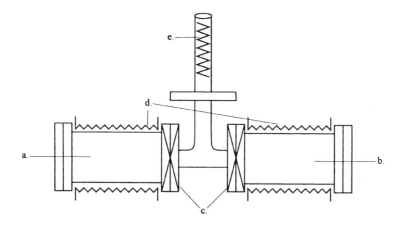

Figure 6. Valved double oven for high-vapor-pressure molecules (e.g., As_4, P_4, etc.): (a,b) externally heated reservoirs for arsenic and phosphorus; (c) controlled leak valves; (d) heater windings for the external resevoirs; (e) thermal filament to crack tetrameric molecules into dimers.

to be dissociated thermally; i.e., P_4 can be thermally dissociated efficiently to P_2 molecules which are more reactive than the tetramer. A number of authors have shown that III–V epitaxial films have lower levels of electrically active impurities when dimeric rather than tetrameric nonmetal beams are used.[82,83] This probably results from a lower vacancy concentration because of the greater reactivity of the dimer molecule; however, the complete mechanism is not entirely understood. Figure 6 shows one configuration for a valved, cracker source that can be used to mix the beams from two elements, say P and As, to grow a ternary or quaternary compound, such as $GaAs_xP_{1-x}$.

Source ovens are also available commercially as off-the-shelf units to supply atomic hydrogen, atomic nitrogen, and a variety of other dissociated gaseous species. Atomic H is used in cleaning the substrate and in providing an adsorbate that can act as a surfactant to facilitate epitaxy. Atomic N is used in the growth of nitrides since molecular N_2 is unreactive. These excited atom sources typically use an ECR plasma to couple RF energy to the gaseous species to induce dissociation.[84–86]

5.1.5. Film Characterization

While a single chamber was used for analysis and growth in the earliest MBE systems, the modern systems limit the analysis in the growth chamber to only those probes that provide real-time information about the growth process, such as RHEED[87] or optical reflectometry.[88] It is also useful to have some means for measuring the flux of atoms from the source ovens, since the growth rate may not be a good indication of the flux. The simplest way of determining beam flux is with an ionizing detector such as an ion gauge or mass spectrometer that can be placed directly in the beam flux. While an ion gauge is compact and can easily be positioned in the beam, a mass spectrometer has the big advantage of distinguishing between various vapor species, and, in addition, can be used to detect problems with the vacuum. For beams that condense on cooled surfaces, such as Si, accurate flux measurements can be made with a quartz crystal microbalance; however, uncertainties in the condensation coefficient as a function of the elevated tempera-ture of the substrate require using some other calibration methods.[75] Atomic absorption by the beams provides a direct measure of the beam flux. Pinsukanjana et al.[89] have described such a system that they believe to be accurate to ~2.5%. Various other optical techniques have been used to observe or infer the growth rate of films, including infrared reflectometry[90,91] and ellipsometry.[92,93] However, growth rate measurements are most easily obtained from RHEED observations, which we will describe.

RHEED is probably the single most important analytical tool for the MBE film grower for the real-time observation of the crystal structure of the growing film. Nearly all commercial MBE systems today include an electron gun and phosphor screen for displaying a RHEED pattern of the substrate to provide an indication of the ordering of the surface. With home-built equipment, the cost of adding a RHEED system is negligible compared to the cost of the system. The appearance of a RHEED pattern not only shows when the surface oxide is removed (since the oxide is amorphous and gives rise to a diffuse diffraction pattern) but also shows the improvement in the surface ordering that occurs with the subsequent annealing. Clean semiconductor surfaces are reconstructed into a geometric configuration that minimizes the energy of localized bonding electrons at the surface; this reconstruc-tion is evident in the RHEED patterns as diffraction features positioned between the bulk, or "integral order" diffraction spots. The fractional order beams indicate that the surface unit cell is a multiple of the bulk spacing, e.g., the well-known Si (111) 7×7 structure. The presence of these "fractional order features" provides a qualitative measure of the long range ordering of the surface.

The most remarkable application of RHEED information, however, has come from the inference of the mechanism of film growth obtained from the time dependence of the intensity of the diffraction features. On a properly cleaned and annealed surface, which has been smoothed by the growth of a few tens of monolayers of material followed by annealing to improve the ordering, the intensity of the diffraction features oscillates in time under certain conditions of growth, with a period of oscillation that corresponds to the monolayer growth time. (A monolayer on the (100) face of a crystal of GaAs is taken to be a bilayer of Ga atoms plus As atoms.) The periodic oscillations are taken as strong evidence for a growth mecha-nism in which the growth of each atom layer is largely completed before the next begins, i.e., a layer-by-layer growth mechanism.[94] Figure 7 shows the intensity of

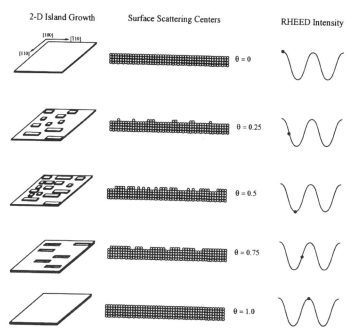

Figure 7. Schematic depiction of two-dimensional island growth with the corresponding RHEED intensities. θ denotes the fractional coverage of the surface by the next layer, and the solid dot on the intensity curve indicates the relative magnitude of the specular beam intensity at the coverage θ.

the specular RHEED beam as a function of time along with a schematic of the condition of the surface. The arriving atoms first nucleate in 2-D islands on the smooth surface. Subsequently, arriving atoms can migrate to the existing step edges to complete the monolayer and return the surface to a smooth condition. Thus, the surface cycles between being atomically smooth and rough, with a period corresponding to the time to complete a monolayer of growth. The rougher surface causes more diffuse scattering of the RHEED beam, leading to a lower intensity of the diffracted beams. The correspondence between the RHEED oscillation period and the monolayer growth rate has been clearly established by empirical measurements of film thickness after growth.[95,96] Thus the RHEED oscillations provide a precise method of measuring growth rates in real time. While RHEED oscillations were first observed in the growth of GaAs layers by MBE, a large body of literature now exists demonstrating the same effect in other materials, including metals,[97,98] and by a variety of growth processes, including various types of CBE, ALE, and CVD. The general growth mechanism that produces oscillations is clearly widespread for a large variety of materials, although the particulars may indeed not be identical.

Besides providing a measure of the growth rate of a film, the RHEED pattern also provides useful information about the geometry of the surface, including the roughness. Soon after Cho introduced RHEED into GaAs MBE systems,[58] it was shown that the etched GaAs surface is relatively rough, but becomes much smoother with growth.[47] RHEED patterns clearly show the improvement of surface geometry, since the smoother surface has diffraction features that are streaked normal to the surface due to the more two-dimensional character of an atomically flat surface. The improvement of the surface occurs because the arriving atoms or molecules are mobile on the heated substrate and predominantly wind up in lattice sites at step edges. Thus, there is a strong tendency for terraces to enlarge by the accumulation of material at the edges.[49]

RHEED also serves as a method for identifying the chemical nature of the surface. Semiconductor surfaces typically reconstruct to minimize dangling bond densities at the surface. In the case of compound semiconductors, where the crystals are often composed of alternating layers of metal or nonmetal atoms, the low-index planes can be predominantly metallic or nonmetallic. This is evident in the diffraction pattern because the reconstructed structure changes as the surface composition changes. A well-known example is the GaAs(100) surface, where the As-rich reconstruction, which is normally observed during the growth of GaAs under an excess of As, is described as the 2×4 structure. However, the Ga-rich structure that forms at a lower As-to-Ga flux ratio is described as a 4×2 structure. Other structures have been observed on this surface which depend on the growth temperature and the ratio of As to Ga in the arriving flux;[98,99] however, the transition from one structure to another, such as from the principal 2×4 As-rich structure to the Ga-rich structure, can be observed from the RHEED pattern to occur in <1 s when the As beam is interrupted during MBE growth, indicating that the transition is the result of submonolayer changes in surface concentrations. The significant point here is that surface structure is often a good indicator of the stoichiometry of the top layer, and this can be an important parameter for determining growth conditions.

Thus, the RHEED system is an excellent technique for monitoring growth rate, for qualitatively measuring surface topography, and for monitoring the surface structure. To an extent, the structure can provide a measure of the surface composition. Also important to the user is that the RHEED system is relatively insensitive to the pressure in the growth chamber, and responds only to changes in surface structure; furthermore, the geometry of electron gun, substrate, and phosphor screen is such that there is no part of the RHEED system positioned in front of the sample to block the molecular beams from the surface.

There are a few potential problems with the use of RHEED in a deposition system. The phosphor screen used to display the diffraction pattern can gradually become coated with the deposited material, but it is not too difficult to clean the glass window and replace the phosphor. Glass windows are presently available commercially that can be kept at an elevated temperature to reduce the condensation of film material. For film growth where film purity and crystalline perfection must be optimized, it is undesirable to flood the growing surface with energetic electrons because beam-induced cracking of residual gases can take place. Thus, it is generally preferable not to use the RHEED for an extended time on the actual surface of the growing film. Many crystal growers use a small auxiliary substrate to establish the beam flux and growth rate prior to actually growing the desired device structure on a large substrate.

One significant requirement for the effective use of RHEED is for the electron beam to be incident at a precise azimuthal angle with respect to the crystal orientation, which requires the substrate holder to have at least some angular adjustment to allow alignment of the RHEED beam along a low-index direction in the crystal; furthermore, this orientation must be maintained during the measurement of the diffraction features such as intensity oscillations. However, it is often highly desirable to rotate the substrate during growth to obtain a uniform film thickness during deposition on large area substrates, and rotation is obviously incompatible with the need for a fixed RHEED azimuth. Some experimenters have gotten around this limitation by gating the RHEED beam and detector synchronously with the angular position of the substrate holder.

An alternative to RHEED, obviously, is to use a thickness measurement that is not dependent on the angular position of the substrate, and which will not cause degradation of the film. Optical methods that do not depend on the angular position of the substrate are becoming more widely used for the measurement of deposition rates. One of these techniques relies on the interference in IR reflectivity when a film is deposited on a substrate with different dielectric constant. In the simplest version, the IR pyrometer used to measure the substrate temperature also measures the periodic variation in emission transmitted through the substrate and film from the heater as the film grows.[90] This variation is caused by interference of reflected and transmitted beams at the film interfaces. This method, of course, requires that the film be relatively transparent to the optical wavelengths detected by the pyrometer. It also requires that the pyrometer window be kept free from deposits that would attenuate the optical intensity. A clever way to avoid deposits on the window is to use a Si surface as a mirror to reflect the optical emission from the substrate and yet avoid a direct path to the window for atoms or molecules desorbing from the substrate.[100] The reflectivity of the Si seems to be little affected by a thin film of deposited material from the growth, and the longer indirect path to the window prevents coatings from forming. Heating the optical windows is also an effective way to keep them free of coatings.[89] Quartz glass windows can be heated to temperatures at which little condensation of high-vapor-pressure species, (e.g., As and P) occurs.

Other workers have used spectroscopic ellipsometry (SE) to obtain real-time information about the surface composition, optical properties, and growth rate of films in an MBE system.[101,102] SE measures the reflectivity of the perpendicular and parallel polarized components of a polarized, incident light beam over a range of wavelengths to determine the complex index of refraction and layer thickness of a transparent film on a reflective substrate. The index of refraction can be related to the composition using the effective medium approximation; thus, SE provides significantly different information than does RHEED. Furthermore, if a substrate holder is used that can rotate precisely in the plane parallel to the substrate surface, it is possible to obtain information while the substrate rotates to improve the growth uniformity. One limitation of using SE is the need to prevent condensation on the optical windows. For high-vapor-pressure materials, this can be achieved by using quartz windows heated to elevated temperatures.

Maracas et al.[102] have used SE to monitor beam flux ratios. After depositing a known amount of Ga on GaAs (determined by the growth rate), they then determined the length of time required for the As beam to convert the Ga into epitaxial GaAs from measurements of the dielectric function as it returned to the values for GaAs. They overcame several experimental challenges, especially the problem of the design of a substrate manipulator with sufficient mechanical stability to allow making measurements during growth when the substrate was heated.

In situ ellipsometry has also been used in the analysis of epitaxial Si deposited by PECVD.[103] These results showed a distinct change in nucleation behavior as the substrate temperature was varied between 600–700°C. The changes were due to the presence of O and H on the surface at low temperature with loss of H at the higher temperature.

Even though the vapors in the chamber used for epitaxial growth can produce deposits of the film constituents that may form insulating films on the electron optics of analytical tools such as AES, there is an ongoing interest for developing additional tools to include in the growth chamber to obtain more real-time information about growth. Chambers et al.[104] described a system that uses the RHEED beam to excite Auger electrons from the growing film. The electrons are analyzed in a small, high-throughput spectrometer that does not block the substrate from the molecular beams. The system described[104] is used to grow epitaxial oxide films and contains electron-beam-heated sources to provide metal beams (Mo and Cr) and an ECR source to provide O atoms. The system also has quartz crystal microbalances (QCM) to measure beam fluxes; however, the in situ AES capability showed clearly that the QCM gave an incorrect measure of film composition because not all of the incident flux was incorporated into the growing film on the 750°C substrate. AES analysis is also used in this system for determining the composition of mixed metal oxides during growth. The system contains a separate analysis chamber with both XPS and X-ray diffraction capability.[104] The message is quite clear: as the material systems become more complex, more in situ analytic capability becomes essential.

5.1.6. Deposition Uniformity

The deposition uniformity will most likely be very good if the substrate size is only a few millimeters across and the distance to the source ovens is many centimeters. Deposition uniformity becomes a significant problem when large-area substrates are used, such as semiconductor wafers that may be 200 mm in diameter or larger. Geometric factors that act to limit the beam flux at the edges of the wafer include the increased distance from the source, the reduced apparent area of the source as viewed from off-axis regions of the substrate, and focusing of the beams along the crucible axis of the source crucibles. This latter factor is correctable to some degree, and better crucible design has not only improved the deposition uniformity significantly but has also reduced the changes in beam geometry as the crucible contents are depleted. The distribution of flux from typical crucible designs has been discussed.[105,106]

To obtain the minimum variation in film uniformity across the substrate, rotation of the substrate during growth is necessary. Optimized crucible design and substrate rotation result in thickness variations that can be as small as a few percent across 75-mm wafers.[107] With a rotating substrate and multiple source ovens, it is important that the rotation speed be fast enough to prevent compositional fluctuations in the film. We have occasionally observed TEM images of films grown in our laboratory which showed the unintentional formation of a superlattice structure rather than the desired uniform ternary alloy when both InAs and GaAs were deposited simultaneously at a deposition rate of ~1 monolayer/s with a rotation speed of <0.2 rps. In our apparatus the In and Ga sources are located on opposite sides of the substrate, and, thus, the deposition ratio for the two metals changes considerably across the 3-inch substrate wafer at any instant. We observed that the rotation speed should be at least 1 rps to obtain a homogeneous film. In other words, the rotation speed must be somewhat greater than the growth rate to even out the geometrical differences in flux from sources located at different positions off the axis of rotation of the wafer. While extremely uniform films are of most interest to semiconductor device applications, significant compositional variation is possible across even a relatively small substrate when beams are arriving from different positions around the periphery of the substrate, and a rotating substrate holder should be considered if multicomponent materials are to be deposited.

5.1.7. MBE Growth Mechanisms

In this section we describe in rather general terms the prevailing wisdom about film growth mechanisms using MBE, since this is an important consideration in using epitaxial growth to improve the substrate. Our discussion applies particularly to MBE films, where the UHV environment is compatible with the tools used to examine the growth process. Because similar behavior (e.g., oscillation in the intensity of RHEED features) has been exhibited by many of the modified types of film deposition, such as GSMBE and UHV CVD, it is very tempting to think that with the addition of some surface chemical reactions, these processes are fundamentally similar to standard MBE growth.

After the thermal removal of surface oxide, an etched substrate surface is typically rough on an atomic scale, as shown by a spotty RHEED pattern and by TEM observations that indicate rough features as much as 10 nm above the surrounding flat areas. The degree of roughness is very much a function of the polishing treatment and subsequent annealing in UHV. Once epitaxial growth begins, however, the surface rapidly becomes much smoother and the RHEED pattern shows this smoothing by developing streaked diffracted features.[108]

This smoothing of the surface was predicted by Frank and van der Merwe[109] based on a model that involved the surface migration of atoms and molecules over short terraces on a rough surface and their incorporation into the lattice at step edges. Wider terraces have a larger collection area for vapor species and thus have bounding step edges that advance more rapidly than those edges adjacent to narrow terraces which collect fewer adatoms. The consequence of this step growth is that terraces tend to become similar in size and the surface becomes smoother. Eventu-

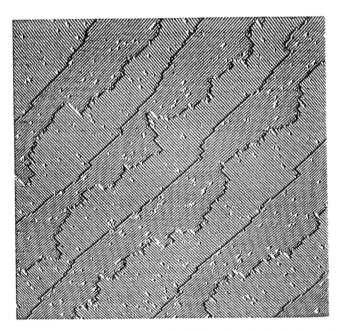

Figure 8. Scanning tunneling microscope image of a Si(100) surface showing the type A step edges (Si dimers parallel to steps) and type B steps (Si dimers perpendicular to steps). Upper portion of surface at lower right, with downward tilt to upper left.[110]

ally the surface develops into a nearly uniform array of terraces, as can be seen in the STM image of a Si(100) surface in Fig. 8.[110] An interesting feature of the Si(100) step edges is the way in which the edges alternate from being relatively smooth to relatively jagged as the layers alternate between a reconstruction with Si dimer chains lying parallel to one with the chains perpendicular to the step.[111] Tsao has reviewed implications of this basic growth model in detail.[49]

Once the surface is smoothed, as described above, further growth can proceed in either of two modes depending on the nature of the surface and the mean-free path of atoms and molecules on the surface. If the terrace width is comparable to or less than the diffusion length of atoms and the substrate temperature is sufficiently high, it is possible for the atoms arriving on the surface to diffuse to step edges without interacting with other atoms, and growth occurs by the steady movement of steps across the surface; i.e., growth will be in the "step flow" regime. Step flow will occur on stepped, vicinal surfaces similar to that in Fig. 9, which are slightly off the axis of principal planes, at temperatures that are high enough to provide good surface mobility.

The other growth mode occurs when the terraces are wider than the diffusion length. In this case, 2-D nucleation occurs on the terraces, which leads to periodic roughening and smoothing of the surface as each monolayer fills in again. This situation produces periodic variations in the RHEED intensities, as discussed previously (Fig. 7). It is obvious that increasing the surface mobility (by, for example, increasing the substrate temperature) can lead to a transition from the 2D nucleation mode to the step flow mode, which is indicated by a disappearance of the RHEED oscillations. Joyce et al. have used measurements of this type to infer

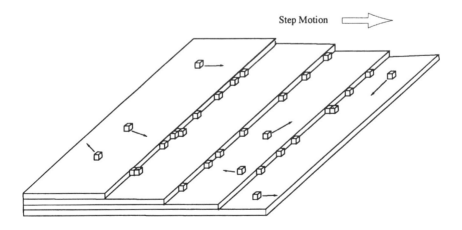

Figure 9. Schematic of the step flow growth mode process.

the kinetic parameters and diffusion length of Ga atoms on vicinal surfaces whose terrace length can be calculated from the crystal orientation.[112] Alternatively, lowering the temperature can produce the onset of RHEED oscillations. Clearly, surface mobility is a key element for determining which growth mode is dominant.

Both of the growth mode mechanisms, as described above, as well as the formation of nuclei and their subsequent enlargement ("ripening"), have been observed very clearly in elegant STM studies on Si surfaces.[113,114] Nuclei on the (100) plane were elongated chains, reflecting the anisotropy in surface reconstruction and mobility on this surface.

5.1.8. Beam Flux and Stoichiometry

A key aspect in the epitaxial growth of multielement films is the influence of beam flux on the stoichiometry of the compound, i.e., maintaining equal numbers of metal and nonmetal atoms in the film. This is particularly important for semiconductor materials, where deviations from stoichiometry can have a large effect on carrier concentration.

5.1.8.1. Elemental Growth. The simplest growth system is one with only a single component, and Si, because of its technological importance, has been studied extensively. While solid Si has been occasionally used as a Si vapor source, the vapor pressure is so low at the melting point, $\sim 4 \times 10^{-4}$ Torr, that the deposition rate is too low to deposit more than a few monolayers. Therefore, the Si vapor in Si MBE is normally obtained from an electron-beam-heated source crucible, to obtain a large enough flux at the substrate for a useful growth rate. Thus, the term "beam" is used very loosely in this instance, since the Si atoms simply evaporate radially outward from the hot zone of the evaporator. Because of the low vapor pressure of Si at the typical substrate growth temperature of ~600°C, the condensation coefficient of the Si vapor is near unity with very little reevaporation.

The vapor pressure of Ge is approximately two orders of magnitude greater than that of Si at the same temperature, but it is still too low at the Ge melting point to use a subliming source. It is possible to derive a sufficient flux of Ge from a molten charge in a heated crucible to grow an elemental film at very slow growth rates, but it is much better to use an electron beam heater for the source to avoid contamination due to slow deposition. Again, the condensation coefficient on the substrate is essentially unity.

For elemental materials, the stoichiometry is obviously not an issue. The beam flux must be high enough to provide a reasonable growth rate, yet not so high that disorder and roughening of the interface occur. That is, the beam flux should be much higher than the rate of contamination from residual gases, yet not so high that growth kinetic limitations affect the film morphology.

5.1.8.2.   Compound Growth.   When the films have more than one constituent, the problem of compositional control becomes an issue. We consider three possible cases, depending on the vapor pressure of the constituents.

In the first case, if both constituents have low vapor pressures so that the adsorbed species have a long surface lifetime before reevaporation, then the problem for the crystal grower is simply to control the flux of both kinds of atoms and molecules to produce the desired ratio in the film. In other words, the film composition is determined by the arrival rate of atoms at the surface. An example of this case is the deposition of the alloy Si_xGe_{1-x}, where there is complete miscibility in the solid phase and reevaporation of either element is minimal. The challenge then lies in precise temperature control of the sources. As indicated above, a 1° temperature change produces roughly 2% variation in beam flux in sources using the evaporated elements.

In the second case, one constituent has a higher vapor pressure as an element than in the compound. For example, elemental As has a vapor pressure at the typical GaAs growth temperature of 550°C that is roughly seven orders of magnitude larger than the vapor pressure of As in equilibrium with GaAs. As we have indicated, a nominally stoichiometric film can be obtained by supplying an excess flux of As (typically as As_2 molecules) to react with all of the arriving Ga; the unreacted As will be reevaporated from the surface. (We are ignoring the point defect structure of the film; the exact beam flux ratio of As to Ga can have a significant influence on the concentration of vacancies of either species.)

In the third case, both constituents have higher vapor pressures as elements than in the compound, which is true for most II–VI compounds. For these materials, the first monolayer of either component can adsorb on the substrate, but subsequent layers will be less strongly bound due to the weaker elemental bond, and at elevated temperatures they will simply desorb. By exposing the substrate alternately to each beam long enough to deposit successive monolayers of each component, stoichiometry is achieved because only a monolayer will stick on each exposure, and each monolayer reacts with the preceding one. This process is known as atomic-layer epitaxy (ALE)[115,116] and is useful for producing a film consisting of alternating layers of the constituents; i.e., the ideal configuration of the zinc blende and wurtzite structures in the <100> direction consists of alternating layers of each species. Atomic-layer epitaxy has the advantage that obtaining uniform coverage is not as dependent on surface migration as in MBE, since the components arrive at adsorption sites from the vapor or by rapid migration from a physisorbed layer. As a result, ALE is particularly useful for providing uniform coverage over nonplanar surfaces.

The ALE process can also be used to form materials usually grown by the second case described above. For example, GaAs can be grown by exposing the substrate alternatively to Ga and As beams. Since the shutters on a standard MBE system can usually be programmed to operate automatically, the timing can be adjusted for any desired cycling of beams. It has been observed that the surface mobility of adsorbed Ga atoms is significantly greater when As is not incident; thus, if the shutter timing is such that Ga begins arriving in the absence of an As flux, there is greater smoothing of the surface. This type of growth is known as migration-enhanced epitaxy (MEE),[117] and is reported to improve epitaxial growth at low temperatures, where the reduced mobility of the metal atoms normally leads to roughening of the surface. It may be particularly useful when growing a heterostructure where the different metal species may have widely differing surface mobilities. This would normally lead to a roughening of the interface as soon as the lower-mobility metal is deposited.

There is an experimental problem with any epitaxial process that demands extensive operation of oven shutters. Most shutters are activated by a mechanical linkage to an external driving device such as a solenoid through a flexible vacuum seal, typically a metal bellows. Bellows seals have a finite lifetime after which weld failure is likely to occur with catastrophic vacuum leakage. If a shutter-controlled epitaxy such as ALE is contemplated, where the shutter operation may be required for each monolayer, a durable shutter mechanism should be used, such as a magnetic coupling device, with no flexing of the vacuum wall. Nevertheless, shutter control provides a very precise method of metering the flux of each component arriving at the surface, as long as the beam stability is good. However, as we have previously

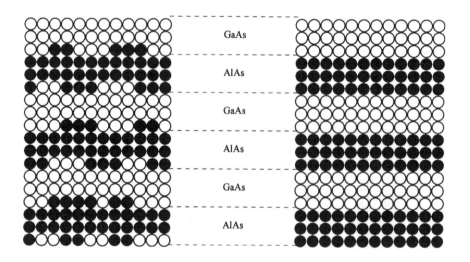

Non-phase locked growth   Phase locked growth

Figure 10. Cross-sectional diagram of superlattices grown with random shutter timing and with shutters timed to change at completion of monolayer growth, i.e., phase-locked epitaxy.

indicated, unless the ovens and shutters are well designed, there can be thermal transients associated with the shutter operation that reduce the precision.

A clever variant of MEE is to phase-lock the shutter operation to the RHEED oscillation period to begin a heterointerface only at the time when the maximum in the RHEED intensity indicates that optimum smoothness of the growing surface has been achieved. For example, in very narrow quantum wells, the variation in well width produced by disorder in a single layer of atoms can produce noticeable effects in the optical properties such as the peak width in the photoluminescent spectrum. In Fig. 10 we show schematically the structure of a quantum well device with interfaces formed when the surfaces are relatively rough and when the surfaces are maximally smooth. Superlattices grown by phase-locked epitaxy (PLE) have shown significant improvement in optical and structural properties over superlattices grown with shutter operation at a random phase of the surface molecular coverage.[118]

5.1.9. Temperature Effects

We have suggested that the role of substrate temperature is twofold. First, the heated substrate provides sufficient surface mobility to allow deposited atoms to migrate to step edges, or at least to the edges of seed nuclei. Second, the elevated temperature of the substrate provides thermal energy to desorb excess unreacted material, at least in the growth of III–V and II–VI materials where one or more elemental species are volatile, and thus, is important in maintaining stoichiometry. In fact, the issue of substrate temperature is rather more complicated than we have so far indicated. We will use the example of GaAs to illustrate some of the effects of temperature on growth. We will also confine the discussion specifically to the (100) crystal plane, noting that the temperature-induced effects will certainly vary somewhat for other planes as a result of the differences in bonding. At high temperatures, >675°C in the case of GaAs, reevaporation of both metal and nonmetal atoms occurs, with the result that the growth rate declines, becoming negative above about 750°C where evaporation of the substrate is greater than growth. The loss of Ga becomes an important issue when attempting to grow a ternary alloy such as AlGaAs, where the greater volatility of Ga over Al can change the film composition at high temperatures. Above about 630°C, the surface species are sufficiently mobile to migrate to step edges, at least for surfaces that are slightly off-axis, and as a result the growth occurs in the step flow mode with little modulation of the RHEED intensities. From 500°C to about 625°C, two-dimensional nucleation on the step terraces occurs because of reduced mobility; as a result the RHEED intensities oscillate as the nuclei form and then enlarge to cover and smooth the surface once again.

Below 450 to 500°C the growth mode is strongly dependent on the topography of the surface. If the initial surface is relatively rough with a spotty, rather than streaked, RHEED pattern, the arriving atoms have insufficient mobility to move to the valleys and the growing surface will remain rough and nonplanar. Below 400°C, it is likely that polycrystalline growth will take place on a rough surface. However, if the surface has been smoothed by careful etching and/or by growth at higher temperatures, then it is possible to reduce the temperature to well below 400°C and still maintain a smooth, planar growth. In fact, on a very smooth surface with a growth rate reduced to perhaps 0.1 monolayer/s, the substrate temperature can be reduced to below 150°C while a planar interface is preserved!

Low-temperature growth of GaAs has become an area of great interest because of the electrical properties of the resulting material. If the growth temperature is reduced to below about 300°C, the vapor pressure of elemental As is lowered and it is possible for excess As to remain on the surface and to become incorporated into the growing film so that the film is no longer stoichiometric, and contains about 1% excess As. Subsequent annealing at higher temperatures causes the excess As to form small nuclei within the film, and these nuclei strongly affect the electrical properties of this low-temperature GaAs (LT-GaAs) by acting as carrier traps. LT-GaAs forms a nonconducting layer because of the carrier trapping; it thus provides good electrical isolation from the substrate while providing a surface that is structurally excellent upon which to deposit further material. These LT-GaAs buffer layers have become widely used in the fabrication of GaAs devices because they provide insulating layers that have the lattice constant of GaAs. In terms of the focus of this chapter, LT-GaAs demonstrates that if the substrate is properly smoothed by epitaxial growth, then the surface mobility is sufficient, even at fairly low temperatures, to allow the arriving species to move into lattice sites; also, in the case of compound growth, nonstoichiometry can be accommodated by the incorporation of excess constituents into metastable point defects while still maintaining a planar epitaxy.

5.1.10. Surfactants

Thus far, we have emphasized the role of surface migration in improving the crystalline quality of epitaxial films. If the surface species are able to migrate to step edges, a planar growth interface is maintained. However, if the temperature is too low, the surface diffusion length may not be great enough to achieve a thermodynamically stable structure, and smooth growth may not result. Recently several authors have explored the use of surfactants to improve surface mobility by incorporating another material such as atomic H, which acts to weaken the bonding to the substrate of the arriving constituents, to enhance their surface mobility. For example, Okada and Harris found that irradiating the surface of GaAs during growth with atomic H permits the growth temperature to be lowered from 580–600°C to as low as 330°C with no loss of structural quality of the epitaxial film.[119] There is much that is not clear about their process; e.g., how much H is adsorbed, and how does it leave the growing film? Presumably the H film is segregated to the surface as the film grows with little incorporation into the film. Nevertheless, the H (derived from a tungsten filament heated in H_2 gas) clearly makes a significant difference in the growth kinetics. Hydrogen is also used as a surfactant in the homoepitaxial growth of Si;[120] however, these authors believe that H on Si *reduces* the surface mobility of the Si. Their evidence for reduced mobility was the much longer persistence of RHEED oscillations on the H-covered surface, indicating a shorter mean-free path for the arriving Si. While in these experiments the surfactant provided better homoepitaxial films at lower temperatures than would normally be observed, surfactants can play an even more important role in the growth of heterostructures, as discussed in the next section.

5.1.11.   Heteroepitaxy, Superlattices, and Quantum Wells

One of the important consequences of 2-D growth is the ability to produce structures consisting of one or more very thin layers of one semiconductor material surrounded by layers of a different material, usually with a wider band gap than the center material. If the thickness of the narrow-band-gap center well is much less than the wavelength of electrons in the material (from a few to tens of atom layers), the electron energy levels are discrete in the well; i.e., they are "quantized" and the structure is described as a quantum well. A structure comprising a periodic sequence of quantum wells is a superlattice. Quantum wells and superlattices are of great interest because the electronic and optical properties are quite different from bulk semiconductor material. Many of the early theoretical predictions of solid-state physics have been precisely confirmed in these quantum well devices that so closely approximate the ideal models used in the simple theory.

Much of the early work on superlattices and quantum wells was based on the material system GaAs–$Ga_xAl_{1-x}As$ because the Al content in the alloy increases the band gap, and yet the two binary compounds have nearly the same lattice constant. The MBE process, at least in principle, makes possible the growth of selected ternary and quaternary III–V alloys, since the composition of the film can be controlled by the relative beam fluxes. Figure 11 shows the lattice constant versus band gap for III–V binary materials and their ternary alloys. It is evident that the lattice constant varies widely for the various alloys. If one is restricted to binary compound substrates, the mismatch between a ternary film and binary substrate can be substantial. Thus, only by the use of strained layers is it possible to have access to the full range of materials and properties offered by the III–V alloy and II–VI families, as well as Si–Ge.

5.1.12.   Strained-Layer Epitaxy

The subject of strain in epitaxial layers is extremely important because of the obvious fact that heteroepitaxial layers are nearly always grown on a substrate with

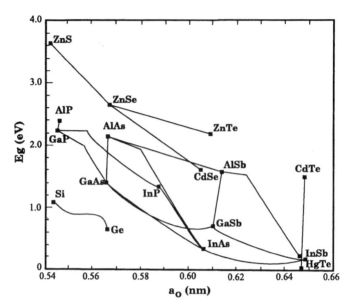

Figure 11. Plot of band gap vs. lattice constant for elemental, III–V, and II–VI semiconductors. The solid squares represent the values for the indicated binary compounds; the lines show the band-gap values for the intermediate ternary alloys.

a different lattice constant. The strongly ordering nature of a clean substrate is clearly shown because the initial growth of a film mismatched to the substrate often occurs with the film adopting the two-dimensional spacing of the substrate; i.e., the film growth is "pseudomorphic" with the substrate. Figure 12 illustrates the difference between commensurate, or lattice-matched, growth, pseudomorphic growth with uniaxial distortion of the film, and relaxed, incommensurate growth. Depending on the amount of mismatch, the distortion of pseudomorphic growth causes increasing strain in the film to an extent that relaxation eventually occurs with the formation of misfit dislocations in the plane of the interface. Frequently, the relaxation is catastrophic with extensive slip and disruption of surface planarity. Calculations have been made to determine the critical thickness of the epilayer at which the strain is sufficient to cause the generation of dislocations. The original theory of Matthews and Blakesley was a straightforward mechanical energy balance and provides a reasonable fit to more recent experimental data.[121] Figure 13, which plots the variation of critical thickness versus percent mismatch for the system of Si_xGe_{1-x} on Si, shows there are definite limitations on the thickness of pseudomorphic growth. The exact value of critical thickness depends on the elastic constants for the film material, so this is *not* a universal curve. People and Bean have discussed

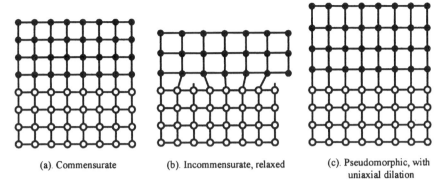

(a). Commensurate   (b). Incommensurate, relaxed   (c). Pseudomorphic, with
                                                                      uniaxial dilation

Figure 12. Cross section of possible growth modes: (a) lattice-matched, commensurate growth; (b) lattice-mismatched, relaxed growth; (c) lattice-mismatched, strained pseudomorphic growth.

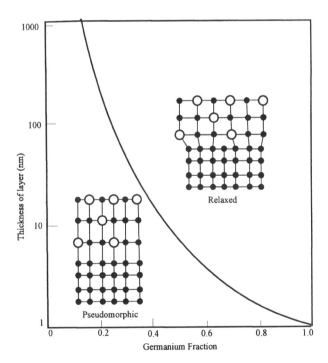

Figure 13. Transition between pseudomorphic and relaxed growth modes for Si_xGe_{1-x} alloy films deposited on Si. The solid line is based on the calculations by Matthews and Blakesley.[121]

the more detailed considerations for strained-layer growth and have modified the curves somewhat, although, in general, the simpler theory provides a good fit to the data.[122]

There are various ways to circumvent the mismatch problem. As indicated in Fig. 12, as long as the film thickness is below the critical thickness, a planar film can be grown on a mismatched substrate. Strain effects will, however, alter the band structure of the film, which may be an important consideration for subsequent experiments. The traditional method of dealing with mismatch has been to grow a graded buffer layer, where the lattice constant of the film is gradually altered from that of the substrate to that of the film. The MBE method is particularly effective for this because the relative beam fluxes of constituents can be varied in a very controlled fashion; thus, it is easy to begin by growing GaAs, for example, and then to direct a gradually increasing In flux toward the substrate by a programmed temperature sequence of the In source to eventually deposit a film of $In_xGa_{1-x}As$, where x can be as large as 0.3 or greater for buffer thicknesses around 0.5 μm. (InP is normally used as a substrate for larger values of x.) Graded buffers, however, do not eliminate stress; they merely distribute it over a larger volume of material, and threading dislocations are often observed in TEM studies of such structures.

Another approach is to introduce a strained superlattice buffer, where alternating layers are strained in tension or compression. This has the advantage that dislocations are often constrained to lie parallel to the interface and thus do not extend into the final layer of interest. The growth of strained superlattices with the alternate layers in compression and tension is an active research area that is much too broad for the scope of this chapter. References 123 and 124 are excellent presentations of the current status of research in this field.

There is an additional problem with strained layers, which is that strain can be relieved by the rejection of the species causing strain; i.e., there can be enhanced bulk diffusion to the surface with the accumulation on the surface of the atoms causing strain. For example, InGaAs quantum wells grown with GaAs barriers on a GaAs substrate show a skewing of the In spatial distribution toward the surface as a result of such strain-enhanced outdiffusion and excess evaporation of the In.[125]

While in some cases, as we stated, the structure of the substrate may force the epilayer into a metastable pseudomorphic configuration, at sufficiently high temperatures the surface layer is quite mobile and, as a result, will attain a thermodynamically stable structure that depends on the energetics of surface and interface and on the lattice strain. This may lead to the formation of islands of the epitaxial material, either immediately (Volmer–Weber growth) or after the growth of a monolayer or so (Stranski–Krastanov growth) rather than the desired layer-by-layer (Frank–van der Merwe) growth. Several investigators[126,127] have shown that the presence of an intermediate layer, a surfactant, can reduce the tendency for island formation while lowering the temperature needed for surface mobility. In the growth of Si–Ge heterostructures, As and Sb have proven useful. With both As and

Sb, only monolayer amounts of the surfactant are required, since the As and Sb are strongly segregated to the growth surface. An As layer is particularly effective in enabling the growth of very thick layers of Ge on Si because strain is relieved not by misfit dislocations but by stacking fault arrays ("V defects"), which do not destroy the planarity of the surface.[126] Unfortunately, As and Sb are doping impurities in the Group IV elements, and sufficient amounts of the surfactants are incorporated into the growing films to alter the doping.

Tin does not act as a dopant in Si–Ge, since it is an isoelectronic element with them. However, Sn has also been found to facilitate the growth of Ge layers on Si and to minimize the segregation of Ge into the subsequent Si layer.[128] It is strongly surface segregated, so only monolayer quantities are needed. A major difficulty with Sn-mediated Ge growth is that the Ge layer thickness cannot exceed about four monolayers to avoid Ge island formation; i.e., the growth mode is clearly Stranski–Kastanov. Nevertheless, the use of surfactant-mediated growth is clearly a very interesting research area that is relatively unexplored.

Recent reports of the use of compliant substrates have indicated another alternative to dealing with strained growth.[129–131] Rather than using a rigid substrate, which produces stress in the growing film, one employs a thin-film substrate that is able to deform to match the lattice constant of the film. One way to obtain such a substrate takes advantage of the different chemical reactivities of GaAs and AlAs. GaAs and AlAs, which have only a small mismatch (~0.15%), have been favored for electronic devices because of the ease in growing thin layers of either material on the other. In addition, AlAs is chemically much more reactive than GaAs, which makes it possible to detach very thin films of GaAs from an AlAs substrate (or, more likely, from a thin layer of AlAs on a GaAs substrate) by chemical etching. Thus, it is possible to grow a film of GaAs 10 nm thick on a AlAs layer, bond the GaAs to a second substrate by a flexible adhesive, such as liquid In, and subsequently detach the thin GaAs from the original substrate by chemical-etching the AlAs in dilute HCl. The resulting structure consists of a very thin film of GaAs bonded to a supporting substrate by a compliant material, In. The GaAs is thin enough and only weakly attached to the supporting structure so that it can deform to match the structure of an epitaxial film of different lattice constant deposited on it. (Note that the In layer will be liquid at the substrate temperature used for deposition.) This approach has proven very promising in the growth of InGaAs, where layers with a thickness as much as eight times the critical thickness have been grown without significant misfit dislocations.[130] An alternative approach which does not use a liquid metal solder obtains a flexible attachment of the GaAs to the supporting structure by direct bonding of the GaAs film to a GaAs supporting crystal. The bonding is accomplished by heating the film and supporting crystal in contact in a hydrogen atmosphere. If the crystal orientation of the film is rotated slightly with respect to the crystallographic orientation of the substrate prior to heating, strong bonding occurs only in small, isolated islands. After the GaAs film is detached from the AlAs substrate by chemical etching, it is only weakly bonded to the GaAs substrate at discrete locations and is therefore able to elastically conform to the spacing of a mismatched film.[131] It is claimed that this approach provides a nearly universal substrate, but further experiments are clearly necessary to confirm this.

5.2. Vapor-Phase Epitaxy

There are several other epitaxial techniques used most extensively for industrial production of epitaxial devices. We provide a rather brief description because these processes are more complex and are less likely to be as useful for *in situ* substrate preparation as is MBE or its variants. Vapor-phase epitaxy (VPE) and the related process, MOCVD, discussed later, are chemical deposition processes as opposed to the physical process of MBE. Vapor-phase epitaxy is a natural development from CVD of Si on Si substrates. The chloride process has been extensively used in industry:

$$SiCl_4(g) + 2H_2(g) \leftrightarrow Si(s) + 4HCl(g) \tag{10}$$

Reaction (10) is typically carried out at ~1200°C and is reversible, which means that if the carrier gas, N_2 for example, which initially enters the chamber contains some HCl, the Si surface will be etched to clean the substrate prior to growth. The VPE process has been used from the early days of Si device processing to prepare thin layers of silicon with properties superior to the bulk substrate material; a typical application has been to grow lightly doped Si on a heavily doped substrate, which is impossible to produce by a diffusion process.

The CVD process is typically carried out at pressures from a few Torr to atmospheric pressure. This has the advantage of simplifying the apparatus and increasing the growth rate; however, the complex fluid flow of reactant gases at these pressures is difficult to model. Furthermore, a large number of intermediate molecular species are actually involved in the reaction. Thus, CVD is mechanistically much more complex than the UHV growth processes. These reactions have been reviewed by many authors because of their commercial importance; see, for example, Refs. 132 and 133.

Meyerson and co-workers have developed a UHV CVD process that has enabled them to fabricate Si–Ge heterostructure devices with exceptional performance. These devices depend on the electronic properties of Si–Ge alloy quantum wells contained within Si barriers. This work illustrates that CVD is indeed compatible with the UHV environment and that it is possible to grow state-of-the-art films in a system without the complications of standard MBE.[134,135]

5.3. Metal-Organic Chemical Vapor Deposition

Metal-organic chemical vapor deposition (MOCVD) is a process that combines some of the best features of MBE and CVD for the growth of III–V semiconductors.[136] Rather than relying on an elemental source to provide the metal constituent, MOCVD utilizes a volatile organic compound to provide a vapor source that does not require the high-temperature cell used in MBE. Typically volatile compounds such as trimethyl gallium are used. This reduces the thermal load in the vacuum system, provides larger fluxes for faster growth, and since the metal–organic is introduced from a source outside the vacuum chamber, there is no need to open the system for replenishing the supply of metal. The MOCVD process also has made the growth of ternary and quaternary materials easier, since it is straightforward to control the partial pressures of several metal sources. In a variation of MOCVD, vapors of nonmetal compounds are also used; for example, AsH_3 has been used as a source of As in the growth of GaAs. With beam sources that rarely need replenishing and an interlock system that allows substrate entry and exit from the growth chamber without impairing the vacuum, the MOCVD system can operate for very long periods without downtime.

While MOCVD was first used in the late 1960s with systems operating at atmospheric pressure that were essentially identical to CVD systems of the period, present systems are much more closely related to MBE systems and operate at reduced pressure, or even at UHV in a molecular beam mode. It has been possible to include RHEED display in UHV growth systems, and the RHEED oscillations during MOCVD growth are nearly identical to those observed in MBE and suggest a similar layer-by-layer growth mechanism.

Hydrocarbons and other organic vapors are normally avoided in UHV systems because of the likelihood of decomposition and the difficulty in removing C contamination from clean surfaces. Since MOCVD utilizes various organic vapors, it is not surprising that early MOCVD films were rather heavily contaminated with C. It has since been found that the C contamination can be essentially eliminated by adjustment of the Group V beam intensity. As a result, MOCVD can be used to produce very high quality III–V films, with the advantage of high throughput due to the faster growth rates possible.

5.4. Liquid-Phase Epitaxy

Liquid-phase epitaxy (LPE) [137] involves the growth of an epitaxial film on a substrate that is placed in contact with a saturated solution of the film material. When the solution is cooled very slightly it becomes supersaturated and growth occurs on the substrate. Liquid-phase epitaxy is essentially an equilibrium process, where knowledge of the thermodynamic parameters allows a prediction of the composition of the film. Cleaning of the substrate surface is accomplished by raising the temperature of the solution briefly to make it unsaturated with respect to the substrate material, thus, causing some etching of the substrate prior to growth. Most likely LPE will be useful for the preparation of substrates of controlled composition, since it takes place in an environment that would appear incompatible with typical surface analytical tools, and would thus not seem a good candidate for a process to prepare a surface for *in situ* studies. The advantages of LPE are (a) simplicity, because the growth chamber is typically glass and no complex pumping or valving is involved since growth occurs at or near atmospheric pressure; and (b) the composition of ternary alloys is determined by thermodynamics. However, the disadvantages of LPE have considerably decreased its use: (a) small substrates are used and growth on whole wafers is not normally possible; (b) film thickness is very difficult to control since the growth rate is a sensitive function of solution temperature; and (c) it is difficult to maintain the film composition constant since only small volumes of solution are used. While LPE played a very significant role in the development of III–V devices in the 1960s, and is still used occasionally because of the optical quality of the layers produced, the film uniformity and thickness are not sufficiently under control for modern device fabrication.

5.5. Solid-Phase Epitaxy

Solid-phase epitaxy (SPE) describes the regrowth of a polycrystalline or amorphous film on a crystalline substrate at an elevated temperature. The most common example of SPE is the regrowth of a semiconductor that has been ion implanted or sputtered to an extent that the surface has been disordered by the ion flux. Annealing can produce regrowth from the undamaged region outward to the surface. Other instances can involve using scanning laser beams or other optical sources to carry out a rapid localized annealing of the surface of a wafer on which a noncrystalline layer has been deposited. In most cases, SPE is a homoepitaxial process used to recrystallize the surface of a film placed on a crystalline substrate. The main limitation for surface preparation would seem to be the time constraint to regrow a layer of a thickness of more than a few tens of atom layers. It should be remembered that ion bombardment and subsequent annealing, in effect a form of SPE, were used in the early work on clean single-crystal surface studies.[138] In recent years ion bombardment has lost favor as a method for cleaning semiconductors because of the evidence for ion-induced damage;[139] nevertheless, ion bombardment and annealing are still commonly used for preparation of other materials, or where the electrical or optical properties of semiconductors are not of concern.

Studies of SPE, particularly on Si, have been made using the full range of surface analysis tools, including LEED,[140] scanning electron microscopy,[141] and scanning tunneling microscopy.[142] These studies have shown the gradual transition from a surface covered with small clusters to a recrystallized, smooth surface through the dissociation of the clusters and smoothing of step edges as the temperature increased from 250 to 600°C.

6. Conclusions: MBE Deposited Thin Films and Surface Science

It is evident from the emphasis in the foregoing discussion that the author's preference for *in situ* surface preparation is by physical processes closely related to MBE. The relative simplicity and UHV compatibility of MBE make it an obvious choice for the preparation and improvement of surfaces to be used in other experiments. The implementation of MBE techniques can range in complexity from the addition of a simple Si filament source used to deposit on a Si surface for STM observations to a complex multichamber system for the preparation of superlattices for optical or electron spectroscopy. These methods are all based on the realization that film growth from the vapor can, under very clean conditions, produce surfaces that are almost atomically ideal. This is possible because the temperature and deposition rate can be kept low in the absence of contamination from the surroundings, and the resulting long surface diffusion lengths produce uniformly stepped surfaces like those in Fig. 8.

Skeptics have sometimes referred to MBE as representing "megabuck epitaxy." It is certainly true that commercial systems designed for large semiconductor wafers, high throughput, and automated operation have become fairly expensive. However, if the intent of the experimenter is simply to generate (or regenerate) a surface for other experiments, it is a simple matter to incorporate one or two source ovens with rotary shutters and to include a rudimentary RHEED system into a vacuum system designed for any of the various electron spectroscopies. As long as very clean vacuum conditions can be maintained during the operation of the source ovens, all the advantages of MBE can be obtained, without some of the more sophisticated hardware that has proven valuable for the development of semiconductor devices. In this way some useful MBE capability can be included quite inexpensively.

On the other hand, there are an increasing number of applications where extensive manipulation and movement of the substrate are required, whether to separate the growth area from the analysis area or to permit a complicated sequence of growth or analysis steps. One particularly important application is that of sample transfer between a MBE growth stage and a mounting stage that provides the mechanical stability required for STM.[143] Research directed toward improved sample handling for MBE applications makes it much easier to design systems with these requirements in mind.

The fortunate combination of technological incentives for unique electronic devices coupled with the advances in vacuum and surface science has made it possible to develop advanced techniques of thin-film deposition that provide remarkable control over film parameters. The importance of the relationship between our understanding of surfaces and the technology of thin-film deposition cannot be overstated. The tools for studying surfaces developed since 1968 have played a crucial role for fabricating epitaxial semiconductor films that are more perfect than their underlying substrates. In particular, the spectroscopic methods for determining surface composition (e.g., AES, XPS, and SIMS), the structural characterization tools, RHEED, and now the various forms of scanning probe microscopies, have been essential for gaining an understanding of how the various deposition methods influence the films they produce.

The wealth of structural detail provided by the scanning probe techniques, particularly when supplemented by other means of analysis, provides the answers to problems that hitherto required much trial-and-error experimentation. For example, we have referred to the problems in wet-chemical-etching GaAs with the resulting rather rough (on an atomic scale) surfaces. Song et al.,[144] using a combination of XPS and AFM, have shown that an etching solution comprising HCl/H_2O produces a surface flat to one monolayer and that oxides and chlorides are removed from the surface in a few minutes, leaving a nearly stoichiometric composition. Other etchants left the surface rougher while changing the composition more drastically. Fan et al.[145] have examined the smoothness of MBE-grown GaAs(100) surfaces by STM and AFM. Although the films were grown in a separate system, they were passivated for the transfer in air by using a cap layer of As. Their finding was that the As cap layer, which is easily desorbed in the microscope vacuum, provided good protection for the transfer; moreover, the capping, transfer, and subsequent thermal removal of the cap layer did not perceptibly degrade the quality of the surface structure. Ohkouchi et al.[146] have shown that thermal cleaning of InP and InAs results in a structurally flat surface, provided that an As flux is directed at the surface to suppress In droplet formation after the loss of P or As by evaporation. In the case of InP, the outermost layer consists of pseudomorphic InAs. Heinrich et al. have examined the increased concentration of P vacancies produced by the preferential loss of P from InP surfaces at elevated temperatures (although these were cleaved surfaces, not grown by MBE).[147]

All of these examples involve the verification of a previous qualitative understanding based on empirically derived methods of surface preparation that have been in use for years; now, the combination of MBE and high-resolution structural analysis have provided a firm understanding of the precise effectiveness of the treatments. Combining epitaxial techniques that form highly ordered surfaces with scanning probe methods for analysis is particularly fruitful, because the probe analysis answers questions about atomic structure and order, while the nearly perfect epitaxial surfaces provide the STM and AFM with the uniform, planar surfaces that are the most readily interpreted and understood.

References

1. Milton Ohring, *The Materials Science of Thin Films*, Academic Press, Boston (1992).
2. R. Glang, in: *Handbook of Thin Film Technology* (L. I. Maissel and R. Glang, eds.), McGraw-Hill, New York (1970).
3. Donald L. Smith, *Thin-Film Deposition: Principles and Practice*, McGraw-Hill, New York (1995).
4. Stephen A. Campbell, *The Science and Engineering of Microelectronic Fabrication*, Oxford University Press, New York (1996).
5. G. Allan, G. Bastard, N. Boccara, M. Lannoo, and M. Voos (eds.), *Heterojunctions and Semiconductor Superlattices*, Springer-Verlag, Berlin (1986).
6. L. C. Feldman and J. W. Mayer, *Fundamentals of Surface and Thin Film Analysis*, North Holland, Amsterdam (1986).
7. A. W. Czanderna (eds.), *Methods of Surface Analysis*, Elsevier, Amsterdam (1975).
8. S. Dushman, *Scientific Foundations of Vacuum Techniques*, Wiley, New York (1962).
9. T. A. Delchar, *Vacuum Physics and Techniques*, Chapman and Hall, London (1993).
10. *Vacuum Technology: Its Foundations, Formulae and Tables*, Leybold AG, Export, PA (1992).
11. M. H. Hablanian, *High Vacuum Techniques: A Practical Guide*, Marcel Dekker, New York (1990).
12. J. F. O'Hanlon, *User's Guide to Vacuum Technology*, 2nd ed., Wiley, New York (1989).
13. L. Holland, *Vacuum Deposition of Thin Films*, Wiley, New York (1956).
14. R. E. Honig, *RCA Rev.* **23**, 567 (1962).
15. L. I. Maissel, in: *Handbook of Thin Film Technology* (L. I. Maissel and R. Glang, eds.), McGraw-Hill, New York (1970).
16. W. R. Runyan and K. E. Bean, *Semiconductor Integrated Circuit Processing Technology*, Addison-Wesley, Reading, MA (1990).
17. J. L. Vossen and J. J. Cuomo, in: *Thin Film Processes* (J. L. Vossen and W. Kern, eds.), Academic Press, New York (1978).
18. W. D. Westwood, in: *Microelectronic Materials and Processes* (R. A. Levy, ed.), Kluwer Academic, Dordrecht (1989).
19. G. K. Wehner, *Phys. Rev.* **102**, 690 (1956).
20. H. H. Anderson and H. L. Bay, in: *Sputtering by Particle Ion Bombardment* (I. R. Behrisch, ed.), Springer-Verlag, Berlin (1981).
21. D. B. Fraser, in: *Thin Film Processes* (J. L. Vossen and W. Kern, eds.), Academic Press, New York (1978), p. 115.
22. J. A. Thornton and A. S. Penfold in: *Thin Film Processes* (J. L. Vossen and W. Kern, eds.), Academic Press, New York (1978).
23. Ref. 1, p. 114.
24. W. Tsai, J. Fair, and D. Hodul, *J. Electrochem Soc.* **139**, 2004 (1992).
25. J. M. Molarius and M. Orpana, in: *Issues in Semiconductor Materials and Processing Technologies* (S. Coffa, F. Priolo, E. Rimini, and J. M. Poate, eds.), Kluwer, Dordrecht (1991).
26. J. M. E Harper, in: *Thin Film Processes* (J. L. Vossen and W. Kern, eds.), Academic Press, New York (1978), p. 175.
27. J. Puretz and L. W Swanson, *J. Vac. Sci. Technol. B* **10**, 2695 (1992).
28. A. Sherman, *Chemical Vapor Deposition for Microelectronics: Principles, Technology, and Applications*, Noyes, Park Ridge, NJ (1987).
29. A. C. Adams, in: *VLSI Technology* (S. M. Sze, ed.), McGraw-Hill, New York (1985).
30. J. E. J. Schmitz, *Chemical Vapor Deposition of Tungsten and Tungsten Silicides*, Noyes, Park Ridge, NJ (1992).
31. W. Kern, in: *Microelectronic Materials and Processes* (R. A. Levy, ed.), Kluwer Academic, Dordrecht (1989).

32. W. Kern and V. S. Ban, in: *Thin Film Processes* (J. L. Vossen and W. Kern, eds.), Academic Press, New York (1978).
33. G. Lucovsky, P. D. Richard, D.V. Tsu, S.Y. Lin, and R.J. Markunas, *J. Vac. Sci. Technol. A* **4**, 681 (1986).
34. V. Herak and D.J. Thomson, *J. Appl. Phys.* **67**, 6347 (1990).
35. L. Royer, *Bull. Soc. Fr. Mineral Cristallogr.* **51**, 7(1928).
36. J. N. Smith, H. Saltsburg, and R. L. Palmer, *J. Chem. Phys.* **49**, 1287 (1968).
37. R. L. Palmer, J. N. Smith, H. Saltsburg, and D. R. O'Keefe, *J. Chem. Phys.* **53**, 1666 (1970).
38. C. T. Foxon and B. A. Joyce, *Sur. Sci.* **50**, 434 (1975).
39. J. R. Arthur, *Sur. Sci.* **43**, 449 (1974).
40. J. R. Arthur and T. R. Brown, *J. Vac. Sci. Technol.* **12**, 200 (1975).
41. J. Y. Tsao, T. M. Brennan, and B. E. Hammons, *J. Cryst. Growth* **111**, 125 (1991).
42. C. Pearson, M. Krueger, R. Curtis, B. Borovsky, X. Shi, and E. Ganz, *J. Vac. Sci. Technol. A* **13**, 1506 (1995).
43. M. Tomitori, K. Watanabe, M. Kobayashi, and O. Nishikawa, *J. Vac. Sci. Technol. B* **12**, 2022 (1994).
44. T. Hasegawa, M. Kohno, S. Hosaka, and S. Hosoki, *J. Vac. Sci. Technol. B* **12**, 2078 (1994).
45. F. C. Frank and J. H. van der Merwe, *Proc. Roy. Soc. A* **189**, 205 (1949).
46. E. Bauer, *Z. Kristallogr.* **110**, 291 (1983).
47. A. Y. Cho and J. R. Arthur, *Prog. Solid-State Chem.* **10**, 157 (1975).
48. B. A. Joyce, *Rep. Prog. Phys.* **48**, 1637 (1985).
49. J. Y. Tsao, *Materials Fundamentals of Molecular Beam Epitaxy*, Academic Press, Boston (1992).
50. M. A. Herman and H. Sitter, *Molecular Beam Epitaxy*, Springer-Verlag, Berlin (1988).
51. C. W. Tu and J. S. Harris, Jr. (eds.), *Molecular Beam Epitaxy 1990*, North Holland, Amsterdam (1991).
52. K. Ploog, in: *Crystals/Growth, Properties, and Applications* (L. F Boschke, ed.), Springer-Verlag, Heidelberg (1979).
53. M. B. Panish and H. Temkin, *Gas Source Molecular Beam Epitaxy*, Springer-Verlag, Berlin (1993).
54. J. R. Arthur, *J. Appl. Phys.* **39**, 4032 (1968).
55. J. R. Arthur, in *Proceedings of Conference on Structure and Chemistry of Solid Surfaces* (G. A. Somorjai, ed.), Wiley, New York (1969).
56. C. T. Foxon, M. R. Boudry and B. A. Joyce, *Surf. Sci.* **44**, 69 (1974).
57. C. T. Foxon, J. A. Harvey, and B. A. Joyce, *J. Phys. Chem. Solids* **34**, 1693 (1973).
58. A. Y. Cho, *J. Appl. Phys.* **41**, 2780 (1970).
59. Y. W. Mo, R. Kariotis, D. E. Savage, and M. G. Lagally, *Surf. Sci.* **219**, L551 (1989).
60. M. G. Lagally, R. Kariotis, G. S. Swartzentruber, and Y. W. Mo, *Ultramicroscopy* **31**, 87 (1989).
61. Y. W. Mo, R. Kariotis, B. S. Swartzentruber, M. B. Webb, and M.G. Lagally, *J. Vac. Sci. Technol. A* **8**, 201 (1990).
62. Commercial versions are available from Epi MBE Products Group, VG Semicon, and Riber.
63. S. L. Wright, R. F. Marks, and A. E. Goldberg, *J. Vac. Sci. Technol. B* **6**, 842 (1988).
64. D. E. Mars and J. N. Miller, *J. Vac. Sci. Technol. B* **4**, 571 (1986).
65. A. J. SpringThorpe and A. Majeed, *J. Vac. Sci. Technol. B* **8**, 266 (1990).
66. T. H. Chiu, W. T. Tsang, and J. A. Ditzenberger, *J. Vac. Sci. Technol. B* **4**, 600 (1986).
67. M. Nouaoura, L. Lassabatere, N. Bertru, and J. Bonnet, *J. Vac. Sci. Technol. B* **13**, 83 (1995).
68. S. Strite, M. Kamp, and H. P. Meier, *J. Vac. Sci. Technol. B* **13**, 290 (1995).
69. W. K. Liu and M. B. Santos, *J. Vac. Sci. Technol. B* **14**, 647 (1996).
70. W. K. Liu, W. T. Yuen, and R. A. Stradling, *J. Vac. Sci. Technol. B* **13**, 1539 (1995).
71. B. J. Garc'a, C. Fontaine, and A. Munoz Yague, *J. Vac. Sci. Technol. B* **13**, 281 (1995).
72. T. Kosugi, H. Ishii, and Y. Arita, *J. Vac. Sci. Technol. A* **15**, 127 (1997).

73. B. G. Orr, C. W. Snyder and M. Johnson, *Rev. Sci. Instr.* **62**, 1400 (1991).

74. C. R. Whitehouse, S. J. Barnett, D. E. J. Soley, J. Quarrel, S. J. Aldridge, A. G. Cullis, M. T. Emeny, A. D. Johnson, G. F. Clarke, W. Lamb, B. K. Tanner, S. Cottrell, B. Lunn, C. Hogg, and W. Hagston, *Rev. Sci. Instr.* **63**, 634 (1992).

75. P. Link, G. Grobbel, M. Worz, S. Bauer, H. Berger, W. Gebhardt, J. J. Paggel, and K. Horn, *J. Vac. Sci. Technol. A* **13**, 11 (1995).

76. M. C. Miller and J.R. Arthur, U.S. Patent #4,426,569.

77. P. A. Maki, S. C. Palmateer, A. R. Calawa, and B. R. Lee, *J. Vac. Sci. Technol. B* **4**, 564 (1985).

78. N. Chand and S. N. G. Chu, *J. Cryst. Growth* **104**, 485 (1990).

79. G. D. Pettit, J. M. Woodall, S. L. Wright, P. D. Kirchner, and J. L. Freeouf, *J. Vac. Sci. Technol. B* **2**, 241 (1984).

80. J. N. Miller, *J. Vac. Sci. Technol. B* **10**, 803 (1991).

81. D. G. Schlom, W. S. Lee, T. Ma, and J. S. Harris, Jr., *J. Vac. Sci. Technol. B* **7**, 296 (1989).

82. J. M. Ballingall and C. E. C. Wood, *Appl. Phys. Lett.* **41**, 947 (1982).

83. M. C. Holland and C. R. Stanley, *J. Vac. Sci. Technol. B* **14**, 2305 (1996).

84. R. J. Hauenstein, D. A. Collins, X. P. Cai, M. L. O'Steen, and T. C. McGill, *Appl. Phys. Lett.* **66**, 2861 (1995).

85. H. Morkoc, S. Strite, G. B. Gao, M. E. Lin, B. Sverdlov, and M. Burns, *J. Appl. Phys.* **76**, 1363 (1994).

86. K. Kim, M. C. Yoo, K. H. Shim, and J. T. Verdeyen, *J. Vac. Sci. Technol. B* **13**, 796 (1995).

87. P. K. Larsen and P. J. Dobson (eds.), *Reflection High-Energy Electron Diffraction and Reflection Electron Imaging of Surfaces*, Plenum Press, New York (1988).

88. W. G. Breiland and K. P. Killeen, *J. Appl. Phys.* **78**, 6726 (1995).

89. P. Pinsukanjana, A. Jackson, J. Tofte, K. Maranowski, S. Campbell, J. English, S. Chalmers, L. Coldren, and A. Gossard, *J. Vac. Sci. Technol. B* **14**, 2147 (1996).

90. A. J. SpringThorpe, T. P Humphreys, A. Majeed, and W. T. Moore, *Appl. Phys. Lett.* **55**, 2138 (1989).

91. R. N. Sacks, R. M. Sieg, and S. A. Ringel, *J. Vac. Sci. Technol. B* **14**, 2157 (1996).

92. D. E. Aspnes, W. E. Quinn and S. Gregory, *Appl. Phys. Lett.* **56**, 2569 (1990).

93. D. E. Aspnes, *Surf. Sci.* **307**, 1017 (1995).

94. B. A. Joyce, J. H. Neave, J. Zhang, and P. J. Dobson, in: *Reflection High-Energy Electron Diffraction and Reflection Electron Imaging of Surfaces* (P. K. Larsen and P. J. Dobson, eds.), Plenum Press, New York (1988).

95. J. H. Neave, B. A. Joyce, P. J. Dobson, and N. Norton, *Appl. Phys. A* **31**, 1 (1983).

96. J. M. Van Hove, C. S. Lent, P. R. Pukite, and P. I. Cohen, *J. Vac. Sci. Technol. B* **1**, 741 (1983).

97. D. E. Chambliss and K. E. Johnson, *J. Vac. Sci. Technol. A* **13**, 1522 (1995).

98. C. M. Gilmore and J. A. Sprague, *J. Vac. Sci. Technol. A* **13**, 1160 (1995).

99. C. T. Foxon and B. A. Joyce, *Surf. Sci.* **64**, 293 (1977).

100. A. J. SpringThorpe and A. Majeed, *J. Vac. Sci. Technol. B* **8**, 266 (1990).

101. W. M Duncan, M. J. Bevan, and H. D. Shih, *J. Vac. Sci. Technol. A* **15**, 216 (1997).

102. G. N. Maracas, C. H. Kuo, S. Anand, R. Droopad, G. R. L. Sohie, and T. Levola, *J. Vac. Sci. Technol. A* **13**, 727 (1995).

103. M. Li, Y. Z. Hu, E. A. Irene, L. Liu, K. N. Christensen, and D. M. Maher, *J. Vac. Sci. Technol. B* **13**, 105 (1995).

104. S. A. Chambers, T. T. Tran, and T. A. Hileman, *J. Vac. Sci. Technol. A* **13**, 83 (1995).

105. J. A. Curless, *J. Vac. Sci. Technol. B* **3**, 531 (1985).

106. Z. R. Wasilewski, G. C. Aers, A. J. SpringThorpe, and C. J. Miner, *J. Cryst. Growth* **111**, 70 (1991).

107. EPI Application Note 2/96, EPI MBE Products Group (1996).

108. M. G. Lagally, D. E. Savage, and M. C. Tringides, in: *Reflection High-Energy Electron Diffraction and Reflection Electron Imaging of Surfaces* (P. K. Larsen and P. J. Dobson, eds.), Plenum Press, New York (1988) p. 139.

109. F. C. Frank and J. H. van der Merwe, *Proc. R. Soc. A* **198**, 205 (1949); **200**, 125 (1949).
110. I am grateful to Prof. M. G. Lagally for providing Fig. 8.
111. R. Becker and R. Wolkow, in: *Semiconductor Surfaces in Scanning Tunneling Microscopy* (J. A. Stroscio and W. J. Kaiser, eds.), Academic Press, Boston (1993), Chap. 5.
112. J. H. Neave, P. J. Dobson, B. A. Joyce, and J. Zhang, *Appl. Phys. Lett.* **47**, 100 (1985).
113. R. J. Hamers, U. K. Kohler, and J. E. Demuth, *J. Vac. Sci. Technol. A* **8**, 195 (1990).
114. Y. W. Mo,, B. S. Swartzentruber, R. Kariotis, M. B. Webb, and M. G. Lagally, *Phys. Rev. Lett.* **63**, 2393 (1989).
115. T. Suntola and J Hyvaerinen, *Ann. Rev. Mater. Sci.* **15**, 177 (1985).
116. C. H. L. Goodman and M. V. Pessa, *J. Appl. Phys.* **60**, 65 (1986).
117. J. Nishizawa, H. Abe, and T. Kurabayashi, *J. Electrochem. Soc.* **132**, 1197 (1985).
118. T. Sakamoto, H. Funabashi, K. Ohta, T. Nakagawa, J. J. Kawai, T. Kojima, and Y. Bando, *Superlattices Microstructures* **1**, 347 (1985).
119. Y. Okada and J. S. Harris, Jr., *J. Vac. Sci. Technol. B* **14**, 1725 (1996).
120. K. Sakamoto, H. Matsuhata, K. Miki, and T. Sakamoto, *J. Cryst. Growth* **157**, 295 (1995).
121. J. H. Matthews and A. E. Blakesley, *J. Cryst. Growth* **27**, 118 (1974).
122. R. People and J. C. Bean, *Appl. Phys. Lett.* **47**, 322 (1986).
123. T. P. Piersall (ed.), *Strained-Layer Superlattices: Physics*, Academic Press, Boston (1991).
124. T. P. Piersall (ed.), *Strained-Layer Superlattices: Materials Science and Technology*, Academic Press, Boston (1991).
125. J.-P. Reithmaier, H. Riechert, and H. Schlotterer, *J. Cryst. Growth* **111**, 407 (1991).
126. M. Copel, M. C. Reuter, M. Horn von Hoegen, and R. M. Tromp, *Phys. Rev. B* **42**, 682 (1990).
127. F. K. LeGoues, M. Copel, and R. Tromp, *Phys. Rev. Lett.* **63**, 1826 (1098).
128. X. W. Lin, Z. Liliental-Weber, J. Washburn, E. R. Weber, A. Sasaki, A. Wakahara, and T. Hasegawa, *J. Vac. Sci. Technol. B* **13**, 1805 (1995).
129. C. Carter-Coman, A. S. Brown, R. Bicknall-Tassius, N. M. Jokerst, and M. Allen, *Appl. Phys. Lett.* **69**, 257 (1996).
130. C. Carter-Coman, R. Bicknall-Tassius, A. S. Brown, and N. M. Jokerst, *Appl. Phys. Lett.* **70**, 1754 (1997).
131. F. E. Ejeckom, Y. H. Lo, S. Subramanian, H. Q. Hou, and B. E. Hammons, *Appl. Phys. Lett.* **70**, 1685 (1997).
132. Y. -M. Houng, *Chemical Beam Epitaxy*, *Crit. Rev. Solid State Mater. Sci.* **17**, 277 (1992).
133. T. Y. Hsieh, K. H. Jung, and D. L. Kwong, *J. Electrochem. Soc.* **138**, 1188 (1991).
134. B. S. Meyerson, *Appl. Phys. Lett.* **48**, 797 (1986).
135. B. S. Meyerson, *IBM J. Res. Dev.* **34**, 806 (1990).
136. G. B. Stringfellow, *Organometallic Vapor-Phase Epitaxy* (Academic Press, Boston (1989).
137. M. B. Panish, *J. Electrochem. Soc.* **127**, 2729 (1980).
138. R. E. Schlier and H. E. Farnsworth, *J. Phys. Chem. Solids* **6**, 271 (1958).
139. H. Temkin, L. R. Harriot, R. A. Hamm, J. Weiner, and M. B. Panish, *Appl. Phys Lett.* **54**, 1463 (1989).
140. T. Shigeta, *Appl. Phys. Lett.* **52**, 619 (1988).
141. S. S. Lau, S. Matteson, J. W. Mayer, P. Revesz, J. Gyulai, J. Roth, T. W. Sigmon, and T. Cass, *Appl. Phys. Lett.* **34**, 76 (1979).
142. K. Uesugi, M. Yoshimura, T. Yao, T. Sato, T. Sueyoshi, and M. Iwatsuki, *J. Vac. Sci. Technol. B* **12**, 2018 (1994).
143. L. J. Whitman, P. M. Thibado, F. Linker, and J. Patrin, *J. Vac. Sci. Technol. B* **14**, 1870 (1996).
144. Z. Song, S. Shogen, M. Kawasaki, and I. Suemune, *J. Vac. Sci. Technol. B* **13**, 77 (1995).
145. Y. Fan, I. Karpov, G. Bratina, L. Sorba, W. Gladfelter, and A. Franciosi, *J. Vac. Sci. Technol. B* **14**, 623 (1996).

146. S. Ohkouchi, N. Ikoma, and I. Tanaka, *J. Vac. Sci. Technol. B* **12**, 2033 (1994).
147. M. Heinrich, Ph. Ebert, M. Simon, K. Urban, and M. G. Lagally, *J. Vac. Sci. Technol. A* **13**, 1714 (1995).

Index